Transportation and Network Analysis: Current Trends

Applied Optimization

Volume 63

Transportation and Network Analysis: Current Trends

Miscellanea in honor of Michael Florian

Edited by

Michel Gendreau

Département d'informatique et de recherche opérationelle (DIRO),
Université de Montréal,
and Centre de recherche sur les transports (CRT)

and

Patrice Marcotte

Département d'informatique et de recherche opérationelle (DIRO),
Université de Montréal,
and Centre de recherche sur les transports (CRT)

KLUWER ACADEMIC PUBLISHERS
DORDRECHT / BOSTON / LONDON

A C.I.P. Catalogue record for this book is available from the Library of Congress.

ISBN 1-4020-0488-5

Published by Kluwer Academic Publishers,
P.O. Box 17, 3300 AA Dordrecht, The Netherlands.

Sold and distributed in North, Central and South America
by Kluwer Academic Publishers,
101 Philip Drive, Norwell, MA 02061, U.S.A.

In all other countries, sold and distributed
by Kluwer Academic Publishers,
P.O. Box 322, 3300 AH Dordrecht, The Netherlands.

Printed on acid-free paper

Printed in the Netherlands.

Contents

Contributing Authors xi

Preface xiii

Préface xv

Michael Florian — Friend and colleague xvii

Michael Florian — Ami et collègue xix

1
Allocation of railroad capacity under competition: a game theoretic approach to track time pricing 1
A. Bassanini, A. La Bella and A. Nastasi
1.1 Introduction 1
1.2 Model description 3
1.3 Customer choice and price competition 5
1.4 Track access and capacity pricing 8
1.5 Schedule requests and optimality issues 11
1.6 Numerical simulations 12
1.7 Conclusions 15

References 16

2
Real Time Simulation of Traffic demand-supply interactions within DynaMIT 19
M. Ben-Akiva, M. Bierlaire, H.N. Koutsopoulos and R. Mishalani
2.1 Introduction 19
2.2 DynaMIT 20
2.3 Demand 22
2.4 Supply 27
2.5 Demand-Supply interaction 32
2.6 Evaluation 33
2.7 Conclusion 35

References 35

3
Implications of Marginal Cost Road Pricing for Urban Travel Choices and User Benefits 37
D. Boyce, K. Balasubramaniam and X. Tian

3.1 Introduction 37
3.2 The Combined Travel Choice Model and Its Implementation 38
3.3 Findings of the Travel Demand Analysis 40
3.4 Analysis of Alternative Pricing Strategies 46
3.5 Conclusions 47

References 47

4
A Dynamic User Equilibrium Model for Traffic Assignment in Urban Areas 49
L. Brotcorne, D. De Wolf, M. Gendreau and M. Labbé

4.1 Introduction 49
4.2 Problem Formulation 51
4.3 Existence of a User Equilibrium 60
4.4 Solution Algorithm 61
4.5 Numerical Results 63
4.6 Conclusion 68

References 68

5
Estimation of travel demand using traffic counts and other data sources 71
E. Cascetta and A.A. Improta

5.1 Static estimation of O-D demand flows from traffic counts 71
5.2 Dynamic estimators of O-D matrices using traffic counts 84
5.3 Estimation of demand models parameters from traffic counts 88

References 92

6
Path Recovery/Reconstruction and Applications in Nonlinear Multimodal Multicommodity Networks 95
T.G. Crainic, G. Dufour, M. Florian and D. Larin

6.1 Introduction 95
6.2 Path Reconstruction 97
6.3 The STAN Modelling Framework and System 100
6.4 Path Analysis Implementation and Applications 103
6.5 Conclusions 107

References 107

7
Linear-in-parameters logit model derived from the Efficiency Principle 109
S. Erlander

7.1 Introduction 109
7.2 Linear-in-parameters logit model 110
7.3 Testing the efficiency assumption 112
7.4 Conclusions 116

References 116

8
A Multi-Class Multi-Mode Variable Demand Network Equilibrium Model with Hierarchical Logit Structures 119
M. Florian, J.H. Wu and S. He

8.1 Introduction 119
8.2 A Brief Literature Review 120
8.3 Model Formulation 121
8.4 The Analysis of the Mathematical Structure of the Model 123
8.5 A Solution Algorithm 125
8.6 Computational Results (STGO Model) 129

References 132

9
A Toll Pricing Framework for Traffic Assignment Problems with Elastic Demand 135
D.W. Hearn and M.B. Yildirim

9.1 Introduction 135
9.2 Traffic Assignment Models with Elastic Demand 136
9.3 Toll Set for the Elastic Demand Case 139
9.4 The Toll Pricing Framework 140
9.5 Summary 144

References 144

10
A Decision Support Methodology for Strategic Traffic Management 147
T. Larsson, J.T. Lundgren, C. Rydergren and M. Patriksson

10.1 Introduction 147
10.2 Formulating the Traffic Management Model 149
10.3 Solution framework 153
10.4 Outline of algorithms 158
10.5 Extensions 161
10.6 Conclusions 163

References 164

11
Column Generation Method for Network Design 165
N. Maculan, M.M. Passini, J.A.M. Brito and A. Lisser

11.1 Mathematical Model 166
11.2 Simplified Model 168
11.3 Column Generation 169
11.4 Getting a Feasible Solution for (P) 175
11.5 Branch-and-Price Method 175
11.6 Numerical Experiments 176

References 178

12
Computing Fixed Points by Averaging 181
T.L. Magnanti and G. Perakis

12.1 Introduction 181
12.2 A General Averaging Framework 182
12.3 An Approximate Averaging Framework 188

References 198

13
Pollution Permits for Spatial Price Networks 199
A. Nagurney

13.1 Introduction 199
13.2 The Spatial Price Network Model with Pollution Permits 200
13.3 Qualitative Properties 212
13.4 The Algorithm 215
13.5 Numerical Examples 216

References 219

14
A new dual algorithm for shortest path reoptimization 221
S. Nguyen, S. Pallottino and M.G. Scutellà

14.1 Reoptimizing shortest paths computations 221
14.2 The shortest path reoptimization problem 222
14.3 A new dual algorithm for shortest path reoptimization 227

References 235

15
Biproportional Matrix Balancing with Upper Bounds 237
H. Spiess

15.1 Introduction 237
15.2 Standard Biproportional Matrix Balancing 238
15.3 Biproportional Matrix Balancing with Upper Bounds 239
15.4 Solution Algorithm 241
15.5 Conclusions 242

References 242

Contributing Authors

KARTHIK BALASUBRAMANIAM
University of Illinois at Chicago
karthikb@seas.upenn.edu

ANNA BASSANINI
Università di Roma "La Sapienza"
bassanini@dis.uniroma1.it

MOSHE BEN-AKIVA
Massachusetts Institute of Technology
mba@mit.edu

MICHEL BIERLAIRE
Massachusetts Institute of Technology
mbi@mit.edu

DAVID BOYCE
University of Illinois at Chicago
dboyce@uic.edu

JOSÉ ANDRÉ M. BRITO
Universidade Federal do Rio de Janeiro
britom@cos.ufrj.br

LUCE BROTCORNE
Université de Valenciennes
luce.brotcorne@univ-valenciennes.fr

ENNIO CASCETTA
University of Naples "Federico II
cascetta@unina.it

TEODOR GABRIEL CRAINIC
Université du Québec à Montréal
theo@crt.umontreal.ca

DANIEL DE WOLF
Université de Lille 3
dewolf@univ-lille3.fr

GINA DUFOUR
INRO Consultants, Inc.
gina@inro.ca

SVEN ERLANDER
Linköping University
sverl@mai.liu.se

MICHAEL FLORIAN
Université de Montréal
mike@crt.umontreal.ca

MICHEL GENDREAU
Université de Montréal
gendreau@iro.umontreal.ca

SHUGUANG HE
INRO Solutions Inc.
shuguang@inro.ca

DONALD W. HEARN
University of Florida, Gainesville
hearn@ise.ufl.edu

ALESSANDRA A. IMPROTA
University of Naples "Federico II"
aimprota@unina.it

HARIS N. KOUTSOPOULOS
Volpe National Transportation Systems
koutsopoulos@volpe.dot.gov

A. LA BELLA
Università di Roma "Tor Vergata"
labella@disp.uniroma2.it

DIANE LARIN
INRO Consultants Inc.
diane@inro.ca

TORBJÖRN LARSSON
Linköping University
tolar@mtah.liu.se

ABDEL LISSER
France Télécom
abdel.lisser@francetelecom.fr

JAN T. LUNDGREN
Linköping University
jalun@mai.liu.se

NELSON MACULAN
Universidade Federal do Rio de Janeiro
maculan@cos.ufrj.br

THOMAS L. MAGNANTI
Massachusetts Institute of Technology
magnanti@mit.edu

RABI MISHALANI
Ohio State University
mishalani.1@osu.edu

ANNA NAGURNEY
University of Massachusetts, Amherst
nagurney@gbfin.umass.edu

ALBERTO NASTASI
Università di Roma "La Sapienza"
nastasi@dis.uniroma1.it

SANG NGUYEN
Université de Montréal
nguyens@iro.umontreal.ca

STEFANO PALLOTTINO
Università di Pisa
pallo@di.unipi.it

MARCOS M. PASSINI
Universidade Federal do Rio de Janeiro
passini@cos.ufrj.br

MICHAEL PATRIKSSON
Chalmers University of Technology
mipat@math.chalmers.se

GEORGIA PERAKIS
Massachusetts Institute of Technology
georgiap@mit.edu

CLAS RYDERGREN
Linköping University
clryd@math.liu.se

MARIA GRAZIA SCUTELLÀ
Università di Pisa
scut@di.unipi.it

HEINZ SPIESS
EMME/2 Support Center
Heinz.Spiess@spiess.ch

XIN TIAN
WilburSmith Associates
xtian@wilbursmith.com

JIA HAO WU
TJKM Consultants Inc.
jhwu@tjkm.com

MEHMET B. YILDIRIM
University of Florida, Gainesville
ybayram@cao.ise.ufl.edu

Preface

MICHEL GENDREAU AND PATRICE MARCOTTE

As an academic, Michael Florian has always stood at the forefront of transportation research. This is reflected in the miscellaneous contributions that make the chapters of this book, which are related in some way or another to Michael's interests in both the theoretical and practical aspects of his field. These interests span the areas of Traffic Assignment, Network Equilibrium, Shortest Paths, Railroad problems, Demand models, Variational Inequalities, Intelligent Transportation Systems, etc. The contributions are briefly outlined below.

BASSANINI, LA BELLA AND NASTASI determine a track pricing policy for railroad companies through the solution of a generalized Nash game. BEN-AKIVA, BIERLAIRE, KOUTSOPOULOS AND MISHALANI discuss simulation-based estimators of the interactions between supply and demand within a real-time transportation system. BOYCE, BALASUBRAMANIAM AND TIAN analyze the impact of marginal cost pricing on urban traffic in the Chicago region. BROTCORNE, DE WOLF, GENDREAU AND LABBÉ present a discrete model of dynamic traffic assignment where flow departure is endogenous and the First-In-First-Out condition is strictly enforced. CASCETTA AND IMPROTA give a rigorous treatment of the problem of estimating travel demand from observed data, both in the static and dynamic cases. CRAINIC, DUFOUR, FLORIAN AND LARIN show how to obtain path information that is consistent with the link information provided by a nonlinear multimodal model. ERLANDER derives the logit model from an efficiency principle rather than from the classical random utility approach. FLORIAN, WU AND HE give a variational inequality formulation of an integrated network equilibrium model where mode choice is driven by a hierarchical logit model. HEARN AND YILDIRIM extend a former model of equilibrium toll pricing to the elastic demand case. LARSSON, LUNDGREN, RYDERGREN AND PATRIKSSON propose a two-stage procedure for improving the functionality of a network while respecting predetermined managerial goals. MACULAN, PASSINI AND DE MOURA BRITO apply column generation to the problem of designing mixed telecommunication networks with rings and meshed circuits. MAGNANTI AND PERAKIS survey the class of 'averaging methods' for computing fixed points of point-to-set mappings. NAGURNEY develops a spatial price equilibrium model of marketable pollution permits, based on the theory of variational inequalities. NGUYEN, PALLOTTINO AND SCUTELLÀ describe a novel dual algorithm for the reoptimization of shortest paths. Finally, SPIESS adapts a biproportional matrix balancing algorithm to the situation where upper bounds are present.

Montréal, April 12, 2001

Préface

MICHEL GENDREAU ET PATRICE MARCOTTE

La recherche de Michael Florian s'est toujours située à la fine pointe des développements dans le domaine du transport, comme en témoignent les contributions au présent livre. En effet, celles-ci sont toutes reliées, d'une façon ou d'une autre, aux recherches fondamentales ou appliquées de Michael et couvrent des domaines aussi variés que l'affectation du trafic, les équilibres de réseaux, les plus courts chemins, les modèles de demande, les inéquations variationnelles, les systèmes intelligents de transport, le transport ferroviaire, etc. Les contributions contenues dans le présent livre sont brièvement décrites ci-dessous.

BASSANINI, LA BELLA ET NASTASI proposent un modèle de théorie des jeux pour déterminer une politique de tarification des rails de chemin de fer. BEN-AKIVA, BIERLAIRE, KOUTSOPOULOS ET MISHALANI étudient, par le biais d'une simulation, les interactions offre-demande dans un système de transport en temps réel. BOYCE, BALASUBRAMANIAM ET TIAN analysent l'impact d'une politique de tarification au coût marginal sur le trafic dans l'agglomération de Chicago. BROTCORNE, DE WOLF, GENDREAU ET LABBÉ présentent un modèle discret pour l'affectation dynamique du trafic où les taux de départ sont endogènes et où la condition 'Premier arrivé Premier servi' est imposée dans les contraintes mêmes du modèle. CASCETTA ET IMPROTA traitent de façon rigoureuse le problème d'estimation de la demande de transport à partir de données observées, à la fois dans un contexte statique et un contexte dynamique. CRAINIC, DUFOUR, FLORIAN ET LARIN indiquent comment récupérer des flots de chemin compatibles avec les flots d'arcs obtenus à partir d'un modèle non linéaire multimodal. ERLANDER dérive le modèle logit à partir d'un principe d'efficience plutôt que par l'approche classique basée sur la théorie de l'utilité aléatoire. FLORIAN, WU ET HE donnent une formulation variationnelle d'un modèle intégré d'équilibre de réseau où les choix modaux des usagers découlent d'un modèle logit hiérarchique. HEARN ET YILDIRIM généralisent un modèle de tarification à l'équilibre au cas où la demande est élastique. LARSSON, LUNDGREN, RYDERGREN ET PATRIKSSON proposent une procédure à deux étapes permettant d'améliorer la performance d'un réseau de transport tout en respectant des contraintes fixées par l'administrateur du réseau. MACULAN, PASSINI ET DE MOURA BRITO appliquent la technique dite de 'génération de colonnes' à la conception de réseaux de télécommunication dont la topologie contient à la fois des anneaux et des treillis. MAGNANTI ET PERAKIS présentent une rétrospective de la classe des algorithmes de 'pondération des itérés' pour le calcul de points fixes de multifonctions. NAGURNEY développe un modèle d'équilibre spatial, basé sur la théorie des inéquations variationnelles, pour le marché des permis de pollution. NGUYEN, PALLOTTINO ET SCUTELLÀ décrivent un nouvel algorithme dual pour la réoptimisation des chemins les plus courts. Finalement, SPIESS adapte un algorithme biproportionnel pour l'équilibrage de matrices à des problèmes incluant des contraintes de bornes supérieures sur les variables.

Montréal, 12 avril 2001

Michael Florian
Friend and colleague

Michael Florian started his career in operations research by working in industry as O.R. analyst for three years. In 1969, he completed his Ph.D. in Operations Research at Columbia University and shortly afterwards joined the faculty of the Département d'informatique of Université de Montréal. His early research interests focused mainly on machine scheduling, an area in which he made several important contributions in the early 70's, and on various network flow problems.

In December 1971, Université de Montréal created the "Centre de recherche sur les transports" (C.R.T., Centre for Research on Transportation), a multi-disciplinary research unit devoted to the study of transportation systems. Mike was involved from the outset with this initiative and, in 1973, he became Director of C.R.T. The following years were ones of intense activity: Mike, his colleagues and their students developed models and algorithms to tackle important problems in transportation planning, in particular in the area of traffic equilibrium. Mike's contribution during this period went far beyond performing top-quality academic research; besides providing scientific leadership for C.R.T., making it one of the best and best-known transportation research centers in the world, he paid great attention to two key factors: the supervision of graduate students and the transfer of models and techniques developed in universities to practitioners. In fact Mike, along with a few colleagues from C.R.T., was among the first in academia to promote the creation of spin-off companies to commercially distribute the results of academic research. While this idea has now become common-place, it was extremely innovative in the context of the 70's where the realm of academia was thought to be limited to universities, scientific journals and perhaps some consulting. In 1976, Mike founded INRO Consultants Inc., a company whose purpose was to develop and support industrial grade software based on research performed at C.R.T.

After stepping down as C.R.T. Director in 1979, Mike continued his various research, teaching and transfer activities at a relentless pace. Among other things, he led two important cooperation projects with Chile and Brazil that brought up new important transportation planning problems. He was also one of the leaders of a major project, funded by the government of Quebec, that was aimed at expanding C.R.T. by bringing in new graduates as post-doctoral researchers, thus securing the future development of the Centre. In 1994, Mike became the Director of C.R.T.'s newly created Laboratory on Intelligent Transportation Systems whose objective was to promote research in this critical new area. He held that position until 1999.

During his career, Mike has covered all major facets, academic as well as non-academic, of transportation research and planning. He has published over 70 articles in refereed journals and over 40 in conference proceedings. He has been Editor, Associate Editor or member of the Editorial Board of several scientific journals, including Operations Research, Transportation Science and Transportation Research B. He has been invited as keynote speaker in numerous meetings. He has acted as consultant on topics related to transportation planning for more than 60 organizations around the world. He has carried out professional development seminars for many organizations. EMME/2 and STAN, two software packages that he helped develop and that are distributed by INRO Consultants Inc. are now being used in 61 countries.

Mike's outstanding contribution to the advancement of transportation and operations research has been recognized in many ways. Among other distinctions, he was awarded in 1988 the Merit Award of the Canadian Operational Research Society and the Prize of the R & D Council of Transportation of Quebec; he was elected in 1990 to the Royal Society of Canada; in 1998, he received the prestigious "Robert D. Herman Lifetime Achievement Award" of the Transportation Science Section of INFORMS, the Institute for Operations Research and Management Sciences of America, and Linköping University granted him in 2000 the degree of "Doctor Honoris Causa". INRO Consultants Inc., which he founded and led for the last 25 years, was awarded in 1998 the Omond Solandt Award of the Canadian Operational Research Society for excellence in operations research.

MICHEL GENDREAU AND PATRICE MARCOTTE
MONTRÉAL, APRIL 15, 2001

Michael Florian
Ami et collègue

Michael Florian entreprend une carrière en recherche opérationnelle comme analyste dans l'industrie puis, peu après l'obtention d'un doctorat de l'Université Columbia (New York), est engagé comme professeur au département d'informatique de l'Université de Montréal. Au début de sa carrière universitaire, sa recherche porte principalement sur les problèmes d'ordonnancement, un domaine auquel il contribue de façon significative dans les années soixante-dix, ainsi que sur divers problèmes de réseaux de transport.

En décembre 1971, l'Université de Montréal crée le Centre de recherche sur les transports (C.R.T.), une unité de recherche multidisciplinaire consacrée à l'étude des systèmes de transport. Mike y est associé dès le début, et en prend la direction en 1973. En collaboration avec des collègues et des étudiants, il développe de nombreux modèles et algorithmes permettant de résoudre des problèmes de planification des transports en général, et d'équilibre de trafic en particulier. Au cours de cette période, la contribution de Mike ne se limite pas à la publication de travaux de recherche de haut niveau; en plus de d'assurer la direction scientifique du C.R.T. et d'en faire l'un des meilleurs centres au monde dans son domaine, Mike accorde une importance toute particulière à la supervision d'étudiants aux cycles supérieurs d'une part, et au transfert technologique de modèles développés dans l'université vers les utilisateurs du 'monde réel' d'autre part. En fait Mike et quelques collègues du C.R.T. sont parmi les premiers à faire la promotion d'entreprises vouées à la commercialisation de la recherche académique. Cette idée était particulièrement innovatrice dans le contexte des années 70, alors que les résultats tangibles de la recherche universitaire semblaient se limiter à des publications scientifiques et peut-être un peu de consultation professionnelle. En 1976, Mike crée Les Consultants INRO, une compagnie dont le but est de développer des logiciels de qualité industrielle basés sur des travaux de recherche entrepris au C.R.T.

Après avoir quitté la direction du C.R.T. en 1979, les activités de recherche et de transfert technologique se multiplient. Mike dirige deux important projets de coopération scientifique avec le Chili et le Brésil. En parallèle, il participe à l'élaboration d'un projet majeur, financé par le gouvernement du Québec, qui vise à intégrer au C.R.T. des chercheurs post-doctoraux et assure ainsi le développement à long terme du Centre. De 1994 à 1999, il dirige le tout nouveau Laboratoire sur les Systèmes Intelligents de Transport, un regroupement de chercheurs qui se consacre à la recherche dans ce domaine stratégique.

Au cours d'une carrière qui est loin d'être terminée, Mike couvre tous les aspects majeurs de la recherche en planification des transports. Il est l'auteur de plus de 70 articles dans des revues avec comité de lecture et de 40 articles dans des compte-rendus de conférences. Il est ou a été rédacteur, rédacteur adjoint ou membre du comité de rédaction de plusieurs revues scientifiques, dont Operations Research, Transportation Science et Transportation Research B. En plus d'avoir été invité comme conférencier dans les congrès et les universités les plus prestigieux, ses services de consultant ont été requis par plus de 60 entreprises de par le monde. Les logiciels EMME/2 et STAN, développés au C.R.T., sont distribués par Les Consultants INRO dans plus de 60 pays, du Mexique à l'Australie, en passant par l'Italie.

La contribution exceptionnelle de Mike à l'avancement du transport et de la recherche opérationnelle a été maintes fois reconnue. En 1988, il se voit décerné le Prix du mérite de la Société Canadienne de Recherche Opérationnelle ainsi que le prix du Conseil pour la Recherche et le Développement du Transport du Québec; en 1990, il est élu membre de la Société Royale du Canada et reçoit le 'Robert D. Herman Lifetime Achievement Award' de la section 'Transport' d'INFORMS ('Institute for Operations Research and Management Sciences of America'); en 1998, la compagnie Les Consultants INRO, dont il assure toujours la direction, obtient le prix d'excellence en recherche opérationnelle 'Omond Solandt', octroyé par la Société Canadienne de Recherche Opérationnelle; en 2000, l'université de Linköping lui octroie un doctorat 'Honoris Causa'.

MICHEL GENDREAU ET PATRICE MARCOTTE
MONTRÉAL, 15 AVRIL 2001

1 ALLOCATION OF RAILROAD CAPACITY UNDER COMPETITION: A GAME THEORETIC APPROACH TO TRACK TIME PRICING

A. Bassanini

A. La Bella

A. Nastasi

Abstract: The reorganization of the European railway sector following the application of Directive 440 requires devising an infrastructure access mechanism for competing transport operators. This paper proposes a market-based approach to railroad track allocation and capacity pricing, formulating a three-stage game-theoretic model where transport operators request their preferred schedules to the infrastructure manager and set the final prices for the transport services on the basis of actual schedules and access tariffs. The latter are simultaneously computed by a non discriminatory mechanism which maximizes the value of the timetable of each operator. Access tariffs are based on the congestion degree each train imposes on the system.

The model is validated by numerical simulations showing the impact of congestion externalities on access tariffs, final service prices and operators' profits.

1.1 INTRODUCTION

The European railway industry is in the midst of a process of restructuring and commercialization (Bowers, 1996; Brooks, 1995; Nash, 1993) triggered by directive 91/440, which came into force in January 1993.

The directive was designed *inter alia* to liberalize the market for providing rail services and expresses a need for increased competition within the sector, with the

M. Gendreau and P. Marcotte (eds.), Transportation and Network Analysis: Current Trends, 1–17.

expectation that this would result in both improved commercial attitudes and increased quality in service provision.

Up to now, the European model for restructuring has normally involved the separation of infrastructure from operations. In general, the railway industry in Europe may be considered to comprise three types of organization, namely the infrastructure manager or infrastructure company (IC), transport operators (TOs) and a regulatory authority. The latter supervises the interactions between the subjects operating in the sector, ensuring the respect of the principles of fairness and non-discrimination sanctioned by the directives, so as to promote an adequate degree of competition within the railway industry.

The duties of the IC vary across countries, reflecting differences in pre-existing market structures and regulatory objectives. However, the IC is often responsible for assigning railway capacity among the competing TOs. The new structure of the sector therefore requires defining a track access mechanism dealing with track allocation and access tarification for transport services.

In allocating track capacity, the IC has to take into account various aspects. First, geographical demand repartition often determines a high intensity of traffic on some particular lines; moreover, customer needs create cyclical variations in demand, with peak periods when available capacity is unable to satisfy demand on one hand, and periods of under-utilization of the infrastructure on the other. This determines congestion problems which must be dealt with in order to ensure the efficient exploitation of the available capacity. Second, and relatedly, priority criteria must be set for the services using the network so as to guarantee that high-valued services have priority over lower-valued ones. Third, coordination among services offered on different lines must be achieved. Fourth, minimum services often have to be granted to final users even on less profitable lines; this raises an issue concerning the amount of State subsidies to be granted to railways.

The tariff system must enhance the efficiency of infrastructure use, avoiding undue discrimination among TOs and ensuring the financial equilibrium of the IC. The tariff system can then take into account technical data regarding prices and costs, commercial data on the transport market and data that allow a better use of available capacity, such as congestion levels. In fact, a considerable part of infrastructure costs is related to congestion, which negatively affects the quality of the services provided.

The evolution trend of the railway sector implies the necessity to adopt "scientific" management methods and modern decision support techniques in order to raise competitiveness and optimize operations logistics. A key issue here is that of the timetable, which is a "compressed" description in time and space of all the trains using the system and thus represents the way access rights are allocated. In the decentralized scenario, conflicts in demand for track access must be solved in order to maximize the total value of a timetable, which also depends on the degree of coordination achieved among the services provided on the whole network.

In this context the existing models for timetable formation are of little practical use since the related literature (Carey, 1994; Kraay and Harker, 1995; Odijk, 1996) assumes a vertically integrated structure with a single decision-maker; on the contrary, the new scenario requires considering the peculiar form of oligopolistic interdependence intrinsic to the emerging scenario. On the other hand, the literature regarding congestion tolls estimation (Levin, 1981; Vickrey, 1969) and peak-load pricing (Crew and Kleindorfer, 1987; Morrison, 1983), either ignores the problem's imperfect competitive nature or treats spatial aspects separately from the temporal ones.

This paper makes an attempt at jointly considering spatial and temporal aspects in a context characterized by oligopolistic interdependence.

In the proposed game-theoretic model, the IC implements a non-discriminatory track allocation and pricing mechanism which maximizes the value of the effective schedule of each TO compatibly with the others' requests (Harker and Hong, 1994). This mechanism differs from traditional track assignment algorithms since it allows to decentralize the allocation of railroad tracks among different operators competing for network access without unduly discriminating between firms. The mechanism is applied sequentially to the rail lines belonging to the network in order of decreasing intensity of traffic, thereby taking into account the need of coordinating services offered on different lines.

The model will be developed for passenger services; however, it can also be extended to freight transportation.

The paper is structured as follows. Section 1.2 contains a description of the model, with assumptions and notation. In Section 1.3 demand-related aspects are analyzed with reference to price competition among rail transport operators. In Section 1.4 the track access mechanism is formulated; then, an analysis of the derivation of track prices is presented. In Section 1.5 a discussion of the possible outcomes of the model is provided, while Section 1.6 presents some numerical simulations. Section 1.7 concludes the paper.

1.2 MODEL DESCRIPTION

We propose a model where competing TOs request their preferred tracks to the IC and fix final service prices according to actual schedules and access tariffs. The IC is assumed to be responsible for managing the timetable, allocating train paths and providing new infrastructure, while the TOs use the rail infrastructure to run train services, paying network access tariffs to the IC. We assume that the IC would also assume the role of the regulator; the latter therefore does not appear in the model.

The competition among the TOs and the interactions between the latter and the IC can be described by a three-stage general model. In the first stage, each TO submits to the IC his track requests. TOs behave non-cooperatively and wish to obtain effective schedules as close as possible to the ideal ones. In the second stage the IC determines the effective timetable and the related tariffs for all the TOs. In the third stage the TOs set the prices of the services for the customers, on the basis of the timetable and the tariffs computed by the IC. Each TO maximizes his profit by adjusting prices, while considering consumer choice.

We assume that the model is characterized by complete information. This means that each firm has all the relevant information about other firms, their available strategies and the potential outcomes of the model. These assumptions might be restrictive, but are justified by assuming a context where TOs have a sufficiently good knowledge of cost and demand conditions for transport services.

The three-stage sequential structure of the game imposes that its solution be derived through a procedure similar to backward induction, from the third stage up to the first two. In other words, the relevant equilibrium notion is the subgame perfect equilibrium (Selten, 1975). In accordance to this solution procedure, in the following sections we will describe the model beginning from the third stage and proceeding back to the first two.

The allocation mechanism we propose determines the global timetable by minimizing the deviation from the requested schedules for each transport operator, and produces congestion-based infrastructure user fees. The mechanism is applied sequentially to the rail lines in order of decreasing intensity of traffic. In other words, the IC first solves the capacity allocation problem on the main traffic line. The mechanism is then applied to lines of decreasing importance in order to take into account the need of coordinating connections between stations on different lines.

The pricing principle we have adopted is justified by considering that trains' delay imposes a time cost that builds up to a large part of the total cost of the system. Thus, a train with a higher priority level will be granted better schedule adherence; on branch lines, this would mean better coordination with the services offered on the main line. However, being assigned tracks closer to the requested ones will generally imply paying higher tariffs.

Moreover, it is worth noticing that the model allows to analyze the impact of congestion on infrastructure access tariffs as well as final prices and operators' profits.

1.2.1 *Notation*

Let us define the following notation:

- K : set of all TOs;
- k : TO $k \in K$;
- $K \backslash k$: set of all TOs, excluding k;
- t_k : train of TO k;
- T_k : set of trains of TO k;
- $T_k \backslash t_k$: set of trains of TO k, excluding t_k;
- $T_K \backslash T_k$: set of trains of all TOs, excluding k;
- I : set of main stations;
- I_k : set of all stations in the schedule of train t_k;
- i : a departure station, $i \in I$;
- j : an arrival station, $j \in I$;
- IA_{jt_k} : requested arrival time at j for train t_k of TO k;
- ID_{it_k} : requested departure time from i for train t_k;
- A_{jt_k} : effective arrival time in j of train t_k;
- D_{it_k} : effective departure time from i of train t_k;
- $A_{t_k} = (\ldots, A_{jt_k}, \ldots)^T$; $D_{t_k} = (\ldots, D_{it_k}, \ldots)^T$;

- $A_k = (A_{t_k} \mid t_k \in T_k)^T$; $D_k = (D_{t_k} \mid t_k \in T_k)^T$;
- $A = (\ldots, A_k, \ldots)$; $D = (\ldots, D_k, \ldots)$;
- $x_k = (A_k, D_k)^T$; $x = (\ldots, x_k, \ldots)$;
- $A_{jT_k \backslash t_k}$: effective arrival times at j of the set of trains $T_k \backslash t_k$;
- $D_{iT_k \backslash t_k}$: effective departure times from i of the set of trains $T_k \backslash t_k$;
- $A_{jT_K \backslash t_k}$: effective arrival times at j of the set of trains $T_K \backslash t_k$;
- $D_{iT_K \backslash t_k}$: effective departure times from i of the set of trains $T_K \backslash t_k$;
- $A_{jT_{K \backslash k}}$: effective arrival times at j of the set of trains $T_K \backslash T_k$;
- $D_{iT_{K \backslash k}}$: effective departure times from i of the set of trains $T_K \backslash T_k$;
- $w_{D_{it_k}}$: monetary value of the deviation from the requested departure time from i for t_k;
- $w_{A_{jt_k}}$: monetary value of the deviation from the requested arrival time at j for t_k;
- DW_{it_k} : minimum dwelling time of train t_k at station i;
- D_i' : consumer's ideal departure time from i;
- $\Phi_{t_k;(i,j)}$: demand for train t_k between i and j;
- $\Phi_{t_k} = \left(\Phi_{t_k;(i,j)} \mid i, j \in I_k\right)$;
- h_{t_k} : tariff that TO k must pay for operating train t_k;
- $p_{t_k;(i,j)}$: unit price for the service provided by train t_k between i and j;
- $p_{t_k} = \left(p_{t_k;(i,j)} \mid i, j \in I_k\right)$; $p_k = (p_{t_k} \mid t_k \in T_k)^T$; $p = (\ldots, p_k, \ldots)$;
- C_k : other costs of TO k, except for tariffs;
- Π_k : profit of TO k.

1.3 CUSTOMER CHOICE AND PRICE COMPETITION

The TOs have the freedom to choose their ideal train schedules and the final prices for the services they provide. The schedule determines a large fraction of an operator's cost as well as the services he can offer. Schedule decisions should therefore be made keeping into account demand data derived by aggregating individual travelers' preferences.

Following well-known models of consumer choice for transit services (McFadden, 1981; Ben Akiva and Lerman, 1985; Dobson and Lederer, 1993), we assume a random utility model such that each individual traveler's demand is given by a logit function. A traveler's demand for a train is a function of the total cost of using that train; this cost has three components: 1) the cost of departing at a time that differs from the customer's most preferred departure time; 2) the cost associated with travel duration and 3) the actual fare. The total demand for each train is the sum of all travelers' demand for the train.

In order to simplify the analysis, we assume that there is only one class of customers; this hypothesis is not realistic, but is made to reduce the complexity of the problem[1].

Consider L cities on an east-west rail line and the demand for rail transport between them. Assume that cities are assigned progressive numbers from west to east, i.e. the first station of the line is indicated as 1, the second as 2 and so on (the last would therefore be L).

For sake of simplicity, from now on we will consider only trains running from west to east. The extension to the general case - with trains running in both directions - is straightforward.

In our case, let the $\frac{L(L-1)}{2}$ origin-destination pairs of cities be denoted by $O-D$. For each pair $(o,d) \in O-D$ ($d > o$ always holding due to our assumptions), consider transport service between the cities as provided by train t_k run by transport operator k; the operator may not be the only one providing transport between the cities.

Note that train t_k traveling from o to d may also serve customers' demand for any $O-D$ pair (i,j) on the same rail line, provided that i coincides with or is east of o and j coincides or is west of d. This should be considered when estimating transport demand for train t_k; therefore, we will first derive the expression for transport demand between the generic $O-D$ pair (i,j) served by t_k and then accordingly evaluate demand between o and d.

For any i,j and $D_i' \in [0,24)$ let $\beta(i,j,D_i')$ denote the density of customers whose most preferred departure time from i to j is at time D_i'. It is assumed that this function is piecewise constant on $[0,24)$. That is, there are $G = G(i,j)$ constant density intervals over $[0,24)$; let $\beta_g(i,j)$ be the density of the gth interval, $g = 0,\ldots,G-1$ for travel between i and j. Assume for each origin-destination pair that customers' travel preferences and densities are identical for each day. If this assumption is violated, one can redefine the period of analysis so that densities are identical for each period.

Assume that a customer's utility function is linear and separable in various attributes of routes and money is the numeraire. If a customer whose ideal departure time from i is D_i' instead leaves at time D_i, the passenger's utility declines by $\varpi|D_i - D_i'|$, for some $\varpi > 0$. The term $|D_i - D_i'|$ is interpreted to be the difference in hours on a 24-hour clock between D_i and D_i'. For example, the difference between 21 and 1 is 4. Customers also suffer a reduction in utility as a function of travel duration, in the amount of ν monetary units per minute for the duration of the trip. Finally, passengers paying μ for traveling have their utility reduced by μ^2.

Assume that, for origin-destination pair (i,j), demand for train t_k by passengers having ideal departure time D_i' is given by a logit function (Dobson and Lederer, 1993). The demand for a train also depends on the degree of intermodal competition, in terms of the availability and the quality of alternative modes of transport. In order to keep into account these effects on the demand for train t_k, we introduce in the formulation a parameter χ, representing the probability of choosing an alternative transport mode.

Then, if D_i' is in the gth interval of β, demand for train t_k for customers with ideal departure time D_i' is:

$$\Phi_{t_k;(i,j)}\left(D_i'\right) = \frac{\beta_g(i,j)\exp\left[-\alpha\left(\varpi|D_i' - D_{it_k}| + \nu A_{jt_k} + p_{t_k;(i,j)}\right)\right]}{\sum\limits_{k'\in K}\sum\limits_{t_{k'}\in T_k(i,j)} \exp\left[-\alpha\left(\varpi|D_i' - D_{it_{k'}}| + \nu A_{jt_{k'}} + p_{t_{k'};(i,j)}\right)\right] + \chi} \tag{1.1}$$

where $T_k(i,j)$ is the set of trains of TO k that serve $O-D$ pair (i,j).

Demand for train t_k going from o to d can therefore be written as:

$$\begin{aligned}\Phi_{t_k} = & \sum_{j=o+1,\ldots,d} \Phi_{t_k;(o,j)}\left(D_o'\right) + \sum_{j=o+2,\ldots,d} \Phi_{t_k;(o+1,j)}\left(D_{o+1}'\right) + \ldots \\ & + \sum_{j=d-1,d} \Phi_{t_k;(d-2,j)}\left(D_{d-2}'\right) + \Phi_{t_k;(d-1,d)}\left(D_{d-1}'\right)\end{aligned} \tag{1.2}$$

This formulation, in order to be used for numerical simulations (see Section 1.6) has previously been tested on demand data provided by the Italian railway company FS. We were thus able to obtain estimates for all relevant parameters.

We also assume each train t_k has capacity M_{t_k}. This capacity constraint must be satisfied at each station, where passengers can get on or off the train. Of course, the TOs would try to determine the optimal capacity of each train in order to satisfy the expected demand. Therefore, trains' capacity constraints are not likely to be binding. Moreover, cars could be added to trains or taken out at each station, so that the capacity of the train could vary along its route; this would obviously not vary the basic form of the capacity constraint.

We thus obtain $(d-o)$ constraints of the form:

$$\sum_{j=o+1,\ldots,d} \Phi_{t_k;(o,j)}\left(D_o'\right) \leq M_{t_k} \text{ at station } o \tag{1.3}$$

$$\vdots$$

$$\sum_{j=r+1,\ldots,d} \left[\Phi_{t_k;(o,j)}\left(D_o'\right) + \cdots + \Phi_{t_k;(r,j)}\left(D_r'\right)\right] \leq M_{t_k} \text{ at station } r \tag{1.4}$$

$$\vdots$$

$$\sum_{i=o,\ldots,d-1} \Phi_{t_k;(i,d)}\left(D_i'\right) \leq M_{t_k} \text{ at station } d-1 \tag{1.5}$$

The above set of constraints may be written in a compact form by defining a matrix S_{t_k} with stopping stations on rows and $O-D$ pairs on columns, such that:

$$s_{t_k} = \begin{cases} 1 \text{ if } j > f \text{ and } i = o, \ldots, f \\ 0 \text{ else} \end{cases} \tag{1.6}$$

Using relationships (1.2) and (1.6) it is possible to maximize firms' profit with respect to final prices. The profit of TO k can be written as:

$$\Pi_k(p) = \left\{\sum_{t_k\in T_k}\sum_{(o,d)\in OD} \left[p_{t_k;(o,d)} \cdot \Phi_{t_k;(o,d)} - h_{t_k;(o,d)}\right]\right\} - C_k \tag{1.7}$$

where p is the vector of service prices. Thus, TO k's profit maximization problem is:

$$\begin{aligned} &\max_{p} \Pi_k(p) \\ &s.t. \;\; S_{t_k}\delta \leq M_{t_k} e \end{aligned} \tag{1.8}$$

where $\delta = \left(\Phi_{t_k;(o,d)} | (o,d) \in O - D\right)$ is a vector whose elements are the demand between each $O - D$ pair and $e = (1, \ldots, 1)^T$ is a $(d - o) \times 1$ (column) vector.

The equilibrium prices vector p_k^* is derived by solving the system of the first order conditions associated to the n profit maximization problems.

1.4 TRACK ACCESS AND CAPACITY PRICING

In the previous section, we illustrated how demand for a specific train is affected by the train's departure time and travel duration. These elements are determined by the IC by implementing the allocation and pricing mechanism to the requests of the TOs, who submit to the IC their ideal schedules - in terms of the desired departure and arrival times in each station of the trains they want to run - and a set of parameters representing the value (in monetary terms) attributed to the deviation of one unit of time from the requested schedule.

The mechanism computes actual schedules for trains and the corresponding tariffs to be paid for network use by maximizing the value of the effective timetable (i.e. the set of all train schedules) for each TO. This is obtained by minimizing the weighted deviation from each TO's requested timetable; the weights are the values the TOs attribute to the deviation from the desired schedule.

The mechanism is sequentially applied to rail lines in order of decreasing intensity of traffic.

The mechanism is non-discriminatory, since the IC maximizes the value of the schedule of each individual TO according to his requests. For this reason, the application of the mechanism on each line is analytically represented as a game whose set of players is composed of the n competing TOs.

1.4.1 Mechanism construction

The peculiar type of interdependence among the TOs is related to congestion-due interactions among trains traveling on the same network. In particular, the impossibility to assign the same track to more than one train, as well as the strong congestion-originated interdependence among interacting tracks make it impossible to determine the set of feasible schedules of a rail operator independently from the others. In other words, each operator cannot determine his own set of feasible schedules without knowing in advance the schedules of the others. A representation of congestion-related effects is therefore needed, together with an adequate analytical structure, keeping into proper account the competitive nature of the allocation and pricing process.

The first requirement is fulfilled by incorporating in the mechanism the elements and the hypotheses of a line model (Harker and Hong, 1990), which computes the delay a train undergoes as a result of its interactions with the other trains on a partially double track rail line, given data about train schedules and distributional information concerning operational uncertainties[3]. The expected running time of a train between stations i and j is computed as the sum of the train's free running time between the two stations and the expected delays due to casual interferences - meets and/or overtakes - with other trains traveling in the system. The priority levels in the delay function grow with the value attributed to the deviation from the requested schedule for the trains and have a direct impact on the expected running time of each train.

The nature of the interaction between transport operators is such that the set of effective feasible schedules of an operator cannot be determined independently from the others' timetable. Therefore, it is not possible to formulate the mechanism as a standard Nash game, as the latter does not allow for interactions among strategy sets. The adequate analytical structure, instead, is that of a *generalized* Nash game.

1.4.2 Problem formulation

The model requires IA_{jt_k}, ID_{it_k}, $w_{D_{it_k}}$, $w_{A_{jt_k}}$, DW_{it_k} as input data, and generates values for A_{jt_k}, D_{it_k}, $A_{jT_k\backslash t_k}$, $D_{iT_k\backslash t_k}$, $A_{jT_{K\backslash k}}$, $D_{iT_{K\backslash k}}$.

The expected running time $E(i,j)_{t_k}$ of a train $t_k \in T_k$ between stations i and j, calculated by the line model, can be written as:

$$E(i,j)_{t_k} = E(i,j)_{t_k}\left(A_{jt_k}, D_{it_k}, A_{jT_k\backslash t_k}, D_{iT_k\backslash t_k}, A_{jT_{K\backslash k}}, D_{iT_{K\backslash k}}\right)$$

When traveling in the railway network, each train t_k must satisfy a whole set of operational constraints. For our purposes, the relevant constraints are those related to the congestion degree in the system; we will thus simplify the analysis by considering two types of constraints.

First, if a train t_k arrives at station j, where passengers can get on or off the train, it must wait for a time at least equal to DW_{jt_k} before leaving at D_{jt_k}. We consider the minimum dwelling time as exogenously given. This type of constraint can then be written as:

$$D_{jt_k} - A_{jt_k} \geq DW_{jt_k} \;\; \forall j \in I_k, t_k \in T_k \tag{1.9}$$

The second type of constraints is the one we will mostly concentrate on, as these constraints are directly related to congestion levels on railway lines.

The arrival time of train t_k at station j must be greater than or equal to the departure time at station i plus the expected running time from i to j:

$$\begin{aligned} A_{jt_k} - D_{it_k} &\geq E(i,j)_{t_k}\left(A_{jt_k}, D_{it_k}, A_{jT_k\backslash t_k}, D_{iT_k\backslash t_k}, A_{jT_{K\backslash k}}, D_{iT_{K\backslash k}}\right) \quad (1.10) \\ \forall i,j &\in I_k, t_k \in T_k \end{aligned}$$

$E(i,j)_{t_k}$ is the most probable running time between stations i and j; therefore, the respect of condition (1.10) ensures the feasibility of the schedule of train t_k.

Let us define: $x_{K\backslash k} = (x_1, \ldots, x_{k-1}, x_{k+1}, \ldots, x_K)$. The set of feasible schedules for TO $k \in K$ is then defined by the set:

$$X_k\left(x_{K\backslash k}\right) = \left\{x_k = (A_k, D_k)^T \mid \text{ constraints (1.9) and (1.10) are satisfied}\right\}$$

The mechanism implemented by the IC, therefore, ensures the respect of the capacity constraints of the rail line as well as the constraints imposed on the travel of each train by the technical characteristics of rail transport.

The mechanism is first applied to the main line, in order to solve the related capacity allocation issues. Then, the problem that has to be solved for each TO k can be written as:

$$\begin{aligned}
\min \quad & u_k\left(x_k, x_{K\backslash k}\right) \\
& = \sum_{t_k \in T_k} \left\{ \sum_{j \in I_k} w_{A_{jt_k}} \left(A_{jt_k} - IA_{jt_k}\right)^2 + \sum_{i \in I_k} w_{D_{it_k}} \left(D_{it_k} - ID_{it_k}\right)^2 \right\} \quad (1.11) \\
s.t. \quad & x_k = \left(A_k, D_k\right)^T \in X_k\left(x_{K\backslash k}\right)
\end{aligned}$$

The first term of objective function (1.11) measures the weighted square of the difference between the requested and the effective arrival time, and the second term the weighted square of the difference between the requested and the effective departure time at each station. However, any convex function could be used as well.

When the mechanism is subsequently applied to branch lines, the TOs could request to be guaranteed a certain degree of coordination between the services they provide on the secondary lines and the trains running on the main line. Consider the case of a TO providing rail services on secondary ("branch") lines branching off from a junction station s situated on the main line. If allowed to do so, the TO would probably require that, for example, his trains leave from s immediately after the arrival of other trains traveling on the main line. In other words, each TO could typically request that the effective departure time of his train from junction station s be "not before" a certain ideal time. The ideal time would presumably be the arrival time in s of the services running on the main line; these times are known to the TOs as a result of the previous application of the mechanism to the main line. This type of requests, submitted by the TOs, can be expressed mathematically through constraints of the form:

$$D_{st_k} \geq ID_{st_k} \text{ (or, respectively, } IA_{st_k} \geq A_{st_k} \text{)} \quad (1.12)$$

Constraints of type (1.12) would add to those of type (1.9) and (1.10) in the formulation of the value maximization problem (1.11) of the schedule of each TO.

1.4.3 Problem solution and tariff derivation

For the determination of the effective timetables and tariffs we exploit some results of variational and quasivariational inequality theory.

In fact, in the presented mechanism we make use of a generalized Nash game formulation; the latter, in turn, can be rewritten as a quasivariational inequality (QVI). The solution set of this QVI contains the solution of a related variational inequality (VI), and in our special case every VI algorithm can be employed in order to calculate a solution to the model; thus, the existence of a QVI solution is established (Harker, 1991). However, the obtained equilibrium is only one out of a set of equilibrium solutions that might exist for the QVI related to the original problem.

In our case, it is therefore possible to find a VI whose solution gives the effective timetable for all the competing TOs. This is particularly expedient since solution algorithms for VIs are more numerous and efficient than those for QVIs.

Once the effective timetable has been obtained, the tariffs the IC should impose in order to ensure the respect of the capacity constraints can be derived from the generalized Nash game formulation. At the equilibrium solution, in fact, the mechanism produces the effective equilibrium arrival and departure times and the dual variables of constraints (1.9) and (1.10). The latter explicitly represents why it is necessary to turn to the generalized Nash equilibrium: TO k should know the schedules of the trains of the others, $T_K \backslash T_k$, in order to determine his own set of feasible schedules, as the expected running time represents the most likely travel time between i and j. The difference between the effective and the expected running time is called slack time. As slack time grows, the dual variable associated to constraint (1.10) decreases; this dual variable is the unit access tariff for the line segment under consideration.

Thus, the TOs request for each train the desired arrival and departure times, then the model produces effective arrival and departure times at each station and tariffs that ensure the respect of constraint (1.10).

In order to demonstrate this, consider that if we first assume that (1.10) holds and solve the resulting equilibrium problem, the solution we obtain will consequently satisfy constraint (1.10) and generate an associated dual price. If we now removed (1.10) from the minimization problem (1.11) of each TO, but we added to the payoff functions u the dual prices times the track occupancy times, we would find that no TO would have an incentive to change. The dual prices, therefore, are exactly the prices that the IC would need to charge the TOs to assure that constraint (1.10) holds in equilibrium.

1.5 SCHEDULE REQUESTS AND OPTIMALITY ISSUES

In this section we analyze some aspects of the model with respect to its solution procedure and final outcome.

It is worth noting that it is possible to determine the subgame perfect equilibrium only when in the first stage of the game the number of choice alternatives (in terms of timetable requests and associated values) for each TO is restricted. In fact, when in the first stage of the game the number of choice alternatives for each TO is very high, it becomes impossible to apply the 'backward induction', due to the fact that the allocation mechanism does not allow for an explicit functional relation to be derived between effective timetable and the tariffs on one side, and track requests on the other. Thus, the timetables and values to be submitted to the IC cannot be endogenously determined by the model and the first stage of the game must be substituted by the exogenous determination of the mentioned elements[4].

On the other hand, the IC could allow each TO to choose his schedule requests on each line from a limited pre-defined set of tracks. This set could be defined, for example, in order to ensure adequate service frequencies on the lines. In this case, it would be possible to enumerate all requests' combinations and apply to each the allocation and pricing mechanism in the second stage. In the third stage there would be a price subgame for each of the effective timetables and tariffs thus obtained. Using a 'backward induction' procedure, therefore, it would become possible to derive the subgame perfect equilibrium. If a sole subgame perfect equilibrium existed, the outcome of the game would be univocally determined. In the case of multiple equilibria, coordination problems among the TOs could arise, in the sense that the combination of the strategies chosen by each TO may not be an equilibrium. These problems, however, could be solved by the IC by opportunely choosing one equilibrium according to a pre-specified criterion.

First, he could select the solution that minimizes the equilibrium price of a particular type of service or the maximum value of the Nash equilibrium prices. Second, he could choose the solution that ensures the maximization of the sum of the profits obtained by the TOs or the equilibrium configuration that ensures the highest degree of coordination of the services on branch lines with those provided on the main line, since this is an important quality factor for the whole rail system.

Finally, the choice criterion could be to minimize the cumulated sum of the weighted deviations from the requested timetables for all the TOs, i.e.:

$$\min \sum_{k\in K} \sum_{t_k\in T_k} \left\{ \sum_{j\in I_k} w_{A_{jt_k}} \left(A^*_{jt_k} - IA_{jt_k}\right)^2 + \sum_{i\in I_k} w_{D_{it_k}} \left(D^*_{it_k} - ID_{it_k}\right)^2 \right\}$$

The latter choice mechanism ensures a repartition as uniform as possible of track demand over the whole rail network, since it is likely to avoid the track demand overlay that could occur in some of the equilibria.

1.6 NUMERICAL SIMULATIONS

The application of the model requires the usage of specific software tools for each of the stages of the model. We have chosen to employ MATLAB (version 4.2c.1), since it entails high-level programming and relatively easy understanding of the procedures. With respect to the QVI problem, the result of (Harker, 1991) shows that any algorithm for a variational inequality, such as the diagonalization algorithm we have chosen, can be employed for solving this special case of quasivariational inequality.

The model has been tested in some simple cases on the main Italian line from Milan to Rome. Demand and cost data have been obtained based on estimates by the Italian national railway company (FS). The constraints related to trains' capacity have been kept into account by assuming an average capacity of 350 passengers per train, according to FS' data on actual train capacity. In the following simulations, trains' capacity constraints are never binding.

We consider only four principal stations on the line, i.e. Rome, Florence, Bologna and Milan. However, the model easily allows to consider a higher number of stations. The simulation has been conducted in the time lag from 7 a.m. to 12 a.m., with a limited number of trains. In order to (partially) reproduce actual congestion levels in the system, we have assumed a partially double track line with one siding every 20 km, while trains' switching time has been set equal to five minutes. The presented results have been obtained on the line segment between Florence and Bologna.

We will compare three cases. The first corresponds to a reference situation (α); in the second case (β) a train is added to the system, and in the third (γ) the priority of one train is also increased. Thus, the latter is the most congested case.

The schedule requests of the TOs on this line segment and the results obtained by applying the allocation and pricing mechanism for case α are represented in the following table. Trains are indicated by a three-digit alphanumerical code where the first digit indicates the operator, the second the train, the third (apex) the case currently analyzed. Priority values and total tariffs for each train are expressed in Euros. The travel direction of each train is from Florence to Bologna if the arrow points right and viceversa.

TRAIN	DIR.	PRIORITY	REQUESTED TIMES		SLACK	TARIFF
$1a^{\alpha}$	$\longrightarrow$	0.6	ID_F 8:50	IA_B 9:40	+10	446
$1b^{\alpha}$	$\longleftarrow$	0.8	ID_B 11:20	IA_F 12:00	0	48
$1c^{\alpha}$	$\longleftarrow$	0.8	ID_B 8:15	IA_F 9:10	+15	0
$2a^{\alpha}$	$\longrightarrow$	0.8	ID_F 9:10	IA_B 10:00	+10	0
$2b^{\alpha}$	$\longleftarrow$	0.5	ID_B 8:15	IA_F 9:05	+10	79

Train 1a pays the highest tariff, although it has provided 10 minutes' slack and has a relatively low priority level. This is due to its high number of interactions with other trains (it interacts with 1c, 2a, 2b), related to the choice of arrival and departure times. This choice has a major effect in determining congestion levels on the track segment considered. In fact, train 2a, although it has been attributed a higher priority level (0.8 against 0.6) has a requested schedule enabling it to interact only marginally with train 1a, and thus pays a lower tariff.

The following table summarizes the results obtained by implementing the allocation and pricing mechanism for case β.

TRAIN	DIR.	PRIORITY	REQUESTED TIMES		SLACK	TARIFF
$1a^{\beta}$	$\longrightarrow$	0.6	ID_F 8:50	IA_B 9:40	+10	1173
$1b^{\beta}$	$\longleftarrow$	0.8	ID_B 11:20	IA_F 12:00	0	48
$1c^{\beta}$	$\longleftarrow$	0.8	ID_B 8:15	IA_F 9:10	+15	0
$2a^{\beta}$	$\longrightarrow$	0.8	ID_F 9:10	IA_B 10:00	+10	127
$2b^{\beta}$	$\longleftarrow$	0.5	ID_B 8:15	IA_F 9:05	+10	79
$2c^{\beta}$	$\longleftarrow$	0.9	ID_B 9:30	IA_F 10:30	+20	0

The overall level of tariffs has increased, since the system is more congested. However, the additional train pays no tariff: this is due to the fact that its schedule comprises slack times (20 minutes) large enough to compensate the congestion it will experience when entering the system. The tariffs of trains 1a and 2a grow considerably, as a result of increased traveling times due to interactions with the new train on the considered line segment.

The results obtained for case γ are represented as in the following table; notice that the priority level of train 1a has been increased.

TRAIN	DIR.	PRIORITY	REQUESTED TIMES		SLACK	TARIFF
$1a^{\gamma}$	$\longrightarrow$	1	ID_F 8:50	IA_B 9:40	+10	1956
$1b^{\gamma}$	$\longleftarrow$	0.8	ID_B 11:20	IA_F 12:00	0	48
$1c^{\gamma}$	$\longleftarrow$	0.8	ID_B 8:15	IA_F 9:10	+15	0
$2a^{\gamma}$	$\longrightarrow$	0.8	ID_F 9:10	IA_B 10:00	+10	127
$2b^{\gamma}$	$\longleftarrow$	0.5	ID_B 8:15	IA_F 9:05	+10	79
$2c^{\gamma}$	$\longleftarrow$	0.9	ID_B 9:30	IA_F 10:30	+20	0

Note that train $1a^{\gamma}$ pays a much higher tariff than in the cases previously examined, since it has a high priority level and there is more congestion in the track segment. Train 2c, although it has increased congestion in the system, has allowed for enough slack in its schedule, thus it pays no tariff. Train 1b, having alone requested a schedule at an uncongested time of the day, always operates in monopoly conditions. Although this train had no slack in its schedule, it pays the same access tariff in all cases.

As expected, tariffs grow with the congestion level; the percent increase is higher for train 1a, since it imposes the highest congestion level on the system.

Congestion levels also have an influence on competition, since both increase with the number of trains travelling on the rail line in the same time band. Changes in competitive pressure would obviously influence the level of service prices for the final users and profits for the operators.

The results obtained in the third stage for cases α and β are summarized in the following tables.

TRAIN	DIR.	PRIORITY	PRICE	PROFIT
$1a^\alpha$	$\longrightarrow$	0.6	12.86	356
$1b^\alpha$	$\longleftarrow$	0.8	14.49	1569
$1c^\alpha$	$\longleftarrow$	0.8	12.50	784
$2a^\alpha$	$\longrightarrow$	0.8	12.93	961
$2b^\alpha$	$\longleftarrow$	0.5	12.52	747

TRAIN	DIR.	PRIORITY	PRICE	PROFIT
$1a^\beta$	$\longrightarrow$	0.6	12.65	-783
$1b^\beta$	$\longleftarrow$	0.8	14.32	1416
$1c^\beta$	$\longleftarrow$	0.8	12.47	694
$2a^\beta$	$\longrightarrow$	0.8	12.98	900
$2b^\beta$	$\longleftarrow$	0.5	12.45	569
$2c^\beta$	$\longleftarrow$	0.9	12.23	8.26

Note that train 1b continues to operate in monopoly conditions and is not much affected by the increase in congestion; thus, it is able to impose higher prices and raise more profits. In general, however, consumers' prices are lower in case β than in case α, thanks to the increase in competitive pressure due to the additional train. Train $1a^\alpha$ even incurs losses due to lowering prices in order to compensate consumers for increased running times. Train $2c^\beta$ raises very low profits due to departing at a time distant to the customers' preferred one.

The introduction of train 2c has increased competition on the segment under consideration, in the traveling direction from Bologna to Florence. For example, the price for train 1c is lowered and its profit decreases; the same happens to train 2b.

The results obtained for case γ can be represented as follows:

TRAIN	DIR.	PRIORITY	PRICE	PROFIT
$1a^\gamma$	$\longrightarrow$	1	12.65	-1628
$1b^\gamma$	$\longleftarrow$	0.8	14.32	1413
$1c^\gamma$	$\longleftarrow$	0.8	12.47	694
$2a^\gamma$	$\longrightarrow$	0.8	12.98	900
$2b^\gamma$	$\longleftarrow$	0.5	12.45	569
$2c^\gamma$	$\longleftarrow$	0.9	12.23	8.26

Notice that increased congestion also influences the level of profits of trains traveling from Florence to Bologna. For example, the profits of train 1a start positive (case α), then they decrease and become negative (case β), and are further lowered in case γ. In fact, the increase in tariffs cannot be compensated by higher prices, since there is a competitor (train $2a^\gamma$) that can even offer a faster trip.

These examples illustrate our conjecture that a higher level of congestion generally implies a higher overall amount of tariffs, while raising competitive pressure forces operators to lower final prices. Taken together, these circumstances would result in an overall decrease of profits. Thus, the model seems to suggest that the system should converge toward an optimal number of trains providing rail services on the available infrastructure, fairly uniformly distributed on the whole network according to demand.

On the other hand, the amount of losses incurred by trains could be an indication of the need to further invest in infrastructures in order to provide additional capacity.

1.7 CONCLUSIONS

The new scenario emerging in the rail transport sector calls for the definition of a mechanism for track allocation among the TOs in competition and an access pricing system.

In this paper we propose a game-theoretic model where the TOs request tracks to the IC and set prices for the transport services they offer with the aim of maximizing their profit, considering demand data. The IC ensures the respect of the capacity constraints of the network, allocates capacity among the TOs and determines congestion-based tariffs for the use of the infrastructure through a mechanism whose intrinsic structure reflects the oligopolistic aspects of the track allocation and pricing problem. The sequential application of the mechanism to lines of decreasing intensity of traffic allows the IC to satisfy the need of coordinating services offered on different lines.

The mechanism ensures the respect of fairness and non-discrimination in the allocation of network capacity, in accordance with Directive 440, which is the basis for the reorganization of the sector.

The model is flexible with respect to market entry; if a new operator requests a track, it is relatively easy to calculate the effects on the operators already present in the system in terms of additional congestion, access tariffs, service prices and profits.

We have developed the model considering passenger transport; however, it can be extended to consider freight transport. First, an appropriate demand function should be employed (see e.g. Winston, 1983); moreover, in this case the dwelling time must be at least sufficient for inbound and outbound inspection, classification, assembly. In this case, a yard model could be incorporated in the allocation and pricing mechanism, so that the dwelling time would become endogenous; moreover, tariffs for yard use could also be derived.

Numerical simulations show that access tariffs grow with the congestion degree imposed on the network and illustrate the impact of congestion externalities on the level of prices for the final services and the profits of the transport operators. Congestion, in turn, depends on schedule choices in various respects.

First, requested schedules comprising large slack times generally resent less the congestion on the network, but also involve long running times, and this has a negative influence over customers' demand for the train.

Second, priority levels influence the proximity of the effective schedule to the requested one. However, raising a train's priority level entails a growth of access tariffs and does not necessarily imply substantially better schedules. Moreover, in the presence of a competitor on the same line, the increase in access tariffs can be only partially compensated by an increase in final prices due to demand considerations, and this could even results in lower profits. Therefore, the model suggests that rail operators have little incentive to misrepresent the value they attribute to schedules.

Another crucial factor is the time band where the requested schedule is located. Asking for a track at a less congested time could entail quasi-monopoly conditions and enable the operator to raise high profits. Thus, the model suggests that charging trains on the basis of congestion could favor efficient capacity use.

It should be noted that in our model negative profits could ensue both from a high level of congestion in the system (and thus from high access tariffs) or from very low demand (scarce revenues). In the former case, negative or very low profits could indicate the need to further invest in the infrastructure in order to provide additional capacity. In the latter case the amount of losses incurred by the operator, calculated by the model, could be an indication for the required level of subsidies to be bestowed on transport operators in order to ensure the financial viability of the considered services.

Notes

1. It is often assumed that there are at least two different groups of passengers with distinct travel-related costs: business and non-business passengers. In this case, the algorithm would need to calculate two different prices for each train. This would double the number of price variables and increase computational time, but not change the model formulation. The presence of two classes of passengers would thus change the optimal solution, but not the solution methodology.

2. Although we have specified that customers have preferences for departure times for trains, the model is easily recast to consider customers with preferences for arrival times. Also, customers preferences for departure (or arrival) time need not be specified by a symmetric function such as $|Di - D_i'|$. An asymmetric time metric where equally early and late departures have different utilities could be easily adopted.

3. A more general line model could be chosen, entailing a better representation of railway operations. For example, we could choose a model for the double track case, which most frequently occurs on European main lines. However, this impacts only the expected running time calculated, while it has no influence on the model's structure or performance. Moreover, assuming a partially double track rail line enables to visualize better the congestion effects which are the focus of our model.

4. This case is not overly restrictive. In fact, in real world operations, TOs will more likely determine their preferred schedules according to various available data, and the IC will apply the allocation and pricing mechanism to these (exogenous) track requests.

References

Ben Akiva, M. and S. Lerman. (1985). *Discrete Choice Analysis: Theory and Applicati on to Predict Travel Demand*, MIT Press, Cambridge, MA.

Bowers, P.H. (1996). Railway Reform in Germany. *Journal of Transport Economics and Policy*, 30:95–102.

Brooks, M. and K. Button. (1995). Separating Transport Track from Operations: a Typology of International Experiences. *International Journal of Transport Economics*, 22:235–260.

Carey, M. (1994). A Model and Strategy for Train Pathing with Choice of Lines, Platforms and Routes. *Transportation Research*, 28B:333–353.

Crew, M.A. and P.R. Kleindorfer. (1987). *The Economics of Public Utility Regulation*. The MIT Press, Cambridge, MA.

Dobson, G. and P.J. Lederer. (1993). Airline Scheduling and Routing in a Hub-and-Spoke System. *Transportation Science*, 27:281–297.

Harker, P.T. (1991). Generalized Nash Games and Quasi-Variational Inequalities. *European Journal of Operations Research*, 54:81–94.

Harker, P.T. and S. Hong. (1990). Two Moments Estimation of the Delay on a Partially Double-Track Rail Line with Scheduled Traffic. *Journal of Transportation Research Forum*, 31:38–49.

Harker, P.T. and S. Hong. (1994). Pricing of Track Time in Railroad Operations: an Internal Market Approach. *Transportation Research*, 28B:197–212.

Kraay, D.R. and P.T. Harker. (1995). Real-Time Scheduling of Freight Railroads. *Transportation Research*, 29B:213–229.

Levin, R.C. (1981). Railroad Rates, Profitability, and Welfare under Deregulation. *Bell Journal of Economics*, 12:1–26.

McFadden, D. (1981). Econometric Models of Probabilistic Choice, in C. Manski and D. McFadden (eds.) *Structural Analysis of Discrete Data with Econometric Applications*, MIT Press, Cambridge, MA.

Morrison, S.A. (1983). Prices and Investment Level for Airport Runways. In T.E. Keeler (ed.) *Research in Transportation Economics*, pp. 103–130.

Nash, C. (1993). Rail Privatisation in Britain. *Journal of Transport Economics and Policy*, 27:317–322.

Odijk, M.A. (1996). A Constraint Generation Algorithm for the Construction of Periodic Railway Timetables. *Transportation Research*, 30B:455–464.

Selten, R. (1975). Re-Examination of the Perfectness Concept for Equilibrium Points in Extensive Games. *International Journal of Game Theory*, 4:25–55.

Vickrey, W. (1969). Congestion Theory and Transportation Investment. *American Economic Review*, 59:251–260.

Winston, C. (1983). The Demand for Freight Transportation: Models and Applications. *Transportation Research* 17A:419–427.

2 REAL TIME SIMULATION OF TRAFFIC DEMAND-SUPPLY INTERACTIONS WITHIN DynaMIT

Moshe Ben-Akiva
Michel Bierlaire
Haris N. Koutsopoulos
Rabi Mishalani

Abstract: DynaMIT is a simulation-based real-time system designed to estimate the current state of a transportation network, predict future traffic conditions, and provide consistent and unbiased information to travelers. To perform these tasks, efficient simulators have been designed to explicitly capture the interactions between transportation demand and supply. The demand reflects both the OD flow patterns and the combination of all the individual decisions of travelers while the supply reflects the transportation network in terms of infrastructure, traffic flow and traffic control. This paper describes the design and specification of these simulators, and discusses their interactions.

2.1 INTRODUCTION

The main role of Dynamic Traffic Management Systems is to improve general traffic conditions using advanced technologies, managed by real-time intelligent software systems. There are two important functions of such systems. In the context of the economic interpretation of transportation, one of them affects transportation supply, while the other influences transportation demand.

Control systems, often called Advanced Traffic Management Systems (ATMS), impose restrictions and constraints on traffic flows. These systems include traffic signal (based on adaptive or pro-active rules), ramp metering, variable speed limit signs and lane use signs. In general, traffic regulations require drivers to comply with

M. Gendreau and P. Marcotte (eds.), Transportation and Network Analysis: Current Trends, 19–36.

these systems. By modifying the capacity of the network, ATMS affect transportation supply.

Information systems, or Advanced Traveler Information Systems (ATIS), provide traffic information and travel recommendations and guidance to drivers aimed at helping them make better decisions. These include radio forecast, web-based or on-board navigation systems and variable message signs. Such systems differ from ATMS in that drivers are not obligated to follow the recommendations of the system. ATIS actions, by influencing drivers' travel decisions, affect transportation demand.

An effective application of these systems must therefore be based on an implicit or explicit simulation of the interaction between demand and supply.

2.2 DynaMIT

DynaMIT is a simulation-based real-time system designed to estimate the current state of a transportation network, predict future traffic conditions, and provide consistent and unbiased information to travelers.

DynaMIT is designed to reside in a Traffic Management Center (TMC). It combines real-time data from a surveillance system (composed, for instance, of loop detectors, probe vehicles, incident detection systems, etc.) with historical data (collected and processed every day) in order to estimate the current state of the network, predict future traffic conditions and provide travel information and guidance through an ATIS.

This travel information is developed according to two main objectives: unbiasedness and consistency. Unbiasedness means that information provided to any traveler is based on the best knowledge of future network conditions that is available. Consistency means that the expected network conditions to be experienced by travelers coincide with the predicted conditions on which the information was based.

Clearly, input and model errors mean that DynaMIT is not a perfect reflection of reality. Within these limitations, if the properties of unbiasedness and consistency hold, then the best travel choice recommended by or inferred from information provided by DynaMIT will not be inferior to other choice alternatives as network conditions develop. This principle of optimality reflects *user optimal* information in the sense that this information is aimed at optimizing the travelers' utilities. The main user criteria considered are travel time and schedule delay, the absolute value of the difference between a traveler's desired and actual arrival time at the destination of interest.

We note that an ATMS, designed to constrain and control the traffic, is naturally driven by *system optimal* objectives, while an ATIS, designed to influence demand, is naturally driven by user optimal objectives. However, it is expected, and it has been shown through simulated scenarios, that the system performance of a transportation network will generally be improved if drivers are better informed and consequently make better decisions.

The overall structure of DynaMIT is illustrated in Figure 2.1. The upper box describes inputs required by DynaMIT. A database, combining a detailed description of the network with historical data (such as time dependent OD matrices, or link travel times), collected and processed off-line every day, is used as a reference representing the "average" or "usual" network state. This database may also contain some description of the driver population, including socioeconomic characteristics (such as

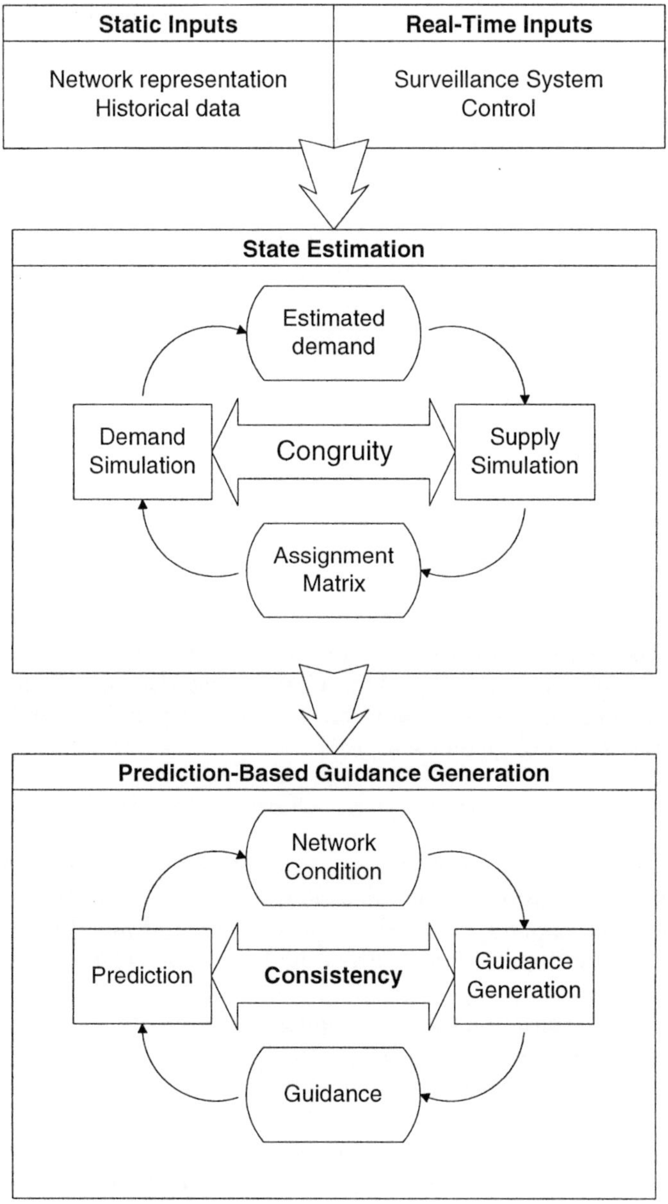

Figure 2.1 Overall structure of DynaMIT

age, gender, income, auto ownership, and trip purpose) obtained from census data and surveys.

Traffic counts from a surveillance system and the settings of the traffic control system (such as traffic signals, ramp meters, and toll plazas) are the source of real-time inputs to DynaMIT. These inputs describe the current conditions of the network.

Traffic counts serve as indirect measurements of the unknown origin-destination (OD) flows. The settings of the control system determine the capacities of the network.

The purpose of the *state estimation* process is to estimate demand levels and traffic conditions on the network given the set of inputs. It is achieved through an iterative simulation of demand-supply interaction designed to reproduce real-time observations from the surveillance system.

The role of the *prediction-based information generation* process is to generate unbiased and consistent traffic information for dissemination to travelers. Information based on predicted network conditions (i.e. anticipatory information) is likely to be more effective than information based on current traffic conditions because it accounts for the evolution of traffic conditions over time which is what travelers will experience.

Anticipatory information is derived from predictions of future conditions, but these conditions will be affected by travelers' reactions to the information they receive. An iterative process that involves demand, supply, and information generation has to take place in order for the system to generate anticipatory information that is unbiased and consistent. A single iteration consists of a trial information strategy, the state prediction (simulating explicitly the demand-supply interaction) under the trial strategy, and an evaluation of the predicted state with respect to unbiasedness and consistency.

In this paper, we describe the demand and supply representation within DynaMIT, and the way they interact for the purposes of estimation and prediction. We refer the reader to Ben-Akiva *et al.* (1998a) for a general description of DynaMIT, to Ben-Akiva *et al.* (1998b) for a description of its prediction capabilities, and to Ben-Akiva *et al.* (1997) for the route guidance aspects.

2.3 DEMAND

Individual travel demand decisions relate to choices of origin, destination, departure time, mode and route. Most of these decisions are made before the trip has begun. Therefore, the term *demand* will mostly refer to *pre-trip* demand in this paper. En-route demand, capturing decisions made while the trip is ongoing, will be explicitly referred to as such.

The demand representation must be compatible both with the available data and the modeling requirements. On the one hand, DynaMIT, as an information generator, must be able to anticipate the response of travelers to the information planned for dissemination (Ben-Akiva *et al.*, 1997). Therefore, a disaggregate representation of demand is required, where each individual is considered with her socio-economic characteristics and access to information. On the other hand, it is infeasible to collect data (historical or real-time) at the individual level for the entire traveling population. Even if some advanced data collection systems (like probe vehicles) may provide more disaggregate information, the availability of such comprehensive disaggregate data is unlikely in the foreseeable future. Therefore, in addition to the disaggregate demand representation, DynaMIT adopts time-dependent origin-destination matrices as a more aggregate representation.

As introduced in Section 2.2, DynaMIT relies on a description of the usual situation on the network and adjusts it in real-time to reproduce the prevailing network state or predict future conditions. Historical time-dependent origin-destination matrices are assumed to be available in that database. This assumption is reasonable, as OD estimation can be performed off-line.

The adjustments performed by DynaMIT to the historical demand capture two phenomena. First, DynaMIT uses disaggregate behavioral models to estimate and predict the response of drivers to real-time information they receive. These are discrete choice models capturing departure time change, mode choice, route choice and route switching (Ben-Akiva and Bierlaire, 1999).

Secondly, there are daily fluctuations that cannot be explicitly captured, but that must be considered. DynaMIT uses OD estimation and prediction models to adjust OD matrices based on real-time traffic counts in order to include these fluctuations.

The combination of these two different models is a key characteristic of DynaMIT's demand simulation. The methodology is described in Bierlaire *et al.* (2000).

In summary, the actual demand $\mathcal{D}$ is represented within DynaMIT as

$$\mathcal{D} = \mathcal{D}_{\text{Hist}} + \Delta\mathcal{D}_{\text{Info}} + \mathcal{D}_{\text{Fluct}} + \varepsilon \tag{2.1}$$

where $\mathcal{D}_{\text{Hist}}$ is the historical demand, $\Delta\mathcal{D}_{\text{Info}}$ represents the influence of the information on drivers behavior. Other factors, that are not directly observable, influence also the demand and cause daily fluctuations. These include modification in the activity pattern of many individuals due, for instance, to occurrence of special events, weather conditions, or personal reasons. These unexplained fluctuations are captured by the random variable $\mathcal{D}_{\text{Fluct}} + \varepsilon$, where the deterministic part $\mathcal{D}_{\text{Fluct}}$ represents mean daily fluctuations, and the error term ε is such that $E[\varepsilon] = 0$.

2.3.1 *Historical information*

The first step of the demand simulation is the disaggregation of the historical OD matrices into a historical population of drivers. Drivers are generated off-line and are stored in a database. Each driver is assigned a vector of socioeconomic characteristics generated by Monte-Carlo simulation based on their distributions within the actual population.

A habitual travel behavior is assigned to each driver. Origin, destination and habitual departure time are directly provided by the historical OD matrix. The habitual mode for all the individuals in the OD matrix is assumed to be private car. A route choice model based on historical travel times is used to determine the habitual route.

2.3.2 *Response to information*

The validity and relevance of DynaMIT's predictions and travel information are critically dependent on its ability to capture travelers' decisions, particularly in response to the information they receive. The demand simulator provides the framework for this procedure, relying on the behavioral models.

Modeling drivers' behavior in an ITS context is an active area of research. Therefore, DynaMIT has been designed to be flexible, and to incorporate new models as they become available. Moreover, behavioral models cannot be generic. They must be calibrated on specific data sets, to take into account specifics of both the population and the information system in place.

The role of these models is to adjust the departure time and the route of each driver receiving information. Also, some drivers may decide to cancel their trip or change transportation mode.

The current version of DynaMIT is mainly based on the models described in Ben-Akiva and Bierlaire (1999), calibrated on synthetic data. These models are used for laboratory evaluation of the system.

Drivers' behavior is categorized based on the information type and whether the choice takes place pre-trip or en-route. The information types are no information (background vehicles), descriptive information, and prescriptive information. If no information is available, then response to guidance behavioral models are not needed, since the drivers will not revisit their habitual travel pattern. Only habitual models are required. In the case of descriptive and prescriptive information, however, both habitual and response to guidance models are necessary. When information is provided, travel decision timings can be classified into two categories: pre-trip and en-route. The habitual decisions in term of departure time and mode choice are available from the historical database. A PS-logit model (Ben-Akiva and Bierlaire, 1999) is used to provide probabilities for selecting each path alternative. The pre-trip decisions based on prescriptive information are captured by a compliance model. The pre-trip decisions based on descriptive information are captured by a choice model, where all combinations of mode, departure time interval and path are included in the choice set. The correlation structure of the choice set can be handled by a nested logit or a probit model. The en-route models share the same structure as the associated pre-trip models, but do not include mode and departure time choices.

2.3.3 OD estimation and prediction

The formulation of the real-time dynamic OD matrix estimation problem is based on a Kalman Filtering framework formulated by Ashok and Ben-Akiva (1993). The basic idea of this approach is to use all the information contained in historical OD data in conjunction with data on traffic counts to generate OD estimates in real-time. Further, each day's estimate is used to update the original historical OD data and hence a learning process ensues. The historical OD matrices that are used in this procedure have already been updated.

Unlike other approaches, this method is based on deviations from historical values. To illustrate how it operates, consider historical OD matrices updated with the previous days' estimates. These OD matrices subsume a wealth of information about the latent factors that affect travel demand and their variations over the course of the day. Such factors are incorporated by including all the prior OD matrix estimations into the real time OD estimation problem. This is achieved by the use of deviations of OD flows from the historical estimates, instead of the actual flows themselves as state variables in the Kalman Filtering methodology. Thus the estimation process indirectly takes into account all the experience gained over many prior estimations and is richer in its structural content. An important input for the Kalman Filtering procedure is the assignment matrix which describes the mapping between OD flows and link flows. The assignment matrix is provided by the supply simulator, as described subsequently.

The basic problem of OD prediction is to compute, in real-time, estimates of future OD flows from the current OD estimates and historical OD flows. The autoregressive process used by the Kalman filtering approach provides a prediction tool, with real-time capabilities, that is consistent with the estimation process. An autoregressive process models the temporal relationship among deviations in OD flows. Unobserved factors that are correlated over time (like weather conditions, unusual events, etc.) give rise to correlation of deviations over time which are reflected by the autoregressive process. More specifically, the autoregressive process is characterized by a set of coefficients describing the effect of the deviations during one time interval on the deviation during another subsequent time interval. These coefficients are determined off-line, using a linear regression model for each OD pair and for each time interval. Predicted deviations are therefore obtained by applying this autoregressive model to the deviations estimated for the current time interval.

The transition equation. Consider a network with n_{OD} OD pairs and n_l link counts. Let us assume the following notation:

- x_h is the vector representing the number of vehicles between each OD pair departing their origins during time interval h and
- x_h^H is the corresponding historical estimate.

In matrix form the transition equation can be expressed as:

$$x_{h+1} - x_{h+1}^H = \sum_{p=h-q'}^{h} f_h^p (x_p - x_p^H) + w_h \tag{2.2}$$

where:

- f_h^p is an $(n_{OD} \times n_{OD})$ matrix of effects of $(x_p - x_p^H)$ on $(x_{h+1} - x_{h+1}^H)$,
- w_h is an $(n_{OD} \times 1)$ vector of random errors, and
- q' is the degree of the autoregressive process, that is the number of past time intervals influencing the current one.

The following assumptions are made about the error vectors:

1. $E[w_h] = 0$,
2. $E[w_h w_l'] = Q_h \delta_{hl}$,

where

- $\delta_{hl} = 1$ if $h = l$, and 0 otherwise, for all h, l, and
- Q_h is an $(n_{OD} \times n_{OD})$ variance-covariance matrix.

The second assumption implies that there is no serial correlation. This is justified because the unobserved factors that could be correlated over time are captured by the historical matrix x_{h+1}^H.

The model in the above form is highly general and assumes dependence of deviations corresponding to one OD pair on deviations corresponding to other OD pairs in prior periods. In practical application this is unnecessarily general and relationships between deviations across different OD pairs may be safely ignored. This simplification is adopted in our implementation.

The measurement equation. Let y_h be a vector representing the link counts measured during interval h. In matrix form the measurement equation, which relates unknown OD flows to the observed link counts, can be stated as follows:

$$y_h - y_h^H = \sum_{p=h-p'}^{h} a_h^p (x_p - x_p^H) + v_h, \tag{2.3}$$

where

- a_h^p is an $(n_l \times n_{OD})$ "assignment" matrix of contributions of x_p to y_h,
- v_h is the vector of measurement errors, and
- p' is the number of time intervals corresponding to the longest trip,
- y_h^H are the link flows obtained by assigning the historical OD flows,

between any O-D pair of the network.

We assume that

1. $E[v_h] = 0$,
2. $E[w_h v_l'] = 0$, $\forall h, l$, i.e. transition and measurement errors are uncorrelated,
3. $E[v_h v_l'] = R_h \delta_{hl}$,

where

- $\delta_{hl} = 1$ if $h = 1$ and 0 otherwise, $\forall h, l$, and
- R_h is an $(n_l \times n_l)$ variance-covariance matrix.

Algorithm. The input to the algorithm comprises from the following matrices (Recall that n_l is the number of links and n_{OD} the number of OD pairs):

- Assignment matrix: a_h^p. There are $p' + 1$ assignment matrices, with dimension $(n_l \times n_{OD})$, per interval h, which are very sparse.
- Historical OD matrix of vehicles departing in interval h,
- Link flows observed in interval h,
- Transition matrices: f_h. There is one transition matrix, with dimension $(n_{OD} \times n_{OD})$, corresponding to each value of $h - q$. These matrices are independent of the actual interval and depend only on $h - q$. Therefore they do not change over time.
- Variance/covariance matrices: Q_h and R_h. There is one pair of these matrices for each time interval h. The dimension of Q_h is $(n_{OD} \times n_{OD})$, while the dimension of R_h is $(n_l \times n_l)$.
- The initial state: x_0, and its variance: $\text{Var}(x_0 - x_0^H)$.

The solution is given by a standard Kalman Filter. We make the following notational simplifications:

$$\begin{aligned} \Phi_h &= f_h^h \\ A_h &= a_h^h \\ Y_h &= y_h - \textstyle\sum_{p=h-p'}^{h-1} a_h^p x_p - a_h^h x_h^H \\ X_h &= x_h - x_h^H \end{aligned} \tag{2.4}$$

Also, the notation $i|j$ indicates an estimate corresponding to interval i based on observations up to and including interval j. Finally, all variables with a hat correspond to an estimate of a random variable. The estimation algorithm proceeds as follows

(i) Initialization Compute

$$\hat{X}_{0|0} = E(X_0) \tag{2.5}$$

and

$$\Sigma_{0|0} = (\text{Var}(X_0))^c \tag{2.6}$$

where superscript c denotes the Cholesky factor. Then, for each time interval h=1,2,3,...

(ii) Variance propagation of the filter

$$\Sigma_{h|h-1} = \left[\Phi_{h-1}\Sigma_{h-1|h-1} \;\; Q_{h-1}^c\right]_{n\times(n+p)}, \tag{2.7}$$

$$H_h = (A_h\Sigma_{h|h-1}\Sigma_{h|h-1}^T A_h^T + R_h)^c, \tag{2.8}$$

and,

$$\Sigma_{h|h} = \Sigma_{h|h-1}\left(I - \Sigma_{h|h-1}^T A^T (H_h^T)^{-1}(H_h + R_h^c)^{-1} A_h \Sigma_{h|h-1}\right). \tag{2.9}$$

(iii) Calculation of the Kalman gain

$$G_h = \Sigma_{h|h-1}\Sigma_{h|h-1}^T A_h^T (H_h^T)^{-1} H_h^{-1}. \tag{2.10}$$

(iv) Measurement update

$$\hat{X}_{h|h} = \hat{X}_{h|h-1} + G_h(Y_h - A_h\hat{X}_{h|h-1}). \tag{2.11}$$

(v) One-step prediction

$$\hat{X}_{h+1|h} = \Phi\hat{X}_{h|h}. \tag{2.12}$$

2.4 SUPPLY

By *supply*, we refer to the transportation network and the ATMS. The network representation consists of static and dynamic components. The static components represent the topology of the network. The dynamic components are designed to capture the traffic dynamics. While the characteristics of the former are fixed during the simulation, the dynamic components can be continuously updated.

The network is divided into links, nodes and loading elements. The nodes correspond to intersections of the actual network, while links represent uni-directional pathways between them. Loading elements represent areas where traffic is generated or attracted. They can be either nodes or loading links and, therefore, they are a generalization of the zone centroid nodes of traditional models.

Each link is divided into *segments* that capture variations of geometry and traffic conditions along the link. While most segments are defined in advance, additional segments can be dynamically created to capture the presence of incidents. Each segment has a capacity constraint at the downstream end, referred to as the *output capacity*. Depending on the nature of the segment, this capacity constraint can be due to the physical characteristics of the road, an incident or a specific control device. Each segment is divided into a moving part and a queuing part. The moving part corresponds to the section of the segment where vehicles can move with some speed. Queuing parts represent vehicles that are queued up. Vehicles are assigned to lanes according to their path leading to their destination. The determination of output capacities is based on recommendations from the Highway Capacity Manual (HCM, 1994). A spill-back occurs on a segment when the downstream segment is blocked. That is, when the length of the queue on this downstream segment is equal to the segment length.

2.4.1 Queuing

The queuing model is illustrated in Figure 2.2. It is based on the delay of the ith vehicle in the queue, given by

$$\frac{i}{c} \tag{2.13}$$

where c is the output capacity of the segment. During a period of time t, there are ct vehicles leaving the queue. If a moving vehicle reaches the end of the queue at time t, its position and, therefore, the position of the end of the queue, will be

$$q(t) = q_0 + l(ct - m) \tag{2.14}$$

where q_0 is the position of the end of the queue at time t=0, l is the average length of vehicles (including headways), c is the output capacity and m the number of moving vehicles between the considered vehicle and the end of the queue at time t=0. That is, m is the number of vehicles that reach the queue before the considered vehicle did. Without loss of generality, it is assumed that the position of the upstream end of the segment is 0.

Note that the model is relevant only when $0 < q(t) < L$, where L is the link's length. The case $q(t) < 0$ will never occur because spill-backs are explicitly modeled. If $q(t) \geq L$, it means that the queue is already dissipated when the vehicles reaches the end of the segment. The queuing model is therefore irrelevant in this context.

2.4.2 Moving

The speed model is based on the assumption that speed is constant on the upstream section of the segment, followed by a deceleration zone covering a downstream section, where the speed of vehicles varies linearly as a function of the position. Figure 2.3 illustrates this model, where v_u is the speed at the upstream end of the segment, v_d is the speed at the downstream end of the segment and L_s is the length of the deceleration zone. L_s depends on the geometry of the segment and traffic conditions.

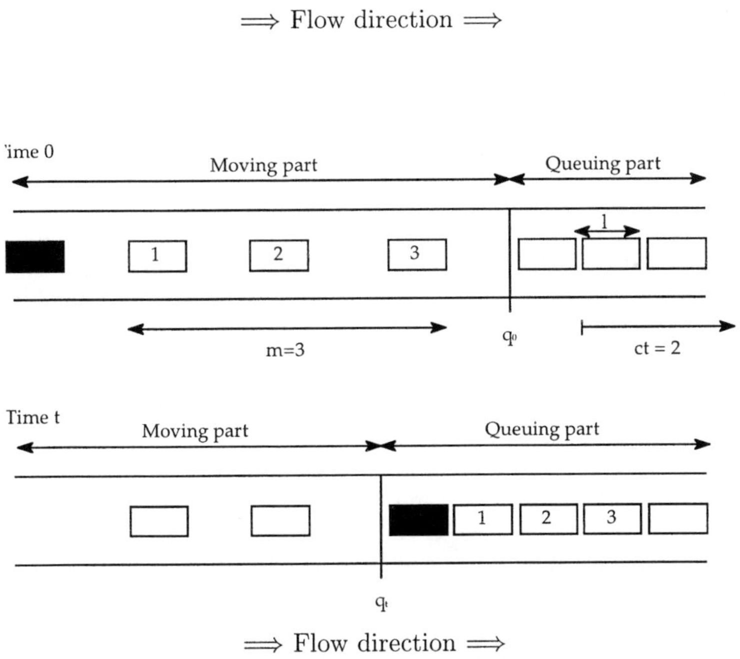

Figure 2.2 Deterministic queuing model

Experiments show that long deceleration zones (in the case of pure deceleration) result in unnecessarily high travel times. The default value in the model is set to 0. This is consistent with empirical evidence indicating that the majority of delays are related to queuing.

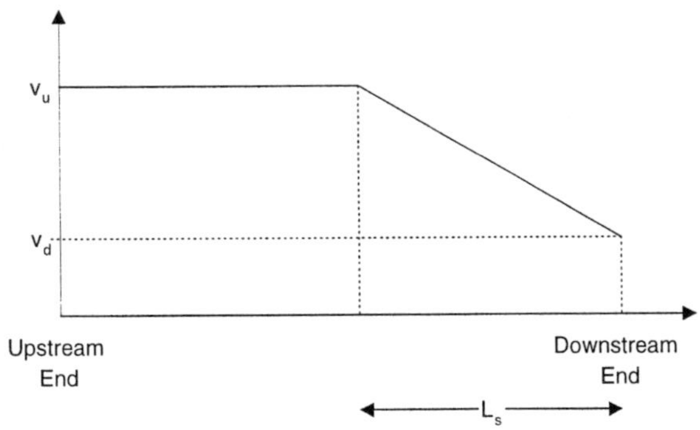

Figure 2.3 Speed profile

Again, without loss of generality, we assume that the position of the upstream end of the segment is 0. Therefore, the downstream end is at position L, where L is the length of the segment. The speed function can then be written as

$$v(x) = \begin{cases} v_u & \text{if} \quad 0 \leq x \leq L - L_s \\ \lambda(x - L) + v_d & \text{if} \quad L - L_s \leq x \leq L \end{cases} \tag{2.15}$$

where

$$\lambda = \frac{v_d - v_u}{L_s}. \tag{2.16}$$

The speed v_u at the upstream end of a segment is a function of the average density on the moving part of the segment. An example of the relationship between the density and the speed is given by

$$v_u = v_0 \left[1 - \left(\frac{K}{K_{\text{jam}}} \right)^{\beta} \right]^{\alpha}, \tag{2.17}$$

where v_0 is the free-flow speed on the segment, K is the density, K_{jam} is the jam density, and α and β are parameters.

When there is no queue on the segment, the speed at the downstream end is affected by the traffic conditions on the downstream segment. Therefore, this quantity is lane-specific. In the presence of a queue, the speed at the downstream end is dictated by the queue dissipation rate, that is, by the output capacity.

2.4.3 *Vehicles Movements*

In the moving part of the segment, vehicle movements are driven by the speed model. The part of the segment with constant speed is straightforward, vehicles being advanced based on that speed. As for the deceleration zone, in the absence of a queue, the vehicle will reach position x at time $t(x)$, assuming that at time $t = 0$ it is at position x_0, given by

$$t(x) = \begin{cases} \frac{1}{\lambda} \log \frac{\lambda x + v_u}{\lambda x_0 + v_u} & \text{if } v_u \neq v_d, \\ \frac{x - x_0}{v_u} & \text{if } v_u = v_d, \end{cases} \tag{2.18}$$

where λ is defined by Equation (2.16).

Assuming that at time t=0, a vehicle is at position x_0, its position at any time t is given by

$$x(t) = \begin{cases} e^{\lambda t}(x_0 + \frac{v_u}{\lambda}) - \frac{v_u}{\lambda} & \text{if } v_u \neq v_d, \\ v_u t + x_0 & \text{if } v_u = v_d. \end{cases} \tag{2.19}$$

where λ is defined by Equation (2.16).

In the presence of a queue the position $x(t)$ of a vehicle at any time t is given by

$$x(t) = e^{\lambda(t)t} \left(x_0 + \frac{v_u}{\lambda(t)} \right) - \frac{v_u}{\lambda(t)} \tag{2.20}$$

where

$$\lambda(t) = \frac{-v_u}{q_0 + l(ct - m)}, \tag{2.21}$$

and q_0, l, c and m are defined as in Equation (2.14).

This model is too complex to be operational. Therefore, we use an approximation: we assume that the model applies only up to a certain time t^*, at which point we consider that the vehicle reaches the queue. After t^*, the pure (linear) queuing model is applied. The value of t^* is given by

$$t^* = \frac{(ml - q_0)\log(e^{-v_u/cl} + \varepsilon)}{v_u + cl\log(e^{-v_u/cl} + \varepsilon)} \tag{2.22}$$

where ε is an approximation parameter. Figure 2.4 shows an example of this approximation for $\varepsilon = 0.1$ and $\varepsilon = 0.001$, where $x(t)$ is the speed model, and $q(t)$ the queuing model.

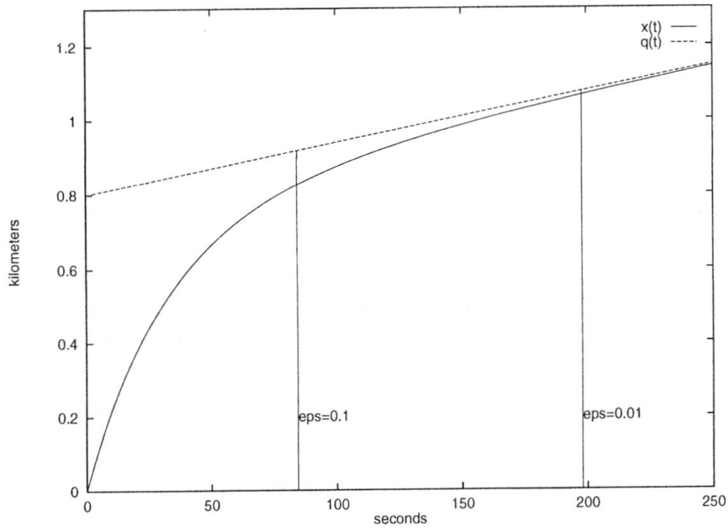

Figure 2.4 Position at time t in the presence of a queue: approximation

Note that if ε goes to 0, t^* goes to infinity. In most cases, this formula can be simplified. When v_u/cl is large enough (for example, greater than 10), we have

$$t^* = \frac{(ml - q_0)\log(\varepsilon)}{v_u + cl\log(\varepsilon)}. \tag{2.23}$$

The position of a vehicle at time t is

$$x(t) = \begin{cases} e^{\lambda(t)t}(x_0 + \frac{v_u}{\lambda(t)}) - \frac{v_u}{\lambda(t)} & \text{if } t < t^*, \\ q_0 + l(ct - m) & \text{if } t \geq t^* \end{cases} \tag{2.24}$$

The vehicle reaches the end of the segment at time:

$$t = \frac{\frac{L-q_0}{l} + m}{c}. \tag{2.25}$$

In the case where $t < t*$, the vehicle reaches the end of the segment before reaching the end of the queue, meaning that the queue is already dissipated.

2.5 DEMAND-SUPPLY INTERACTION

As introduced in Section 2.2, the role of the estimation subsystem is to reproduce a full description of the current state of the network which matches as closely as possible the available real-time data. Designed for real-time operations, DynaMIT cannot calibrate all parameters online. Instead, DynaMIT focuses on the estimation of OD matrices using real-time data.

To perform the state estimation, DynaMIT uses

- historical OD matrices,
- the travel time information and guidance actually provided to the drivers,
- real-time traffic counts from the surveillance system.

As presented in Section 2.3, disaggregate behavioral models are combined with an aggregate OD estimation algorithm to obtain a full description of the pre-trip demand at the individual traveler level for the estimation time interval. That demand is loaded on the network by the supply simulator. Behavioral models are also applied for en-route demand simulation in order to capture the impact of information on drivers already on the network. The resulting network state is then compared to the observed data from the surveillance system.

Unfortunately, there is a missing link in this description. The OD estimation algorithm requires assignment matrices which, as mentioned in Section 2.3, represent the mapping of OD flows into link flows. These matrices can only be obtained from the supply simulator, as they describe the way OD flows have been distributed over the network on the various paths connecting origins with destinations.

Therefore, a fixed point problem is solved, as illustrated in Figure 2.5. The objective here is to find a description of the demand which is congruent with the link flows reflecting the supply. The solution of this fixed point problem is reached through an iterative process where a new assignment matrix is computed at each iteration.

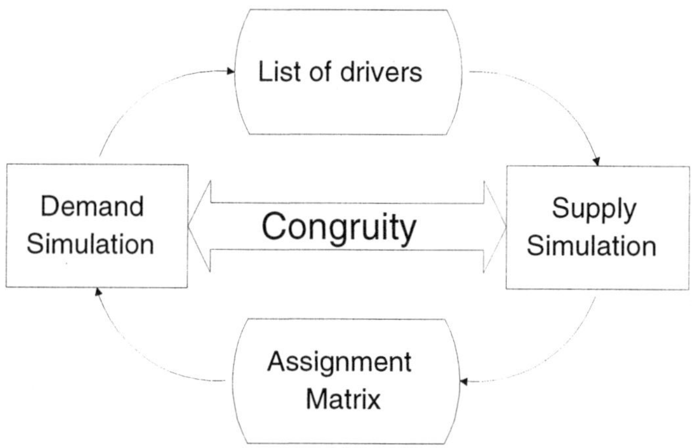

Figure 2.5 Fixed point problem for estimation

As introduced in Section 2.2, the prediction-based information generation module of DynaMIT also consists of a fixed point problem. But its nature is different from the state estimation. Here, the objective is not to match observations (as they are obviously not available), but to generate traveler information and guidance which is consistent with the traffic predictions. This is illustrated in Figure 2.6.

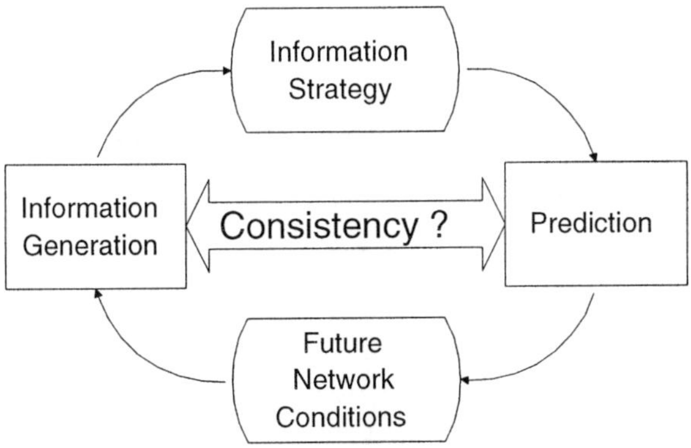

Figure 2.6 Fixed point problem for prediction

The network state most recently estimated is used as a starting point for prediction. Moreover, the most recently disseminated information and guidance is used as a trial strategy. The demand and supply simulators are then used almost in the same way for prediction as they are for estimation. The key difference is for OD matrices. In order to forecast future OD matrices, DynaMIT relies on an autoregressive process (calibrated off-line).

Based on the resulting prediction, a revised travel-time-based information strategy is generated using a combination of the original and new predictions (see Bottom *et al.* (1999) for more details). This revised information strategy is in turn used by the coupled demand and supply simulators to produce a new prediction. Iterations continue until convergence where the final prediction is very similar to the previous prediction the travelers are assumed to react to. This final prediction is the one disseminated to travelers.

2.6 EVALUATION

Computer simulation-based laboratory evaluation is adopted for assessing the performance of DynaMIT prior to field testing (see Ben-Akiva *et al.*, 1996). The laboratory set up used for evaluation of DynaMIT is illustrated in Figure 2.7. It consists of two interacting parts, one of which is DynaMIT, shown in the right box of the diagram. The other part, shown in the left box, is designed to serve as a full representation of the "true" real world and consists of MITSIM (Yang and Koutsopoulos, 1997) and a travel demand generator. The "true" demand is obtained from the habitual travel demand, combined with the simulation of drivers' response to pre-trip information. That true demand is then loaded on the network using MITSIM, the microscopic traffic simulator. The interface between the two parts of the laboratory is constrained in order to represent the role that a true traffic management center would play in the real world. As shown in the middle of the diagram, this interaction takes place at two points: the surveillance system and ATIS. The surveillance system consists of a set of sensors placed within the MITSIM network. These sensors are able to send various types of data to DynaMIT, such as traffic counts and speeds. Such data allow DynaMIT to estimate the current state of the network and, therefore, have a good starting point for the prediction task.

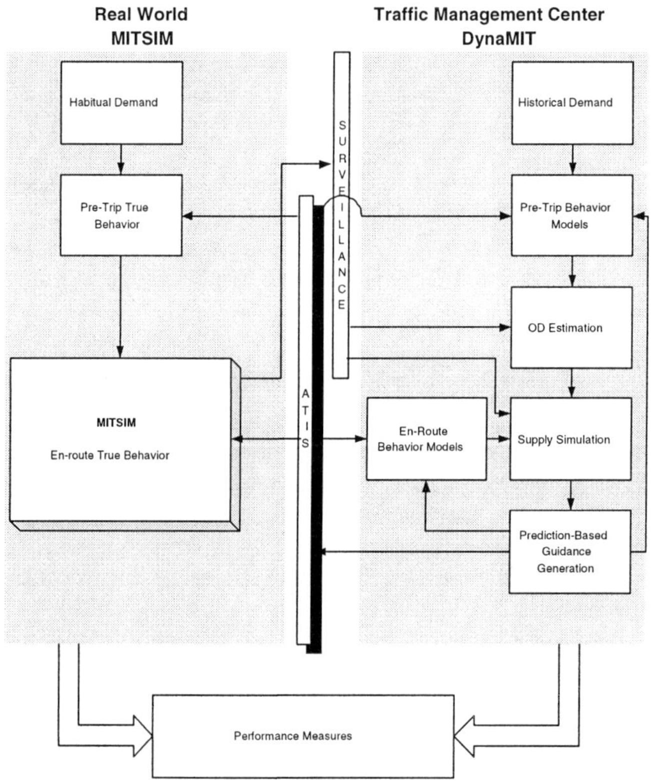

Figure 2.7 MITSIM Laboratory environment

ATIS takes the prediction-based information produced by DynaMIT, and distributes that information to real-world drivers represented in MITSIM at both the pre-trip and en-route stages. The distribution could take place through radio broadcasts, variable message signs within the network and in-vehicle devices. Information and guidance may change the decisions that drivers make regarding departure time, transportation mode and route choice, improving individual utilities and potentially improving the overall traffic flow in the network. Each driver in the real world is given a set of unique behavioral characteristics such as desired travel speed and aggressiveness, as shown in the left box of the diagram. In addition, drivers differ in terms of their access to ATIS and their response to information. Some drivers will elect to ignore any guidance instructions, while others will be more responsive depending on their own current and previous observations of the network. DynaMIT is also explicitly simulating these effects, and a parallel can be drawn between the structures of the two boxes in the diagram. The behavior models in DynaMIT are approximations of the true behavior represented in MITSIM, and several scenarios are considered to test the sensitivity of the system to the error in representing the true behavior. A field test will follow.

2.7 CONCLUSION

DynaMIT is a simulation-based system designed to perform, in real time, estimation of the current state of the network, prediction of future traffic conditions, and generation of guidance strategies. The generated guidance is designed to maximize travelers' utilities.

The main components of DynaMIT include a demand simulator, a supply simulator, and algorithms that capture the interactions between supply and demand in the presence of information. The demand simulator models all travelers' decisions in the presence of information (pre-trip and en-route) and estimates and predicts O-D flows. The supply simulator is based on a mesoscopic representation of the network and speed-density and queuing models to capture traffic dynamics and interactions.

DynaMIT is currently undergoing extensive testing and evaluation using the simulation-based laboratory MITSIMLab.

References

Ashok, K. and M. Ben-Akiva. (1993). Dynamic origin-destination matrix estimation and prediction for real-time traffic management systems. In Daganzo, C., editor, *Transportation and Traffic Theory.* Elsevier Science Publishing Company Inc. Proceedings of the 12th ISTTT.

Ben-Akiva, M. and M. Bierlaire. (1999). Discrete choice methods and their applications to short-term travel decisions. In Hall, R., editor, *Handbook of Transportation Science.* Kluwer.

Ben-Akiva, M., M. Bierlaire, J. Bottom, H.N. Koutsopoulos, and R.G. Mishalani. (1997). Development of a route guidance generation system for real-time application. In *Proceedings of the 8th IFAC Symposium on Transportation Systems*, Chania, Greece.

Ben-Akiva, M., H.N. Koutsopoulos, R. Mishalani, and Q. Yang. (1996). Simulation laboratory for evaluating dynamic traffic management systems. *Journal of Transportation Engineering.*

Ben-Akiva, M.E., M. Bierlaire, H.N. Koutsopoulos, and R.G. Mishalani. (1998a). DynaMIT: a simulation-based system for traffic prediction and guidance generation. Paper presented at TRISTAN III, San Juan, Porto Rico.

Ben-Akiva, M.E., M. Bierlaire, H.N. Koutsopoulos, and R.G. Mishalani. (1998b). DynaMIT: a simulation-based system for traffic prediction. In *Proceedings of the DACCORD Short-Term forecasting workshop.*

Bierlaire, M., R. Mishalani, and M. Ben-Akiva. (2000). General framework for dynamic demand simulation. Technical Report RO-000223, ROSO-DMA-EPFL Swiss Institute of Technology, CH-1015 Lausanne.

Bottom, J., M. Ben-Akiva, M. Bierlaire, I. Chabini, H. Koutsopoulos, and Q. Yang. (1999). Investigation of route guidance generation issues by simulation with DynaMIT. In Ceder, A., editor, *Transportation and Traffic Theory. Proceedings of the 14th ISTTT*, pages 577–600. Pergamon.

HCM (1994). Highway capacity manual. Transportation Research Board, Special Report 209.

Yang, Q. and H. Koutsopoulos. (1997). A microscopic traffic simulator for evaluation of dynamic traffic management systems. *Transportation Research C*, 4(3):113–129.

3 IMPLICATIONS OF MARGINAL COST ROAD PRICING FOR URBAN TRAVEL CHOICES AND USER BENEFITS

David Boyce
Karthik Balasubramaniam
Xin Tian

Abstract: A model that combines origin-destination, mode and route choice into a single, consistent formulation and solution method is used to analyze the effects of marginal cost pricing of roads in the Chicago region. The solution of the system-optimal version of the model corresponds to minimizing the total cost of travel, subject to behavioral and definitional constraints. These results are compared to the standard user-optimal version of the same model and interpreted in graphs showing the effects of variations in total travel demand during the morning peak period, as well as the effect of alternative road pricing strategies.

3.1 INTRODUCTION

Interest in road pricing as a means of suppressing travel demand during congested periods of the day continues to grow throughout the world. The notion of charging each traveler for the external costs imposed on others through their use of the roadway originated 80 years ago (Pigou, 1920) and has been championed by economists ever since (e.g. Downs, 1992). Despite extensive analytical investigations and descriptive explanations, there are few, if any, large-scale studies of the effects of marginal cost road pricing on urban travel choices. In this chapter, we offer such a study based on our *combined travel choice model* of the Chicago region. In this combined model, choices of origin-destination, mode and auto route are represented in an integrated manner.

M. Gendreau and P. Marcotte (eds.), Transportation and Network Analysis: Current Trends, 37–48.

The application of marginal cost road pricing is equivalent to solving a network equilibrium model in which the objective is to minimize the total cost of travel, subject to constraints necessary to make the solution behaviorally realistic and definitionally correct. This formulation corresponds to the system-optimal (second) principle of Wardrop (1952). By comparing the predicted travel choices from this model with the standard user-optimal, variable demand model, we examine the effects of road pricing on origin-destination and mode choices, as well as route choices. Using a measure of user benefits proposed by Williams (1976), we then estimate the total benefits of road pricing. The study is performed on an aggregated network of the Chicago region known as the Sketch Planning Network.

The chapter is organized as follows. The next section briefly describes the system-optimal and user-optimal combined models, and their relation to marginal cost road pricing. The data and parameter values used in the study are also described. The effects of road pricing are investigated in two ways. First, a sensitivity analysis on total demand is performed. The response of the model to variations of ± 10% in total demand are compared to results from a similar sensitivity analysis on the generalized cost parameter pertaining to auto monetary costs. Then, the system-optimal solution is compared to other simpler pricing strategies. A discussion of the findings and some thoughts about future research conclude the paper.

3.2 THE COMBINED TRAVEL CHOICE MODEL AND ITS IMPLEMENTATION

The combined origin-destination, mode and user-optimal (UO) route choice model applied in this study stems from the formulation of Evans (1976), and is a specific version of the variable demand, network equilibrium model originally proposed by Beckmann et al. (1956). A detailed description of this approach may be found in Boyce and Daskin (1997). In the specific implementation used here, the generalized travel costs by auto are defined as a weighted sum of auto travel time and operating costs, including tolls, parking fees and fuel consumption. Auto travel times on road network links are increasing functions of each link's own flow. Auto fuel consumption on a link is a weighted function of the link's length and travel time. Hence, both travel time and fuel consumption are affected by link flows, and more generally by congestion on the network. By defining the objective function of the combined model in the usual way as the sum of the integrals of these functions plus the total cost of other fixed costs related to auto and transit travel, the resulting optimality conditions state that used auto routes from each origin to each destination have equal generalized travel costs and no unused route has a lower cost. At the optimal solution, the origin-destination-mode choices also are functions of these equilibrium auto generalized costs and fixed transit generalized costs.

In a combined model with system-optimal (SO) route choice, the integrals of travel time are replaced by the total travel cost on the network. The optimality conditions of this problem are that used auto routes from each origin to each destination have equal marginal generalized travel costs and no unused route has a lower marginal cost. At the optimal solution, the origin-destination-mode choices also depend on these equilibrium marginal auto travel costs as well as fixed transit travel costs.

The equivalence of the optimality conditions to marginal costs can be seen as follows. Let $c_a(v_a)$ be the travel cost on link a at flow v_a. Then, $\sum_a c_a(v_a)v_a$ is the total travel cost experienced on all links. Taking the partial derivative of this sum with respect to the route flow h_r, we obtain:

$$M_r = \frac{\partial \sum_a c_a(v_a)v_a}{\partial v_a}\frac{\partial v_a}{\partial h_r} = \sum_a (c_a(v_a) + \frac{\partial c_a(v_a)}{\partial v_a}v_a)\delta_{ra}$$

where M_r is the marginal travel cost on route r, $v_a = \sum_r h_r\delta_{ra}$ = sum of the route flows using link a, and $\delta_{ra} = 1$ if link a belongs to route r, and 0 otherwise. Therefore, the term in the optimality condition of the system-optimal route choice problem corresponding to the total travel cost is the marginal travel cost of route r.

The user-optimal and system-optimal combined models may be solved by a generalization of the algorithm proposed by Evans. Taking care that the problems are solved to the same level of convergence, the resulting origin-destination-mode flows and link flows may be summarized to obtain comparisons of the two solutions. In this study, all problems were solved to a relative gap of 0.005, which constitutes fairly fine convergence for this algorithm and problem size.

These models were solved for the Sketch Planning Network of the Chicago region, an aggregation of the 40,000 link, 1790 zone network used for travel forecasting by the Chicago Area Transportation Study (CATS). The Sketch Network has 2,950 links, 933 nodes and 387 zones. Of these links, 1818 represent arterial roads, 358 represent highways (freeways and tollways) and the remainder are centroid connectors. Travel times on links are determined by the widely used BPR function.

Transit travel times and costs for each origin-destination pair with transit service consist of in-vehicle and out-of-vehicle travel times and fares over the minimum cost route by transit, including walk or auto access. These values were aggregated from an analysis performed by CATS on a transit network for the 1790 zone region. The Chicago CBD and the City of Chicago are well served by rapid transit, commuter rail and bus service. The remainder of the region is primarily served by commuter rail, as well as suburban bus service.

The period of study is restricted to the morning peak period from 7:00 - 9:00 am. Model inputs also include the estimated total number of trips for all purposes from each origin and to each destination for the year 1990. According to CATS's trip generation model, which is the basis of these estimates, there are about 1.38 million trips per hour in the morning peak period. Cook County, which contains the City of Chicago, includes about 61% of all trip origins and 64% of all trip destinations. To simplify the analysis, all trips were grouped into a single trip purpose. A fixed matrix representing origin-destination flows of truck trips was also assigned to the minimum cost routes.

The parameter values used in both models were estimated by a method developed by Tian (1999). The values estimated from CATS's 1990 Household Travel Survey are shown in Table I. These estimated parameter values imply a relatively high overall value of time of \$19.10 per hour for auto person trips, as compared to \$2.56 per hour for transit trips.

In order to determine the effects of marginal cost road pricing over a reasonable range of total travel demand, the models are solved for three cases: the base case of 1.38 million trips per hour; a less congested case corresponding to 10% fewer trips (1.24 million trips per hour); a more congested case corresponding to 10% additional trips (1.52 million trips per hour). The proportion of trips departing from each origin

Table 3.1 Parameter Values

Variable	Auto	Transit
In-vehicle time	0.1541 gcu/min	0.0404 gcu/min
Out-of-vehicle time	0.05861 gcu/min	0.09684 gcu/min
Monetary cost/fare	0.003873 gcu/cent	0.01492 gcu/cent
Transit bias	–	-1.65685 gcu

gcu: generalized cost unit

zone and arriving at each destination zone is held fixed. The model solutions from these three cases form the basis for the figures shown in the next section.

In addition to performing this sensitivity analysis with respect to total travel demand, a second sensitivity analysis was performed on the base case by varying the auto monetary cost parameter by ± 20%. The role of this parameter in the models is to represent the effect of monetary costs, including tolls in the SO case, in relation to in-vehicle and out-of-vehicle travel times. This variation corresponds to increasing/decreasing the auto value of time by ± \$3.82. The objective of this second analysis is to provide a benchmark for interpreting the results of the travel demand analysis.

3.3 FINDINGS OF THE TRAVEL DEMAND ANALYSIS

In the solutions of the user-optimal and system-optimal models with variable demand, travel choices may be expected to differ substantially. System-optimal travelers tend to use more circuitous routes as their minimum marginal cost routes, as compared with user-optimal travelers. Because the number of routes is fairly large, and since the route choices are not unique in either case, we make no attempt to summarize the route choice results. However, the computed travel times and costs from these route choices are the basis for the results shown below. Likewise, road tolls may be expected to shift some travelers to transit modes, as well as encourage some travelers to choose lower cost, nearer destinations. All three effects, route, mode and origin-destination choice, are working simultaneously and consistently in this model.

The effects of road pricing on selected measures of origin-destination and mode choice are shown in Figures 3.1, 3.2 and 3.3 in terms of trip length, in-vehicle travel time and monetary cost. Trip lengths for auto are based on over-the-road trip distances, whereas trip lengths for transit are based on straight-line interzonal distances.

Figure 3.1 shows the mean trip lengths from the UO model are about 9.3 miles for auto and 7.9 miles for transit. Road pricing reduces the mean SO auto trip length to 8.7 miles, or about a 6% decrease, and increases the mean transit trip length to 8.3 miles. As the total level of trip-making changes by ± 10%, the trip lengths change by about ± 2% in response to increasing congestion in the UO case, and by nearly ± 3% in response to both congestion and tolls in the SO case. The change in mean auto trip length from a ± 20% variation in the auto monetary cost parameter is only

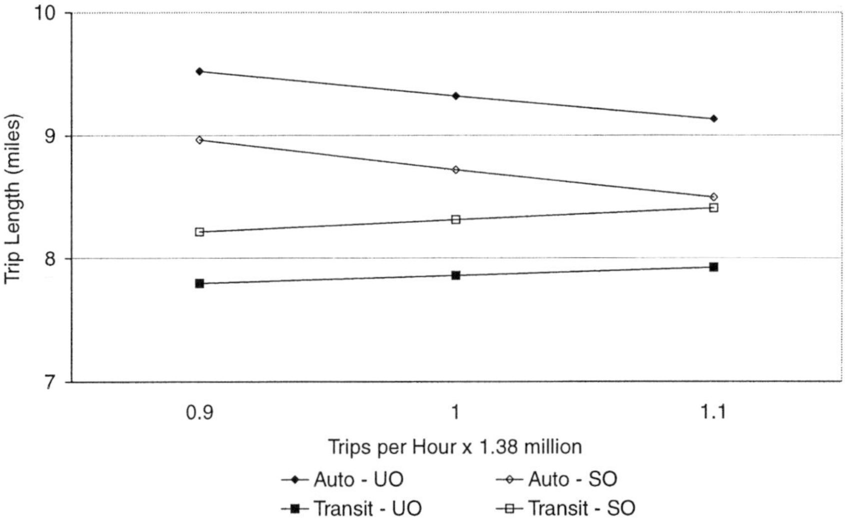

Figure 3.1 Auto and transit trip length for SO and UO combined models

± 0.2% for the UO model and ± 0.05% for the SO model. By comparison, then, the changes in trip length from variations in demand are relatively substantial.

The estimated mean in-vehicle travel times for auto and transit are shown in Figure 3.2: about 14.4 minutes for auto and 26.8 minutes for transit. Introduction of road pricing is predicted to reduce the mean auto travel time to 12.9 minutes and increase the mean transit time to 28.0 minutes. These changes are clearly the result of longer auto trips shifting to transit to avoid road tolls.

Examination of the UO results for a ± 10% variation in total demand reveals that mean auto travel times are predicted to change only about ± 0.4%, while mean transit travel times change by ± 0.6%. This result, in itself, is quite interesting. The model predicts mean auto travel times remain relatively constant compared to mean trip length. Comparable stability in mean auto travel times has also been observed in survey data (Boyce, 1999). In contrast, road pricing alters mean auto travel times by ± 1.8%. The change in mean auto travel time from a ± 20% change in the value of the auto monetary cost parameter is ± 0.6% for the UO model and only ± 0.2% for the SO model, again suggesting that the travel time variations are relatively substantial.

Figure 3.3 shows the mean auto monetary costs ($0.39) and the mean transit fare ($1.80) for the UO case. For autos, these costs include existing tolls and CBD parking fees as well as the cost of fuel. For the SO case, mean transit fares increase slightly to $1.86, but the mean auto costs increase to $2.06, since these costs now include the SO road tolls. Here, we see for the first time the magnitude of the prices implied by marginal cost road pricing. As Hearn and Yildirim (this volume) have shown, however, the same system-optimal flows can be achieved at substantially lower prices, or by a combination of prices and subsidies.

The auto mode shares for the combined UO and SO models are shown in Figure 3.4. The transit mode shares are 1.0 minus the auto mode shares. The sampled auto mode share for the region in 1990 for the morning peak period was about 83%; the user-

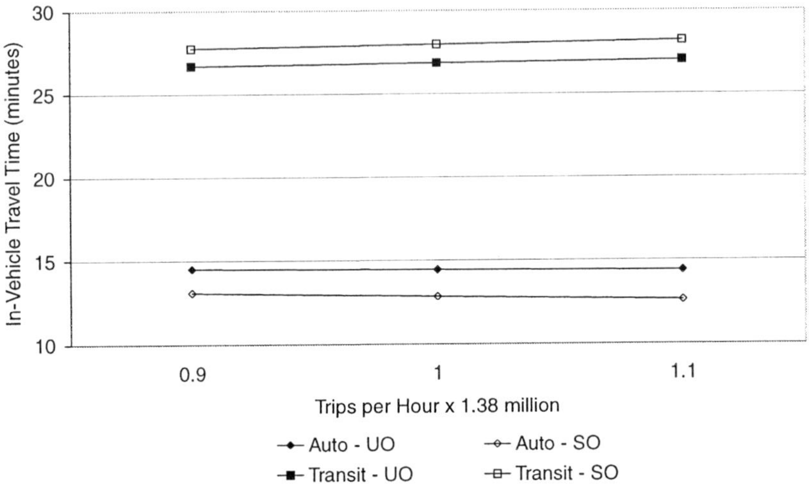

Figure 3.2 Auto and transit in-vehicle travel time for SO and UO combined models

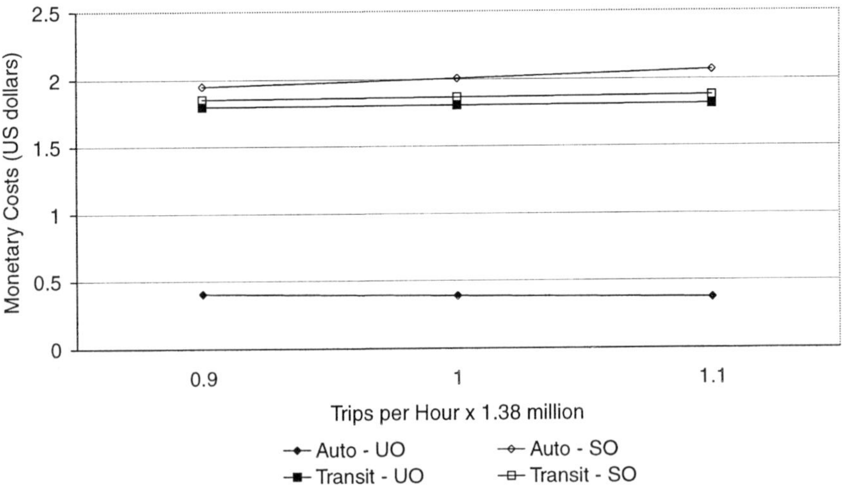

Figure 3.3 Auto and transit monetary cost for SO and UO combined models

optimal model is estimated on this value, as shown by the black circle over the base value of 1. The corresponding UO model estimates of auto mode share by destination are: Chicago CBD - 35%; Cook County including the Chicago CBD - 77%; outlying counties, or rest of the region - 96%. The auto mode shares for the SO model are substantially lower for the Chicago CBD (29%), slightly lower for Cook County (73%), and essentially the same for the outlying counties (95%). The SO auto mode share for the region falls slightly to 81%.

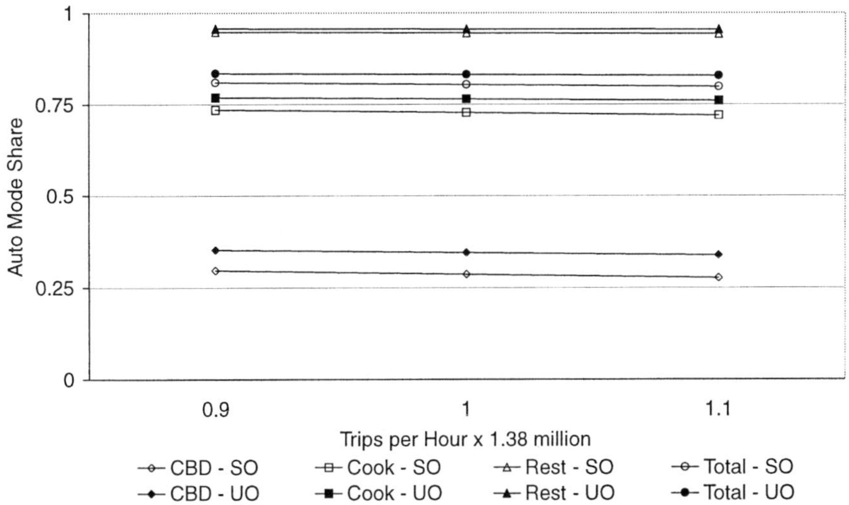

Figure 3.4 Auto mode share for SO and UO combined models

For the two other cases with 10% fewer and additional trips, the results are rather similar. The observant reader may note that the auto mode share with UO route choice varies slightly (± 0.3%) as the total demand changes by ± 10% times the base demand; the effect is slightly larger (± 0.7%) in the SO case. The reason for this decrease is that increased total demand results in higher link and route travel times, which in turn shifts choices to the transit mode.

Because the value of times for auto trips is nearly eight times higher than for transit trips, auto mode choice is heavily influenced by travel time. The effect of varying the auto monetary cost parameter by ± 20% on total UO auto choice is about ± 1%. However, for travel to central business district (CBD) destinations the effect is ± 11%. CBD destinations are more sensitive to this parameter because transit service to these destinations is superior to auto travel. Therefore, mode choices are much more sensitive to variations in the auto monetary cost parameter than to variations in travel demand.

The components of auto monetary costs are shown in Figure 3.5 for the SO case. For all cases, the road tolls are higher for arterial roads than for highways. Mean travel distances, however, are also greater for arterial roads, so it is reasonable that arterial road tolls are higher. For a ± 10% change in travel demand, highway operating costs and tolls fall as demand increases, indicating that an effect of road pricing may be to shift travel from highways to arterials.

Finally, we offer an estimate of the total user benefits of road pricing based on the method of Williams (1976). He derived a net user benefit measure for comparing

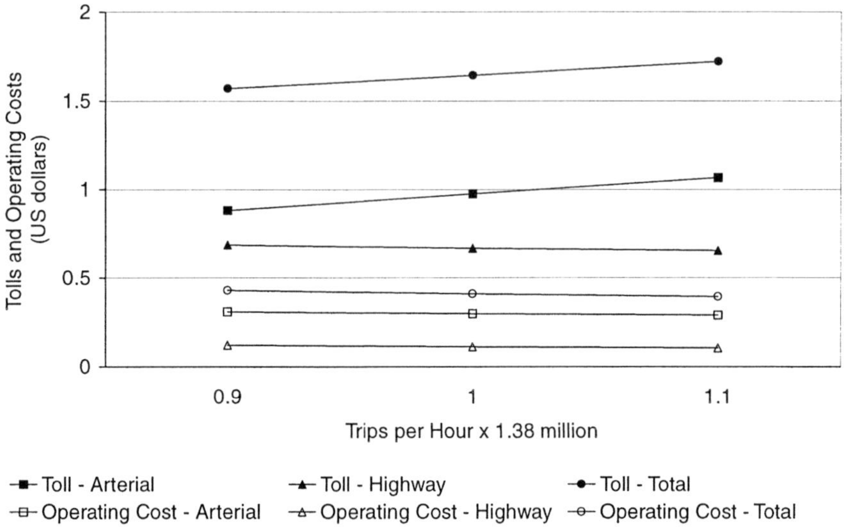

Figure 3.5 Auto tolls and operating cost for the SO combined model

travel on two road networks, as follows:

$$UB = \sum_i O_i(\ln(\frac{a_i^2}{a_i^1})) + \sum_j D_j(\ln(\frac{b_j^2}{b_j^1}))$$

where O_i and a_i are the origins per hour and the balancing factor for origin i, and D_j and b_j are the destinations per hour and the balancing factor for destination j. The superscripts 2 and 1 on the balancing factors refer respectively to the system-optimal and user-optimal balancing factors. These balancing factors assure that the origin and destination constraints are met. This equation was disaggregated to show the user benefits for auto and transit trips separately. The results of this calculation are shown in Figures 3.6 and 3.7.

Figure 3.6 shows that the total net user benefits from road pricing for one morning peak period in 1990 is 263,000 generalized cost units (gcu). About 60% of these benefits accrue to Cook County and the remainder to the outlying counties. The net user benefits are very sensitive to total demand: ± 10% change in demand results in ± 17% change in benefits.

The total amount of road tolls for one morning peak period is $1.59 million, or 609,000 gcu, which is 2.3 times as large as the estimated benefits to all travelers. However, the road tolls collected to achieve SO choices constitute an additional benefit, since these funds are available to finance other societal benefits, including transportation improvements. The total benefits including tolls amount to nearly 900,000 gcu. The estimated benefits, excluding tolls, vary ± 17% with a ± 10% variation in travel demand. The tolls themselves vary ± 13% with the same variation in demand. By comparison, a ± 20% change in the value of the auto monetary cost parameter results in a ± 4% change in user benefits and a ± 3% change in tolls in gcu.

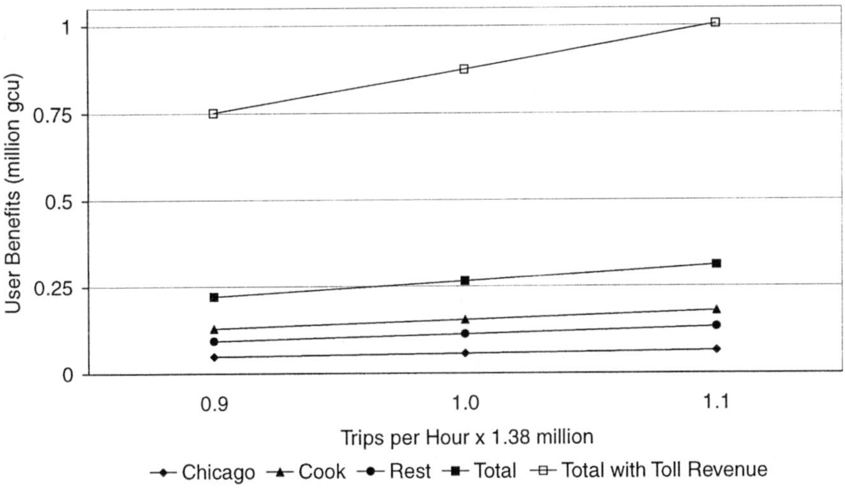

Figure 3.6 Distribution of total user benefits by subregion

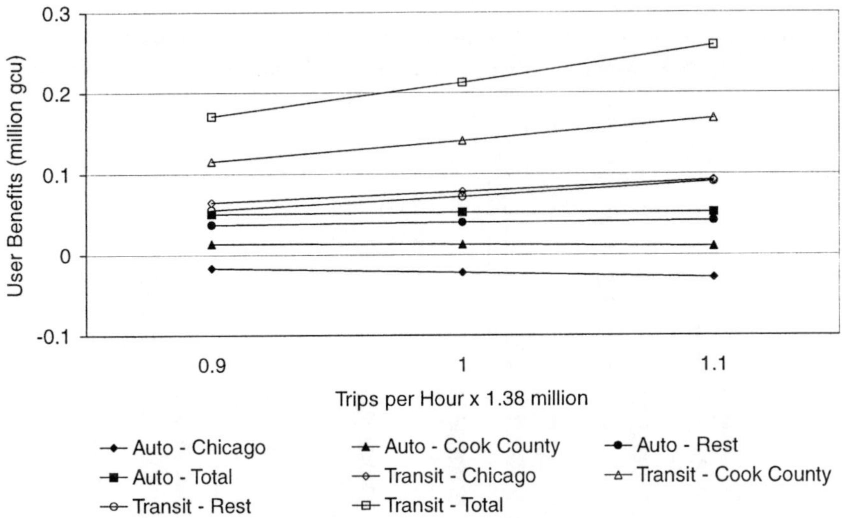

Figure 3.7 Distribution of user benefits by mode and subregion

Figure 3.7 shows the allocation of these benefits by mode: 213,000 gcu for transit and 53,000 gcu for auto. Using the estimated model parameters, these amounts can be transformed into monetary units. Because the value of money is estimated to be different for auto and transit users, the imputed benefit is \$143,000 for transit and \$136,000 for auto, or a total of \$279,000 for 1.38 million travelers or about \$0.20 per trip. The SO toll to achieve these benefits is about \$1.65 per auto interzonal trip, which generates an additional benefit equal in magnitude. Hence, the total daily peak period benefits are \$1,865,000 as compared with daily user tolls of \$1,586,000.

Figure 3.7 suggests the benefits accruing to auto travelers are relatively constant with total demand, whereas benefits accruing to transit users are sensitive to changes in demand. This result holds for the City of Chicago, Cook County and the remainder of the region.

3.4 ANALYSIS OF ALTERNATIVE PRICING STRATEGIES

Until now, the analysis has focused on one road pricing strategy, marginal cost pricing. But marginal cost tolls depend on knowing the peak period vehicle flow on each link, which may make it a difficult strategy to implement. For example, how could travelers make route choices without knowing the tolls in advance? For this reason, we also consider other pricing strategies and compare them with the SO and UO cases.

To examine this issue, we solved the UO model for two other road pricing strategies: a) a flat or constant toll on each road link; b) a doubling of the fuel price. The flat toll on each link was set so that the mean toll per auto traveler is the same as in the SO solution. The doubling of fuel price was motivated, in part, by the fact that gasoline prices in the Chicago Region in June, 2000, were approximately double ($2.20) their level of just one year earlier.

Figure 3.8 compares several measures of system performance for the above three road pricing strategies with the UO case for the base travel demand of 1.38 million trips per hour. The SO case is taken as the basis for computing standardized values for the three alternative cases. The actual values of each measure are shown above the bars.

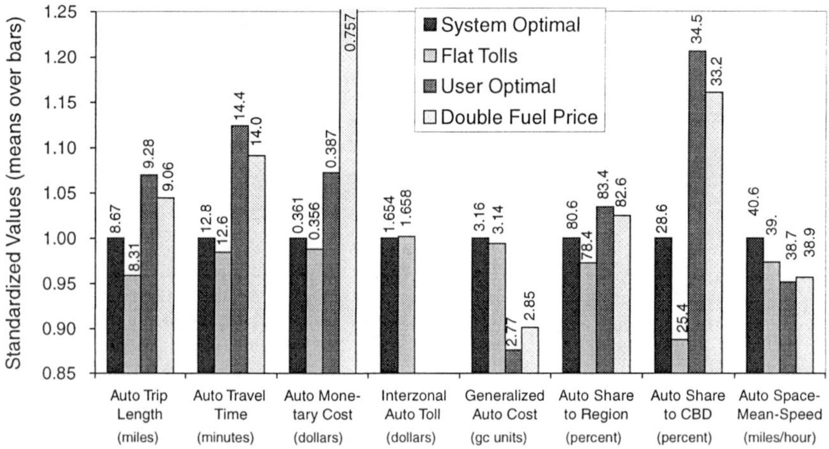

Figure 3.8 Measures of auto system performance

The case of the flat toll on each link can be seen to produce rather similar results to the SO case. The principal differences are nearly 3% fewer trips using auto and 11% fewer trips using auto to travel to the CBD. As a result both mean trip length and mean travel time are somewhat lower.

The UO solution differs most substantially from the SO case: UO trip length is 7% longer than SO trip length; UO travel time is 12% greater, regional auto share is 4% higher and auto share to the CBD is 21% higher than for the SO case. The generalized cost of auto travel is 12% less for the UO case (without tolls) than for the SO case (with tolls), largely because the mean interzonal toll is $1.65.

Doubling of fuel prices in the UO model produces a result rather similar to the UO case. The generalized cost of auto travel with doubled fuel prices is 10% less than the SO cost with tolls. However, auto use is over 2% higher throughout the region and 16% higher for auto travel to the CBD than in the SO case. Since fuel consumption in the model is primarily proportional to travel distance (miles), this solution is also indicative of the case of a toll simply based on travel distance. The result is relatively unattractive as compared with the SO case of marginal cost tolls.

3.5 CONCLUSIONS

Introduction of marginal cost road pricing could reduce congestion to some extent on the road network of the Chicago region by diverting trips to nearer destinations and to transit modes, and by spreading trips over the marginal cost routes so as to reduce the total generalized cost of travel. The cost in terms of tolls would be high for auto travelers; however, these tolls are not unique, and similar benefits might be obtained by a combination of tolls and subsidies.

The benefits obtained by marginal cost pricing are not evenly distributed over the region, just as congestion occurs unevenly. The City of Chicago and the inner suburbs of Cook County would receive more benefits, in part because they enjoy an alternative mode in the form of transit. The estimated total benefits are relatively modest on a per person basis, but the reduction in congested vehicle miles traveled is large. The reduction would likely also result in lower levels of emissions.

A shortcoming of this analysis is that only one time period, the two-hour peak period, is represented. Assuming lower tolls would prevail in off-peak periods, then substantial numbers of travelers might shift their departure times to avoid peak period tolls, thereby also reducing congestion. Adding multiple traveler classes to the model would also improve its validity, since nonwork travelers may be more flexible with regard to their departure times. Representing the income classes of travelers would also serve to identify the effects of tolls by the ability to pay. Hence, through further disaggregation of the combined model, more realistic and interesting results should be obtainable.

Acknowledgments

The support of this research by the U.S. National Science Foundation through the National Institute of Statistical Sciences, Research Triangle Park, NC, is gratefully acknowledged.

References

Beckmann, M., C. B. McGuire and C. B. Winsten (1956) *Studies in the Economics of Transportation.* New Haven: Yale University Press.

Boyce, D. (1999) "Urban Travel and Social Change: American and German Comparisons." Conference on Social Change and Sustainable Transport, Berkeley, CA.

Boyce, D. E. and Daskin, M. S. (1997) "Urban Transportation." ReVelle, C. and McGarity, A. E. eds. *Design and Operation of Civil and Environmental Engineering Systems.* New York: John Wiley & Sons, Inc.

Downs, A. (1992) *Stuck in Traffic.* Washington, D.C.: Brookings Institution.

Evans, S. P. (1976) "Derivation and Analysis of Some Models for Combining Trip Distribution and Assignment." *Transportation Research,* **10**, 37-57.

Pigou, A. C. (1920) *The Economics of Welfare.* London: Macmillan.

Tian, X. (1999) *Parameter Estimation for Single-Class and Multiple-class Combined Travel Choice Models.* Ph.D. Thesis, University of Illinois at Chicago, Chicago, IL.

Wardrop, J. G. (1952) "Some Theoretical Aspects of Road Traffic Research." *Proceedings of the Institution of Civil Engineers,* Part II, **1**, 325-378.

Williams, H. C. W. L. (1976) "Travel Demand Models, Duality Relations and User Benefit Analysis." *Journal of Regional Science,* **16**, 147-165.

4 A DYNAMIC USER EQUILIBRIUM MODEL FOR TRAFFIC ASSIGNMENT IN URBAN AREAS

Luce Brotcorne
Daniel De Wolf
Michel Gendreau
Martine Labbé

Abstract: In this paper, we present a model for dynamic traffic assignment in urban transportation networks. The transportation demand is given by the number of users traveling from the same origin to the same destination and having the same desired arrival time. The model determines for each user both the path and the departure time so that he cannot decrease his disutility by unilaterally changing his departure time or/and his path. The disutility measure is given by a generalized travel time function taking into account total path travel time and schedule delays. Departure time choices are restricted to a finite set of instants, but users with same choice of departure time and same first arc on selected path are loaded uniformly on the network during a time interval whose lower bound is the chosen departure time. A finite dimensional variational inequality formulation of the problem is given and the existence of a dynamic user equilibrium is proved. A heuristic method is proposed to compute a dynamic user equilibrium. Numerical results are provided.

4.1 INTRODUCTION

The growing traffic congestion in urban areas creates an urgent need for better transportation planning. In this context, traffic assignment models can be used to help public authorities assess the impact of various transportation policies such as enlarging areas where bottlenecks occur, work start time flexibility, etc. The representation of traffic variations over time allows dynamic transportation models to tackle real

M. Gendreau and P. Marcotte (eds.), Transportation and Network Analysis: Current Trends, 49–69.

situations. As departure time choice sharply influences traffic congestion, the simultaneous route and departure time models seem to be the more accurate dynamic models. These problems have been treated by using either simulation or analytical methods. The simulation based approaches lead to easier implementations of traffic control strategies and allow studies on the influence of information system techniques in the context of Intelligent Transportation Systems. However, the number of attributes involved are normally quite large and the properties of the solutions remain uncertain. In contrast, with analytical based methods, the optimality conditions can be derived from a preset driver behavioural principle such as utility maximization. Moreover, the sensitivity analysis for different scenarios is easier to perform since the procedure is usually less time consuming than with a simulation based approach.

Many analytical based dynamic user optimal departure time and route choice models have been proposed. According to Friesz *et al.* (1996), they can be clustered according to the mathematical formulation used: optimal control (Ran *et al.* (1992,1994) and Bernstein *et al.* (1993)), variational inequalities (Friesz *et al.* (1993), Wie *et al.* (1993), Ran *et al.* (1996), Gendreau and Bouzaiene-Ayari (1997)), mathematical programming (Janson (1993) and Jayakrishnan (1995)). The presence of time-dependent congestion makes the dynamic equilibrium problem much more difficult than the static one. Thus in most existing models major simplifications are made to make the problem more tractable and the proposed algorithms are generally heuristics.

In this paper, we develop a new dynamic user equilibrium model that includes route choice and departure time decisions on a multiple origin-destination network. The transportation demand is the number of users leaving their origin to reach their destination within a time window around the desired arrival time. The model determines for each user a choice of path and departure time such that he cannot decrease his disutility by unilaterally changing his departure time or/and his path. The disutility integrates travel times and schedule delays. Although the dynamic traffic assignment problem is a continuous phenomenon, one major drawback of continuous models is the necessary discretisation used to solve them. To provide a more realistic description, we consider a finite set of possible departure time variables and continuous traversal times. Users with the same choice of departure time and first arc of path are uniformly loaded on the network during a time interval whose lower bound is the departure time. In order to reduce asymmetric interactions, the path travel time of users of a choice is computed according to the number of users of the choice and the number of users preceding them on the arc. The originality of our model is based on these two models caracteristics. A finite dimensional variational inequality formulation of the problem is given and the existence of a dynamic user equilibrium is proved. A heuristic method that simulates daily changes of departure time and path of users traveling repeatedly on the network is proposed.

The main differences between our model and the previous variational inequality formulations of the dynamic user equilibrium problem are the followings. In Friesz *et al.* (1993) and Ran *et al.* (1996), departure times and travel times are continuous functions of time whereas in Wie *et al.* (1993) both of them are discrete. In Gendreau and Bouzaiene-Ayari (1997), even if a finite set of possible departure time variables and continuous traversal times are considered, the path travel time depends on the total number of users present on the arcs of the path at the entering times.

The paper is organized as follows. In Section 4.2, we describe the formulation of the model as a variational inequality problem. A proof of existence of a dynamic user equilibrium is given in Section 4.3. We present the main lines of the solution algorithm in Section 4.4. Section 4.5 is devoted to numerical results. Finally, some conclusions are drawn in Section 4.6.

4.2 PROBLEM FORMULATION

The **notations** introduced in this paper are summarized in Table 4.1. The transportation network is represented by a graph $G = (\mathcal{N}, \mathcal{A})$ where $\mathcal{N}$ is the node set and $\mathcal{A}$ the arc set. A node represents either an origin or a destination, or an intersection. An arc represents a one-way street. Let $[0, T]$ be the time period of study. The length of this time interval is sufficient to allow all users to complete their trip with some flexibility of departure times. Moreover, we denote by $S = \{0, 1, 2, ...T\} \subset [0, T]$ the finite set of possible departure times for users.

To simplify notations, we define a commodity k as the set of users traveling between the same origin-destination pair $(o(k), d(k))$ and having the same desired arrival time $at(k)$ at destination. The set of all commodities k is denoted by $\mathcal{K}$. The **transportation demand** is given by n^k the total number of users associated with the same commodity, i.e., those with same origin $o(k)$, same destination $d(k)$ and same desired arrival time $at(k)$. Note that users do not generally have to arrive at their destination at a precise moment, but have some latitude to arrive slightly earlier or later. This leads us to define a time window of half-width Δ around the desired arrival time $at(k)$. Users of each commodity $k \in \mathcal{K}$ can be spread among several choices of departure time τ and path p linking origin $o(k)$ to destination $d(k)$. $\mathcal{SP}(k)$ is the set of these choices, $\mathcal{P}(k)$ is the set of all possible paths linking origin $o(k)$ to destination $d(k)$. The **decision variables of the problem** are thus the number of users $n^k_{\tau,p}$ of commodity k with the same choice (τ, p). There is a finite number of possible departure times $\tau \in S$ at any origin node. However, all users with the same departure time τ and same first arc a on the path will have an effective departure time belonging to an interval with lower bound τ. These users will be uniformly loaded on the first arc of the path during the departure time interval. The rate of leaving users is chosen in such a way that all users with same first arc of path and same departure time lower bound τ will enter the arc during the time interval $[\tau, \tau + 1[$. Thus the departure rate varies with the number of users whereas the length of the departure time interval is fixed (See Figure 4.1).

The **network capacity** is specified through travel time functions D_a continuous in time. For all $t \in [0, T]$, $D_a(t)$ represents the traversal time for a user entering arc a at time t. It is computed according to the number of users of the choice to which the entering user belongs and the number of users of other choices preceding him on the arc. Function D_a is defined to satisfy the so-called First-In First-Out (FIFO) property stating that a user cannot arrive earlier at his destination by leaving his origin later. This is a natural assumption in urban congested networks. Let g_a be a non-decreasing piecewise linear function from $\mathbb{R}^+$ into $[0, T]$ that represents the classical "arc performance" function linking the arc traversal time with the number of users present on arc a (i.e., the flow on this arc). According to Branston (1975), a continuous piecewise linear function with two breakpoints seems reasonable to model the arc performance function.

$\mathcal{N}$	set of nodes
i	a node
$\mathcal{A}$	set of arcs
a or (i,j)	an arc
p	a path
$[0,T]$	time period of study
S	finite set of departure times
(τ,p)	a choice of departure time and path
k	a commodity (*i.e.* a set of users associated to a triplet $(o(k), d(k), at(k))$
$o(k)$	origin of commodity k
$d(k)$	destination of commodity k
$at(k)$	desired arrival time of commodity k
$\mathcal{K}$	set of commodities
$\mathcal{P}(k)$	set of possible paths linking origin $o(k)$ to destination $d(k)$
$\mathcal{SP}(k)$	set of choices of path and departure time for commodity k
n^k	total number of users belonging to commodity k
$n^k_{\tau,p}$	number of users of commodity k with choice (τ,p)
$f_a(t)$	load on arc a at time t
$\tilde{f}_a(t)$	effective load on arc a at time t
$x_a(t)$	number of users entering arc a at time t
$y_a(t)$	number of users leaving arc a at time t
$D_a(t)$	traversal time of arc a for some users entering on the arc at time t
$f^k_{a,\tau,p}(t)$	number of users of commodity k with choice (τ,p) present on arc a at time t
$x^k_{a,\tau,p}(t)$	number of users of commodity k with choice (τ,p) entering arc a at time t
$y^k_{a,\tau,p}t)$	number of users of commodity k with choice (τ,p) leaving arc a at time t
$s^k_{\tau,p}(i,t)$	arrival time at node i for users of commodity k with choice (τ,p) leaving the origin at effective time t.
$C^k(\tau,p)$	disutility function corresponding to the choice (τ,p) associated to commodity k
Δ	half-width of the time window around the desired arrival time
α	travel time coefficient in the disutility function
β	early arrival penalty coefficient in the disutility function
γ	late arrival penalty coefficient in the disutility function

Table 4.1 Notations

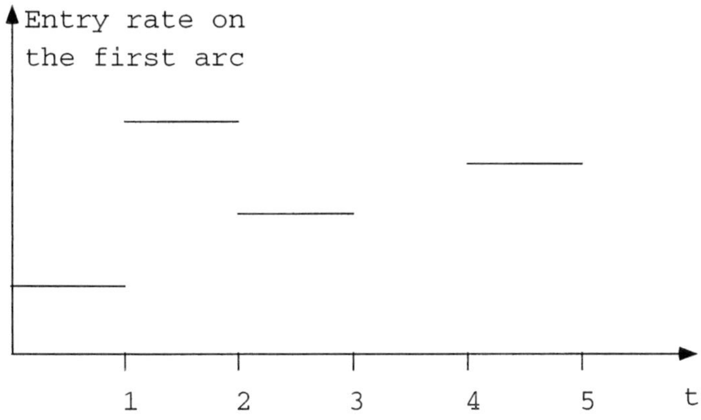

Figure 4.1 Entry on the first arc of a path

The arc traversal time function D_a is then defined as follows:

$$D_a(t) = \begin{cases} g_a(f_a(0)) & \text{if } t = 0, \\ \max\{g_a(f_a(t)); D_a(t-\epsilon) - \epsilon, \forall \epsilon \in [0,t]\} & \text{otherwise,} \end{cases} \tag{4.1}$$

where $f_a(t)$ represents the load (flow) on arc a at time t.

For $t > 0$, $g_a(f_a(t))$ represents the arc performance function evaluated at the load value on an arc at time t. As illustrated in Figure 4.2, a traversal time function given by that term may be unrealistic and may violate the FIFO assumption. Suppose that a large number of users leave arc a in a small interval starting at time t. Then, if the traversal time of this arc was only defined as $D_a(t) = g(f_a(t))$, it would suddenly decrease, as illustrated in Figure 4.2 by the dashed line. Furthermore, it would be possible that a user entering the arc just after t would pass users who have entered it before t. This justifies the presence of the second term in the maximum of (4.1). With such a definition, the arc traversal time of a user who entered just after t will be determined by a line segment of slope -1 starting at point $(t, D_a(t))$. Note that, as long as the function $D_a(t)$ is well defined, it can be evaluated progressively for increasing values of time. The following proposition asserts that the arc traversal time function respects the FIFO queue discipline.

Proposition 1 *Let $g_a(.)$ be a nondecreasing function from $\mathbb{R}^+$ into $[0,T]$. We have that:*

$$\forall t_2 > t_1, t_2 + D_a(t_2) \geq t_1 + D_a(t_1). \tag{4.2}$$

Proof: Without loss of generality, suppose that $t_1 = t_2 - \epsilon$ for $\epsilon > 0$. By definition (4.1), it immediately follows that:

$$D_a(t_2) \geq D_a(t_2 - \epsilon) - \epsilon$$

which is equivalent to (4.2).

Let us now examine the influence of the number of users present on an arc at a given time on the travel time value at that time. To simplify the exposition, consider a single arc a and a single commodity of n users distributed on 2 choices whose lower

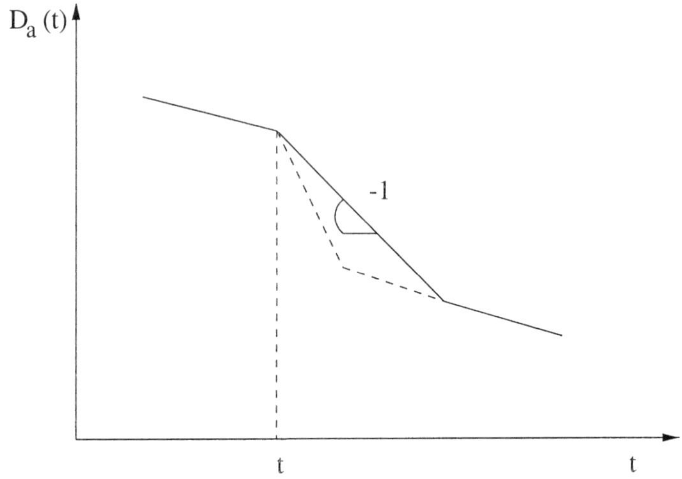

Figure 4.2 Arc traversal time function

bounds of departure time intervals are τ_1 and τ_2 with $\tau_1 < \tau_1 + 1 \leq \tau_2 < \tau_2 + 1$. We make the assumption that the slope of traversal time function is greater or equal to -1 at all times. More precisely, we assume that $D_a(t) = g_a(f_a(t)),\ \forall t \in [0, T]$, and that function g_a is linear. Without loss of generality, we also assume that function D_a has no breakpoints in the interval $]\tau_2, \tau_2 + 1[$. Indeed, if it is not the case, the same reasoning can be done on a sub-interval. To consider the general case where users are scattered among different choices of departure times, let $\underline{t} \in [\tau_2, \tau_2 + 1[$ be the departure time of user x of choice (τ_2, a). Note that for $\underline{t} \in [0, \tau_2[$ the same reasoning can be done except that only the users with choice τ_1 would be considered. Suppose that all users with choice (τ_1, a) have not left the arc a at time $\underline{t}$, more precisely, assume that $\underline{t} < \tau_1 + 1 + D_a(\tau_1 + 1)$. We denote by $f_a^1(\underline{t})$, respectively $f_a^2(\underline{t})$, the number of users with choice (τ_1, a), respectively (τ_2, a), present on arc a at time $\underline{t}$. These values are updated during the course of the algorithm as will be explained in Section 4.4. According to the defining relationship (4.1), the traversal time value for arc a at time $\underline{t}$, $D_a(\underline{t})$, depends on the total number of users present on the arc at that time, i.e., $f_a^1(\underline{t}) + f_a^2(\underline{t})$. This amounts to assuming that $f_a^1(\underline{t}) + f_a^2(\underline{t})$ users of choices (τ_1, a) and (τ_2, a) will be present on arc a during the time interval of presence on arc a of the user of choice (τ_2, a) entering at time $\underline{t}$. This can lead us to overestimate $D_a(\underline{t})$ since the users of choice (τ_1, a) leave the arc long before user x. We should thus rather consider the average number of users of choice (τ_1, a) (computed from time $\underline{t}$ to the last leaving time of users of choice (τ_1, a)) that are present on the arc to compute $D_a(\underline{t})$. This average value is given by:

$$\frac{1}{\tau_1 + 1 + D_a(\tau_1 + 1) - \underline{t}} \int_{\underline{t}}^{\tau_1 + 1 + D_a(\tau_1 + 1)} f_a^1(t)\, dt. \tag{4.3}$$

Since $D_a(\underline{t})$ also depends on the number of users of choice (τ_2, a), we could consider by analogy that only the average number of users of the choice (τ_2, a) (computed from time $\underline{t}$ to the leaving time of the user x of choice (τ_2, a) entering at $\underline{t}$) will be present on the arc:

$$\frac{1}{\underline{t}+D_a(\underline{t})-\underline{t}}\{f_a^2(\underline{t})\{\tau_2+D_a(\tau_2)-\underline{t}\}+\frac{1}{2}f_a^2(\underline{t})\{\underline{t}+D_a(\underline{t})-\tau_2-D_a(\tau_2)\}\},$$

or,

$$\frac{1}{2}f_a^2(\underline{t})+\frac{1}{2}f_a^2(\underline{t})\frac{\tau_2+D_a(\tau_2)-\underline{t}}{D_a(\underline{t})}. \tag{4.4}$$

However, this value requires the knowledge of the travel time value $D_a(\underline{t})$ to compute itself. Therefore, we will approximate the average value (4.4) by $f_a^2(\underline{t})$. When the entry time interval $[\tau_2, \tau_2+1[$ is not too large, this is reasonable since (4.4) is upper-bounded by $f_a^2(\underline{t})$ due to the FIFO hypothesis: $\tau_2 + D_a(\tau_2) \leq \underline{t} + D_a(\underline{t})$.

Finally, to compute $D_a(\underline{t})$ we take into account the time interval by which the user entering at time $\underline{t}$ will be delayed by users of other choices entered before him on the arc. More precisely, the delay induced by users $f_a^1(\underline{t})$ on the travel time of user x of choice (τ_2, a) entering the arc at time $\underline{t} \in [\tau_2, \tau_2+1[$ will depend on their position with respect to x. If the users $f_a^1(\underline{t})$ are far from user x, they will slightly delay him. However, if they are just preceding him, the induced delay will be more significant. To take into account these observations, the load charge (4.3) will be weighted by a ratio between the interval where users $f_a^1(\underline{t})$ and x are both present and the travel time of user x entering at time $\underline{t}$. This ratio is given by:

$$\frac{\tau_1+1+D_a(\tau_1+1)-\underline{t}}{D_a(\underline{t})}.$$

The farther the users of choice (τ_1, a) are from the users x of choice (τ_2, a), the smaller is the numerator $\tau_1+1+D_a(\tau_1+1)-\underline{t}$ and thus the smaller will be the impact of the presence of the users of choice (τ_1, a) on the computation of the traversal time for x.

In conclusion, the contribution of the users of choices (τ_1, a) and (τ_2, a) in the computation of the travel time $D_a(\underline{t})$ is given for $\underline{t} \in [\tau_2, \tau_2+1[$ by

$$\tilde{f}_a(\underline{t}) \;=\; \frac{\tau_1+1+D_a(\tau_1+1)-\underline{t}}{D_a(\underline{t})\{\tau_1+1+D_a(\tau_1+1)-\underline{t}\}}\int_{\underline{t}}^{\tau_1+1+D_a(\tau_1+1)} f_a^1(t)\,dt \;+\; f_a^2(\underline{t}), \tag{4.5}$$

or:

$$\tilde{f}_a(\underline{t}) \;=\; \frac{1}{D_a(\underline{t})}\int_{\underline{t}}^{\tau_1+1+D_a(\tau_1+1)} f_a^1(t)\,dt \;+\; f_a^2(\underline{t}). \tag{4.6}$$

We call this value the *effective load* on arc a at time t and use it, instead of $f_a(t)$ in (4.1). Next, two cases have to be considered to evaluate the contribution of the users of choice (τ_1, a). Note that function $f_a^1(t)$ is linear on $[\tau_1+1, \tau_1+1+D_a(\tau_1+1)[$ in Figure 4.3 and 4.4 due to the linearity assumption on g_a made in this example. First, if the users of the choice (τ_1, a) are still present on the arc at time $\underline{t}$ (i.e., if $\tau_1 + D_a(\tau_1) \geq \underline{t}$) (cf. Figure 4.3), then

$$\begin{aligned}\tilde{f}_a(\underline{t}) \;=\; & \frac{1}{D_a(\underline{t})}\Big\{n_{\tau_1}(\tau_1-\underline{t}) \\ & +\frac{n_{\tau_1}}{2}\left(D_a(\tau_1+1)+1+D_a(\tau_1)\right)\Big\}+f_a^2(\underline{t}).\end{aligned} \tag{4.7}$$

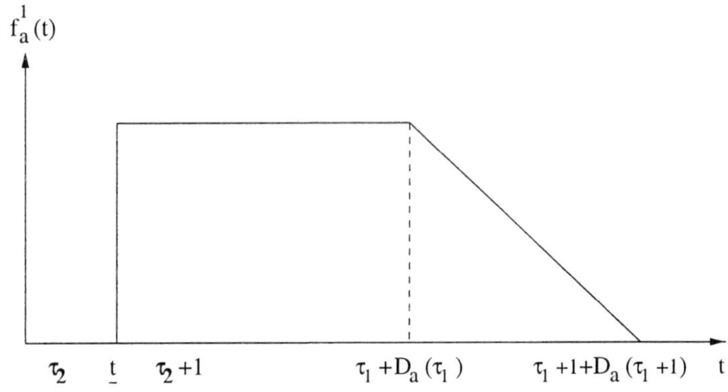

Figure 4.3 Presence of choice τ_1, entry of choice τ_2.

Second, if the users of the choice (τ_1, a) have started to leave the arc at time $\underline{t}$ $(\tau_1 + D_a(\tau_1) < \underline{t}\ )$ (cf. Figure 4.4), then

$$\tilde{f}_a(\underline{t}) = \frac{\tau_1 + 1 + D_a(\tau_1 + 1) - t}{D_a(\underline{t})} \frac{f_a^1(\underline{t})}{2} + f_a^2(\underline{t}). \tag{4.8}$$

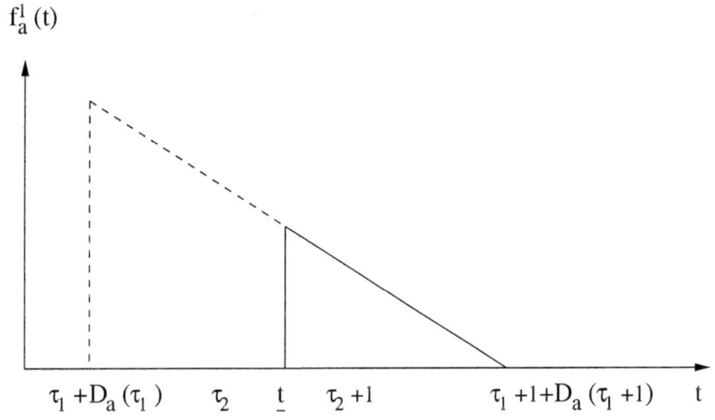

Figure 4.4 Exit of choice τ_1, entry of choice τ_2.

The difficulty with expressions (4.7) and (4.8) comes from the fact that they require the knowledge of $D_a(\underline{t})$ to compute $\tilde{f}_a(\underline{t})$ which is used to compute $D_a(\underline{t})$. To remedy this difficulty, we proceed as follows. Let $\tilde{t}$ be the last instant preceding $\underline{t}$ for which the value of $D_a()$ was computed. $D_a(\underline{t})$ is then approximated by $D_a(\tilde{t})$. Thus $\tilde{f}_a(.)$ is linear, $\forall t \in [\tau_2, \tau_1 + D_a(\tau_1)]$, (see (4.7)) and nonlinear, $\forall t \in [\tau_1 + D_a(\tau_1), \tau_1 + 1 + D_a(\tau_1 + 1)]$ (see (4.8)). As a result, functions $g_a(\tilde{f}_a(\underline{t}))$ and $D_a(\underline{t})$ are also linear and nonlinear on the same intervals. We approximate the nonlinear part of function $g_a(\tilde{f}_a(t))$ on the interval $[\tau_1 + D_a(\tau_1), \tau_1 + 1 + D_a(\tau_1 + 1)]$ by a piecewise linear function with breakpoints defined by $\tau_1 + D_a(\tau_1)$, $\tau_1 + 1 + D_a(\tau_1 + 1)$, corresponding to times associated with the entry or leaving of usersn and abscissa of function g_a

breakpoints. When the entry time interval it is not too large, these approximations are reasonable since the approach described hereafter tends to generate a large number of breakpoints.

On the basis of the reasoning developed in the previous example, we can deduce the value of the arc performance function $g_a(\tilde{f}_a(t))$ at any time. We note $tep^k_{\tau,p}$ and $ted^k_{\tau,p}$ the entry times of the first and last user of the group $n^k_{\tau,p}$ the arc a and $tsp^k_{\tau,p}$ and $tsd^k_{\tau,p}$ the exit times of the first and last user of the same group on arc a. Let t_0 the last breakpoint before $\underline{t}$ of the travel time function $D_a(t)$. The computation of the value $g_a(\tilde{f}_a(\underline{t}))$ is described hereafter.

- Step 1: (Initialization).

  Initialize $\tilde{f}_a(t) = 0, \forall t \in \mathcal{T}$.

- Step 2: (Computation of $\tilde{f}_a(\underline{t})$).

  For all commodities $k \in \mathcal{K}$,

  For all choices $(\tau, p) \in \mathcal{SP}(k)$ such that $n^k_{\tau,p} \geq 0$ and $a \in p$:

  1. Determine $tep^k_{\tau,p}$, $ted^k_{\tau,p}$, $tsp^k_{\tau,p}$ and $tsd^k_{\tau,p}$.
  2. If $\underline{t} \in [tep^k_{\tau,p}, ted^k_{\tau,p}]$, where $[tep^k_{\tau,p}, ted^k_{\tau,p}]$ is the entry time interval for the users with choice (τ, p) :
     - Evaluate the number of users $f^k_{a,\tau,p}(\underline{t})$ of the choice (τ, p) present on arc a at time $\underline{t}$.
     - $\tilde{f} = \tilde{f} + f^k_{a,\tau,p}(\underline{t})$.
  3. If $\underline{t} \in [ted^k_{\tau,p}, tsp^k_{\tau,p}]$, where $[ted^k_{\tau,p}, tsp^k_{\tau,p}]$ is the time interval for which all the users of choice (τ, p) are present:
     - Set $f = n^k_{\tau,p}(tep - \underline{t}) + \frac{1}{2} n^k_{\tau,p}\{D_a(ted) + 1 + D_a(tep)\}$.
     - $\tilde{f} = \tilde{f} + \frac{f}{D_a(t_0)}$.
  4. If $\underline{t} \in [tsp^k_{\tau,p}, tsl^k_{\tau,p}]$, where $[tsp^k_{\tau,p}, tsl^k_{\tau,p}]$ is the exit time interval for the users with choice (τ, p):
     - Determine the number of users $f^k_{a,\tau,p}(\underline{t})$ of the choice (τ, p) present on arc a at time $\underline{t}$.
     - Compute $f = (tsd - \underline{t})\, \frac{f^k_{a,\tau,p}(\underline{t})}{2}$.
     - $\tilde{f} = \tilde{f} + \frac{f}{D_a(t_0)}$.

- Step 3: (Computation of $g_a(\tilde{f}_a(\underline{t}))$).

  Evaluate the piecewise linear function g_a at the load value $\tilde{f}_a(\underline{t})$.

Given the transportation demand and the network capacity, we now model the problem of the determination of the path and departure time choices (τ, p) of users to reach a **user equilibrium**. We first describe the constraints of the model. As mentioned previously, the decision variables of the problem are the number of users $n^k_{\tau,p}$ of each commodity k with choice $(\tau, p) \in \mathcal{SP}(k)$ where p can be defined by the node sequence $p = \{i_0, i_1, ..., i_m\}$ or the arc sequence $p = \{a_1, a_2, ..., a_m\}$ The **variables** used are denoted by $f_a(t), x_a(t)$ and $y_a(t)$ for $a \in \mathcal{A}$. Specifically, $f_a(t)$

represents the total number of users present on arc a at time $t \in [0,T]$; $x_a(t)$ and $y_a(t)$ are the rates at which users, respectively, enter or leave arc a at time t. In the following, they will be referred to as the inflow and exit rates on arc a at time t. These variables are the sum of partial variables for choices (τ, p) associated to commodities k. For all $t \in [0,T]$:

$$\sum_{k \in \mathcal{K}} \sum_{(\tau,p) \in \mathcal{SP}(k)} f^k_{a,\tau,p}(t) = f_a(t),$$

$$\sum_{k \in \mathcal{K}} \sum_{(\tau,p) \in \mathcal{SP}(k)} x^k_{a,\tau,p}(t) = x_a(t),$$

$$\sum_{k \in \mathcal{K}} \sum_{(\tau,p) \in \mathcal{SP}(k)} y^k_{a,\tau,p}(t) = y_a(t).$$

$$\sum_{(\tau,p) \in \mathcal{SP}(k)} n^k_{\tau,p} = n^k \qquad \forall k \in \mathcal{K}, \tag{4.9}$$

$$x^k_{a,\tau,p|a_1=a}(t) = n^k_{\tau,p|a_1=a} \qquad \forall k \in \mathcal{K},\ \forall (\tau,p) \in \mathcal{SP}(k),\ \forall t \in [\tau, \tau+1], \tag{4.10}$$

$$\frac{df_a(t)}{dt} = x_a(t) - y_a(t) \qquad \forall a \in \mathcal{A},\ \forall t \in [0,T], \tag{4.11}$$

$$y^k_{a,\tau,p}(t + D_a(t)) = x^k_{a,\tau,p}(t) \qquad \forall k \in \mathcal{K},\ \forall a \in \mathcal{A},\ \forall t \in [0,T], \tag{4.12}$$

$$y^k_{(i_{l-1},i_l),\tau,p}(t) = x^k_{(i_l,i_{l+1}),\tau,p}(t) \qquad \forall k \in \mathcal{K},\ \forall (\tau,p) \in \mathcal{SP}(k), \quad \forall (i_{l-1}, i_l, i_{l+1}) \in p,\ \forall t \in [0,T]. \tag{4.13}$$

The final constraints are the non-negativity constraints:

$$\begin{aligned} n^k_{\tau,p} &\geq 0, \quad \forall k \in \mathcal{K}, \forall (\tau,p) \in \mathcal{SP}(k), \\ f_a(t) &\geq 0, \quad \forall a \in \mathcal{A}, \forall t \in [0,T], \\ x^k_{a,\tau,p}(t) &\geq 0, \quad \forall k \in \mathcal{K}, \forall (\tau,p) \in \mathcal{SP}(k), \\ &\qquad\ \ \forall a \in p, \forall t \in [0,T], \\ y^k_{a,\tau,p}(t) &\geq 0, \quad \forall k \in \mathcal{K}, \forall (\tau,p) \in \mathcal{SP}(k), \\ &\qquad\ \ \forall a \in p, \forall t \in [0,T]. \end{aligned}$$

The first group of constraints (4.9) ensures that all users of each commodity k will choose a departure time and a path to reach their destination. The second group of constraints (4.10) guarantees a uniform distribution of the users entering the same first arc of path during the same departure time interval. The third group (4.11) specifies that the variation of the number of users present on an arc at a given time is the difference between the inflow and exit rates on this arc at that time. The fourth group (4.12) of constraints guarantees that the exit rate on arc a at time t of users of commodity k with choice (τ, p) is equal to the entry rate of these users on the arcs at time $\underline{t}$ where $\underline{t} = t - D_a(\underline{t})$. Finally, the fifth group of constraints (4.13) is composed of load conservation equation at intermediate nodes. Constraints (4.9) define allocation of users to choice of path and departure time while constraints (4.10), (4.11), (4.12), and (4.13) model the users propagation on the network.

In order to define the dynamic user equilibrium conditions for the dynamic traffic assignment, we define the disutility functions associated with choices (τ, p) for each commodity k. As the users are linearly loaded on the first arc of the path and the traversal time functions are continuous in time, the arrivals at destination are also continuous. Moreover, each user incurs a disutility value corresponding to its effective departure time. The disutility function for a user of a commodity k is a weighted sum of travel time and penalties for arrival outside the time window of half-width Δ around $at(k)$. The total path travel time is defined as the summation of arc traversal times determined at the time users enter the arc. Thus, the model described in this paper is based on the notion of actual path times computed as follows. Let $s^k_{\tau,p}(i,t)$ be the arrival time at node i for the users of $n^k_{\tau,p}$ leaving the first node of path p at effective time t, and let $\{i_1, i_2, ..., i_l\}$ denote the sequence of nodes belonging to path $p \in \mathcal{P}(k)$. The values $s^k_{\tau,p}(i,t)$ are computed recursively as follows.

$$\begin{aligned} s^k_{\tau,p}(i_2,t) &= t + D_{i_1,i_2}(t) \\ s^k_{\tau,p}(i_{m+1},t) &= s^k_{\tau,p}(i_m,t) + D_{i_m i_{m+1}}(s^k_{\tau,p}(i_m,t)), \\ & \quad m = 1, ..., l-1 \end{aligned}$$

and this for all $t \in [\tau, \tau+1[$.

As we suppose a uniform distribution of users, the disutility value of users of a commodity k with choice (τ,p) will be defined as the average disutility value of users computed for the departure time interval. If the discretisation step for departure times is sufficiently small, we can hope that the disutility values of users with the same choice will not be too spread out around this average value.

Then, the disutility function corresponding to the choice (τ,p) associated to commodity k is written as:

$$\begin{aligned} C^k(\tau,p) &= \alpha \int_\tau^{\tau+1} [s^k_{\tau,p}(d(k),t) - t]\, dt \\ &+\beta \int_\tau^{\tau+1} [at(k) - \Delta - s^k_{\tau,p}(d(k),t)]^+\, dt \\ &+\gamma \int_\tau^{\tau+1} [s^k_{\tau,p}(d(k),t) - at(k) - \Delta]^+\, dt, \end{aligned}$$

where

$$[x]^+ = \begin{cases} x & \text{if } x \geq 0, \\ 0 & \text{otherwise.} \end{cases}$$

We assume that the disutility function penalty coefficient associated with the path travel time is greater than the penalty coefficient associated with an early arrival at destination, i.e.:

$$\alpha \geq \beta. \tag{4.14}$$

This assumption means that the users prefer to wait at their destination rather than in their car. Furthermore, empirical studies show that this inequality holds (see, for instance, Small (1982)).

The **dynamic user equilibrium conditions** can then be defined as follows. We denote by $\mathcal{NA}$ is the set of feasible assignments and n the assignments $n^k_{\tau,p}$ vector.

More precisely,

$$\mathcal{NA} = \left\{ n = (..., n^k_{\tau,p}, ...),\ k \in \mathcal{K}, (\tau,p) \in \mathcal{SP}(k) \;\middle|\; \begin{array}{rcl} \sum\limits_{(\tau,p)\in\mathcal{SP}(k)} n^k_{\tau,p} & = & n^k, \\ n^k_{\tau,p} & \geq & 0. \end{array} \right\}.$$

Definition 1 *A traffic assignment $n \in \mathcal{NA}$ is said to be* **a dynamic user equilibrium** *if the two following conditions are satisfied for all commodities $k \in \mathcal{K}$:*

1. *All the choices $(\tau,p) \in \mathcal{SP}(k)$ that are used have the same disutility function value.*
2. *All the choices $(\tau,p) \in \mathcal{SP}(k)$ that are unused, produce a disutility function value greater or equal to those of the used choices.*

There exists a finite number of user equilibrium conditions since the number of path and departure time choices is finite and the disutility value of a set of users with same choice (τ,p) is the average disutility value of the users of the choice.

Mathematically the user equilibrium conditions can be written as follows:

Definition 2 *$n \in \mathcal{NA}$ is a* **dynamic user equilibrium** *if the following conditions are satisfied:*

$$\left\{ \begin{array}{rcll} n^k_{\tau,p}\left[C^k(\tau,p) - C^k_{min}\right] & = & 0 & \forall k \in \mathcal{K},\ \forall(\tau,p) \in \mathcal{SP}(k), \\ C^k(\tau,p) - C^k_{min} & \geq & 0 & \forall k \in \mathcal{K},\ \forall(\tau,p) \in \mathcal{SP}(k), \\ n^k_{\tau,p} & \geq & 0 & \forall k \in \mathcal{K},\ \forall(\tau,p) \in \mathcal{SP}(k), \end{array} \right. \tag{4.15}$$

Finally we can state the dynamic user equilibrium conditions as a variational inequality problem. Let $C(n)$ be the vector of disutility functions $C^k(\tau,p)$, $\forall k \in \mathcal{K}$, $\forall(\tau,p) \in \mathcal{SP}(k)$. We mention here the dependency of this global disutility function on the decision variables $n^k_{\tau,p}$.

Theorem 1 *The simultaneous route-departure equilibrium problem defined in (4.15) is equivalent to the following variational inequality problem: Find a point $\tilde{n} \in \mathcal{NA}$ such that:*

$$C(\tilde{n})^t(n - \tilde{n}) \geq 0 \qquad \forall n \in \mathcal{NA}. \tag{4.16}$$

4.3 EXISTENCE OF A USER EQUILIBRIUM

Theorem 2 *The simultaneous route and departure time dynamic user equilibrium conditions of Definition (2) possess a solution.*

Proof: First observe that the feasible set $\mathcal{NA}$ is non-empty, compact and convex.

Second note that the functions $C^k(\tau,p)$ are continuous functions in variables $n = (...; n^k_{\tau,p}; ...)$. Indeed, the functions $C^k(\tau,p)$ are the integrals of the destination arrival time functions $s^k_{\tau,p}(d(k),t)$. These are continuous functions of the load variables $\tilde{f}_a(t)$ that are continuous in the decision variables $n^k_{\tau,p}$. As $C(n)$ is a continuous function on a convex, compact, non-empty set $\mathcal{NA}$, it follows (see Harker and Pang (1990)) there exists a solution for the variational inequality problem (4.16).

4.4 SOLUTION ALGORITHM

The algorithm is based on an iterative scheme that attemps to capture the day-to-day changes of path and departure time that the users of each commodity make. One iteration of the algorithm can be described as follows. The current state of the network (i.e., the arc traversal times) resulting from an assignment of users of each commodity to several choices of path and departure time is first evaluated (Loading step). Then, on the basis of the new arc traversal times, a new choice of path and departure time is computed for each commodity (Computation of a choice step). Finally, the users of each commodity are distributed among the new choices of path and departure time computed in this iteration, and those used in preceding iterations in order that the used choices have the same disutility value (Reallocation step). This process is repeated until each user has found a satisfying departure time and path. The algorithm is illustrated in Figure 4.5.

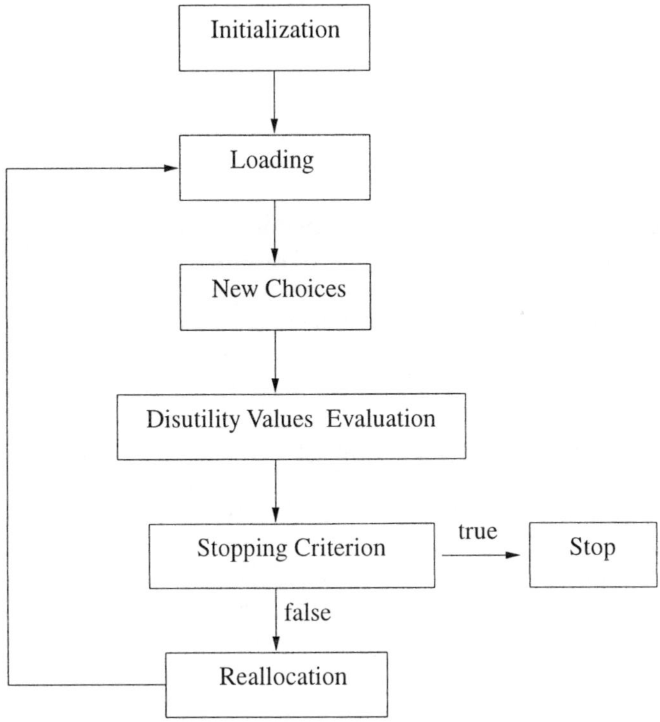

Figure 4.5 Solution Algorithm.

The first step concerns the **initialization** of the process. For each commodity k, we determine the best choice of path and departure time in the static network in which each arc traversal time corresponds to a free flow situation. We denote these paths and times by p and τ respectively. We then assume that all users of each commodity make the same choice (τ, p), namely $n^k_{\tau,p} = n^k$, $\forall k \in \mathcal{K}$.

The second step is the **loading**. The number of users present on each arc ($f_a()$), the effective loads ($\tilde{f}_a()$) and the corresponding traversal times ($D_a()$) are determined according to the current assignment of the users of each commodity k to their choices of departure time and path. The loading is defined to satisfy the users propagation constraints on the network (4.10), (4.11), (4.12), and (4.13). The passing of time is handled by an event drive technique. A detailed description of the loading step is given in Brotcorne (1998)).

On the basis of the traversal times evaluated in the loading, the third step computes a **new choice** of path and departure time (τ, p) for each commodity k. Moreover, it evaluates the **disutility value** $C^k(\tau, p)$ for each used choice associated to each commodity k (i.e, for all (τ, p) for which $n^k_{\tau,p} > 0$ and for the new choices). The algorithm to compute a new choice for a given commodity k searches for the best departure time and path of a single user who is added to the network without altering the link loads. In other words, the arc traversal times are uniquely determined by the effective loads value before adding the new user. To simplify notations, the index k referring to a given commodity is omitted for all variables and the disutility function. The algorithm relies upon the following decomposition principle:

$$\min_{\tau,p} C(\tau, p) = \min_{\tau} \left(\min_{p} C(p|\tau) \right) = \min_{\tau} C^*(\tau).$$

Given a departure time, the problem

$$\min_{p} C(p|\tau)$$

determines the path linking the origin to the destination minimizing the disutility function. The optimal departure time is obtained by minimizing a unidimensional function $C^*(\tau)$. Its analytical expression is unknown, but its value can be computed for any fixed departure time τ by solving the optimal departure time problem.

Let us first consider the optimal path problem. For a given departure time τ from the origin, the marginal user will enter the first arc of his path during the time interval $[\tau, \tau + 1[$ and will encounter the average disutility of the path for departure inside this interval. Similarly to de Palma et al. (1990b), it can be proved that the path with minimal average traversal time will be the one with minimal average disutility value. However, the determination of the path with minimal average traversal time value cannot be solved with classical shortest path algorithms since a sub-path of an optimal one may not be optimal (*c.f.* Brotcorne (1998)). To avoid this difficulty, we modify the assumption according to which the marginal user chooses the path with minimal average path travel time. More precisely, we consider that he chooses the optimal path p^* for a departure from the origin at $\tau + \frac{1}{2}$. Then, we define the value $C(p|\tau)$ as the average disutility of this path p^* for a departure inside $[\tau, \tau + 1[$. It follows from de Palma et al. (1990b) that the optimal path p^* for the marginal user leaving the origin at $\tau + \frac{1}{2}$ can be determined by Dijkstra' s algorithm.

The optimal departure time problem is solved by evaluating the function $C^*(\tau)$ on a subset of departure times $\mathcal{S}' \subset \mathcal{S}$. $\mathcal{S}'$ can, for instance, correspond to the morning rush hour and be determined from lower and upper bounds on travel times of users.

Note that the choice determined with this procedure may not be the minimal average disutility value choice. Moreover, it is possible that its disutility value is not lower that the disutility values of the used choices. In this case, the new choice is not considered.

The fourth step is the **convergence test**. Let C^k_{min} be the smallest disutility value associated with each commodity k. If the following condition is reached for each commodity, the algorithm stops.

$$\max_{k \in \mathcal{K}} \max_{(\tau,p)\,|\,n^k_{\tau,p}>0} \left\{ \frac{C^k(\tau,p) - C^k_{\min}}{C^k_{\min}} \right\} \leq \epsilon. \tag{4.17}$$

The fifth step is a departure time path-swapping process. More precisely, for each commodity k, we swap users from some used choices of path and departure time to the least costly choices. The distribution attempts to equalize the disutility function values of these choices. For each commodity k, the variation of the number of users with choice (τ, p) is given by:

$$n^k_{\tau,p}(\theta^k) = n^k_{\tau,p} + \theta^k dn^k_{\tau,p}.$$

The direction of movement $dn^k_{\tau,p}$ and the step θ^k are determined as follows. Let N^k_{min} be the number of choices for which the minimal value C^k_{min} is reached. The number of users with non-optimal choice will be decreased proportionally to the difference between the disutility function values $C^k(\tau,p)$ of the considered choice, and C^k_{min}. The users taken from the non-minimal choices are then transferred to the least costly ones. The direction of movement for the allocation to the non-optimal choices (τ, p) and the optimal ones $(\underline{\tau, p})$ are thus respectively given by:

$$dn^k_{\tau,p} = C^k_{min} - C^k(\tau,p).$$

$$dn^k_{\underline{\tau},\underline{p}} = -\frac{\sum\limits_{(\tau,p)\neq(\underline{\tau},\underline{p})} dn^k_{\tau,p}}{N^k_{min}}. \tag{4.18}$$

The reallocation step θ^k associated with commodity k is determined by the successive average method (Powell and Sheffi (1982)). More precisely,

$$\theta^k = \min\{B^k, \frac{1}{\ell \bmod L}\}.$$

where ℓ is the iteration counter, L is a fixed number of iterations and B^k the bound on step θ^k to maintain non-negativity of allocations of users to choices. More precisely,

$$B^k = \min_{(\tau,p)\neq(\underline{\tau},\underline{p})} \frac{-dn^k_{\tau,p}}{n^k_{\tau,p}}.$$

Note that the ratio $\frac{1}{\ell \bmod L}$ aims to slow the decrease of step θ^k since it is reduced only after a block of L iterations.

4.5 NUMERICAL RESULTS

We consider a network with 3 links and 2 commodities. These are described in Figure 4.6 and in Tables 4.2, 4.3. Two types of roads are present: roads with high capacity and roads with low capacity.

The arc performance function used in the computation of the arc traversal time is a piecewise linear approximation with 2 breakpoints. It was derived from a function

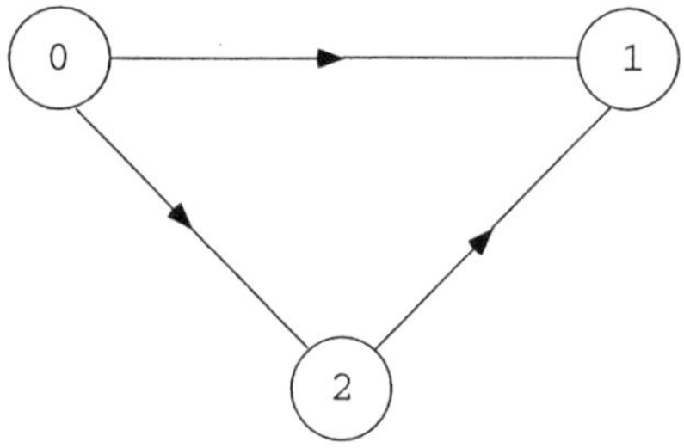

Figure 4.6 A simple network.

Origin	Destination	Length (km)	Congestion density (veh/km)	Free-flow speed (km/h)	Minimal speed (km/h)
0	1	35	100	60	10
0	2	35	200	120	10
1	2	15	100	60	10

Table 4.2 Links Description.

proposed by Jayakrishnan *et al.* (1995) to which the free flow travel time was added. This is a density based link cost function that is monotonically non-decreasing and convex. The parameters used in the disutility function are those calibrated by Small (1982):

$$\begin{aligned} \alpha &= 6.4 \text{ (travel time coefficient)}, \\ \beta &= 3.9 \text{ (penalty for early arrivals coefficient)}, \\ \gamma &= 15.2 \text{ (penalty for late arrivals coefficient)}. \end{aligned}$$

Commodity	Origin	Destination	Desired arrival time
1	0	1	9.00
2	0	2	9.00

Table 4.3 Commodities Description.

All times (departure time and travel time) used in the numerical results are given according to the following time units. One time unit corresponds to 10 minutes. Thus, the desired arrival time which is 9 a.m. is 54 in the time unit system. The solutions for desired time windows with half-width $\Delta = 1$ or 2 time units are described in Tables 4.4 for commodities with 500 users ($n^1 = n^2 = 500$) and in Tables 4.5 for commodities with 600 users ($n^1 = n^2 = 600$). The successive average reallocation

Δ	Commodity	Departure time	Path	Number of users	Average time	Average disutility value	Maximum relative criterion
1	$(0, 1, 54)$	44	0,2,1	4.16	3.30	41.39	0.0016
		45	0,1	64.62	4.86	41.39	
		45	0,2,1	139.66	4.86	41.39	
		46	0,1	34.26	6.31	40.39	
		46	0,2,1	120.76	6.24	41.39	
		47	0,1	23.64	6.47	41.39	
		47	0,2,1	47.22	6.47	41.39	
		48	0,1	12.92	6.36	40.39	
		48	0,2,1	52.76	6.34	41.39	
1	$(0, 2, 54)$	50	0,2	412.66	2.75	18.49	
		52	0,2	87.34	2.56	18.46	
2	$(0, 1, 54)$	45	0,1	40.00	4.34	36.20	0.0113
		45	0,2,1	96.47	4.34	36.20	
		46	0,1	35.06	5.54	36.20	
		46	0,2,1	109.95	5.50	36.20	
		48	0,1	36.11	5.66	36.20	
		48	0,2,1	101.72	5.66	36.20	
		49	0,1	3.37	5.66	36.20	
		50	0,1	15.48	5.40	36.20	
		50	0,2,1	61.84	5.37	36.17	
2	$(0, 2, 54)$	49	0,2	144.84	2.33	15.96	
		50	0,2	82.51	2.52	16.12	
		51	0,2	84.09	2.49	15.94	
		52	0,2	140.57	2.50	16.02	
		53	0,2	47.99	2.40	16.03	

Table 4.4 Solutions obtained for commodities with 500 users.

Δ	Commodity	Departure time	Path	Number of users	Average time	Average disutility value	Maximum relative criterion
1	$(0,1,54)$	44	0,1	22.29	3.97	43.07	0.0091
		44	0,2,1	70.61	4.05	43.27	
		45	0,1	58.81	5.46	42.91	
		45	0,2,1	140.92	5.49	42.96	
		46	0,1	42.15	6.67	43.16	
		46	0,2,1	114.58	6.60	42.88	
		47	0,1	16.03	6.74	43.13	
		47	0,2,1	49.48	6.74	43.12	
		48	0,1	10.67	6.52	42.99	
		48	0,2,1	74.46	6.53	43.27	
1	$(0,2,54)$	49	0,2	134.60	2.31	19.42	
		50	0,2	377.73	2.97	19.37	
		52	0,2	87.67	2.63	19.46	
2	$(0,1,54)$	44	0,1	37.95	4.3	39.99	0.0118
		44	0,2,1	92.63	4.3	39.99	
		45	0,1	49.89	5.78	39.99	
		45	0,2,1	140.39	5.74	40.00	
		46	0,1	20.01	6.25	39.98	
		47	0,1	20.57	6.25	40.01	
		47	0,2,1	123.39	6.25	40.00	
		48	0,1	17.13	6.25	39.99	
		49	0,1	19.43	6.20	40.00	
		49	0,2,1	78.61	6.08	40.23	
2	$(0,2,54)$	49	0,2	196.51	2.48	16.93	
		50	0,2	58.41	2.67	17.10	
		51	0,2	171.44	2.64	16.90	
		52	0,2	97.87	2.66	17.04	
		53	0,2	75.77	2.44	16.93	

Table 4.5 Solutions obtained for commodities with 600 users.

steps are reduced by 1 unit after a block of $L = 50$ iterations. The solutions in Tables 4.4 and 4.5 are the best ones obtained after 6000 iterations, i.e., the ones with the smallest maximum relative criterion (4.17). Their values are given in the last column of Tables 4.4 and 4.5.

For $\Delta = 1, 2$, the users of the first commodity are split on the two possible paths $(0, 1)$ and $(0, 2, 1)$ linking 0 to 1. Even if there is only one path linking 0 to 1, the users of the second commodity are split among different choices of departure times in order to avoid congestion. This emphasizes the importance of the departure time choice in dynamic models. For both arrival time windows, early arrivals or slightly late arrivals at destination are observed for both commodities. For $\Delta = 1$ and $n^1 = n^2 = 500$, the users of commodity 1 with departure times 44 et 45 arrive fairly sharply at the destination, while the users with departure times 46 and 48 arrive respectively slightly early or late. The users of the second commodity experience small early or late penalties for departure times 50 and 52 from the origin. For $\Delta = 2$ and $n^1 = n^2 = 500$, the commodities 1 and 2 users with departure times 45, 46 and 49 arrive early at destination, whereas the users of commodities 1 and 2 with respective departure times 50 and 53 arrive late. For $\Delta = 1$ and $n^1 = n^2 = 600$, the commodity 1 users arrive early for departure times 44 and 45, although the ones with departure times 46 et 48 are respectively slightly early or late. The users of the second commodity arrive early for departure at 49 and slightly early or late for departure times equal to 50 and 52. For $\Delta = 2$ and $n^1 = n^2 = 600$, the users of the first commodity with departure time 44 arrive fairly early, whereas those with departure time 49 arrive slightly late. As for the commodity 2, the users with departure times 49 and 53 arrive slightly early or late.

An increase in incurred penalties occurs when the width of the arrival time window decreases or when the number of commodities increases. Moreover, for a larger arrival time window width, the users are split on a larger number of departure time and path choices. This leads to a decrease of each disutility value.

Finally, note that the described heuristic converges slowly to good solutions. Solutions described in Tables 4.4 and 4.5 are the best ones reached after 6000 iterations, except for the problem defined with 500 commodities users and $\Delta = 1$ for which a solution of quality is determined after 1704 iterations. Note that after 3000 iterations, the maximum relative criterion value for the best known solution was already small for $\Delta = 1$ and reasonably small for $\Delta = 2$. Thus, for commodities with 500 users and $\Delta = 2$, the maximum relative criterion is equal to 0.0208. For commodities with 600 users and $\Delta = 1$, respectively $\Delta = 2$, it was equal to 0.0094 and 0.0143. Computation times on a Sun Ultra Sparc Station is on average equal to 385 seconds for 6000 iterations.

The model proposed in this paper may appear unsuitable for solving real instances size since the computation time required by the algorithm is large even for small instances. However, medium sized network could be considered since the difficulty encountered in the computation of a user equilibrium is mainly related to users interactions. This has been illustrated in our example that was built to point out the congestion effects. Therefore, our approach can be helpful to develop a better understanding of dynamic traffic assignment problems and for studies on medium sized networks in which congestion is frequently observed.

4.6 CONCLUSION

In this paper, we have described a **dynamic user equilibrium model** for traffic assignment in urban transportation networks. Given the number of users traveling from the same origin to the same destination and willing to reach it at the same desired arrival time, the model determines both path and departure time of the users so that none of them can decrease his disutility by unilaterally changing his departure time or his path. The disutility measure is given by a weighted sum of the total path travel time and penalties for arrivals outside a desired time window. For the users of a given choice of path and departure time, the total path travel time is computed according to the number of users of the choice and the number of users preceding them. Moreover, it is defined to preserve the FIFO queue discipline. Users with the same choice of departure time and same first arc of selected path are uniformly loaded on the first arc of their path during a time interval whose lower bound is the departure time.

A finite dimensional variational inequality formulation of the problem was given and the existence of a dynamic user equilibrium was proved. A heuristic method was proposed to compute a dynamic user equilibrium defined by the model. Numerical results on small networks are promising even if the more realistic modeling makes good solutions more difficult to achieve. The results emphasize the importance of the departure time choice for users.

References

Bernstein D., Friesz T.L., Tobin R.L. and Wie B.W. (1993). "A variational control formulation of the simultaneous route and departure-time choice equilibrium problem", *Transportation and Traffic Theory*, 107-126.

Branston D. (1975). "Link Capacity Functions : A Review", *Transportation Research*, 10, 223-236.

Brotcorne L. (1998). *Approches opérationnelles et stratégiques des problèmes de trafic routier*, Doctoral dissertation, Université Libre de Bruxelles.

de Palma A. and Hansen P. (1990). "Optimum departure times for commuters in congested networks", *Annals of Operations Research*, 25, 279-290.

de Palma A., Hansen P. and Labbé M. (1990b). "Commuters paths with penalties for early or later arrival time", *Transportation Science* 24, 276-286.

Friesz T.L., Bernstein D., Smith T.E., Tobin R.L. and Wie B.W. (1993). "A variational inequality formulation of the dynamic network user equilibrium problem", *Operational Research*, 41, 179-191.

Friesz T.L., Bernstein D. and Stough R. (1996), "Dynamic systems, variational inequalities and control theoretic models for predicting time-varying urban network flows", *Transportation Science*, 30, 14-31.

Gendreau M. and Bouzaiene-Ayari B. (1997). "A discrete-time macroscopic dynamic traffic assignment model", Publication CRT 96-31, Centre de recherche sur les transports, Université de Montréal.

Janson B.N. (1993). "Dynamic traffic assignment with schedule delays", presented at: The 72nd Annual Meeting of the Transportation Research Board, Washington DC.

Jayakrishnan R., Wei K. Tsai and Chen A.(1995). "A dynamic traffic assignment model with traffic-flow relationships", *Transportation Research* , 3C, 51-72.

Leblanc L.J., Hegalson R.V.and Boyce D.E. (1985). "Improved efficiency of the Franck-Wolfe algorithm for convex network programs", *Transportation Science*, 19, 445-461.

Harker P.T. and Pang J.-S. (1990). "Finite-Dimensional Variational Inequality and Nonlinear Complementarity Problems: A Survey of Theory, Algorithms and aApplications", *Mathematical Programming*, 48, 161-220.

Powell W. and Sheffi Y. (1982). "The convergence of equilibrium algorithms with predetermined step sizes", *Transportation Science*, 16, 44-55.

Ran B., Boyce D.E and Leblanc L.J. (1992). "Dynamic user optimal departure time and route choice model: a bilevel, optimal-control formulation", Advance Working Paper 12, University of Illinois at Chicago.

Ran B. and Boyce D.E. (1994). "Combined departure time/route choice models", in *Dynamic urban transportation network models, Lecture Notes in Economics and Mathematical Systems* 417, Springer Verlag.

Ran B., Hall R.W. and Boyce D.E. (1996). "A link-based variational inequality model for dynamic departure time/route choice", *Transportation Research* , 30B, 31-46.

Small K. (1982). "The scheduling of consumer activities: work trips", *Am. Econ. Rev.* 82, 467-479.

Smith M.J. (1979). "The existence, uniqueness and stability of traffic equilibria", *Transportation Research*, 13B, 295-304.

Wie B.W., Tobin R.L., Friesz T.L. and Bernstein D. (1995). "A discrete nested cost operator approach to the dynamic network user equilibrium problem", *Transportation Science*, 29, 79-92.

5 ESTIMATION OF TRAVEL DEMAND USING TRAFFIC COUNTS AND OTHER DATA SOURCES

Ennio Cascetta

Alessandra A. Improta

Abstract: Over the last decades, considerable work has been devoted to improve the quality of travel demand estimators by using cheap and easily collectable traffic counts.

In this paper it is first presented a review of the methodology for estimating within-day static O/D demand flows by efficiently combining traffic counts with all other available information, taking into account whether the information is experimental (Classic inference) or "a priori" (Bayesian inference). Within this framework, an analysis of different solution methods is carried out, both in case of link costs known and unknown (congested networks).

Subsequently, it is proposed an extension of previous results to the case of time-varying (within-day dynamic) demand and flows, through within-day dynamic assignment models, by using simultaneous and sequential estimators of O/D matrices.

Finally, the possibility of using aggregate information, namely traffic counts, in order to improve the estimation of demand models parameters is discussed, also considering extensions to the case of joint estimation of O/D flows and demand parameters.

5.1 STATIC ESTIMATION OF O-D DEMAND FLOWS FROM TRAFFIC COUNTS

The problem that will be dealt with in this section is that of "improving" the estimates of the present Origin-Destination demand flows by combining direct and model estimators with the traffic counts carried out on some links of the transportation supply system (network). More formally, the problem can be defined as that of estimating

M. Gendreau and P. Marcotte (eds.), Transportation and Network Analysis: Current Trends, 71–94.

the O-D matrix by combining traffic counts with all other available information in a statistically efficient way.

Estimation models of O-D matrices using traffic counts have received considerable attention in recent years both from the theoretical and from the empirical point of view. This is easily understandable considering the cost and complexity of sampling surveys for demand, as well as the inevitable lack of precision related both to the direct estimation and to the model estimation of O-D flows. On the other hand, the flows measured on some network links (i.e. traffic counts) are a source of information on transportation demand which is cheap and easily obtainable, also by automatic means. Furthermore, in many applications O-D flows estimation is essentially aimed at predicting traffic flows deriving from changes in the supply system (network). In other words, the focus is on estimating and predicting "aggregated" values of the O-D matrix, that is the traffic flows, and it is expected that a matrix capable of reproducing some of such aggregates with sufficient precision will give better predictions in correspondence with network changes.

Many estimation methods have been proposed in the literature for the O-D traffic count based estimation. Some are based on the maximum entropy principle (Van Zuylen and Willumsen, 1980; Willumsen, 1984); others are derived from classical statistical approaches, such as the Generalized Least Square (GLS) estimator (Cascetta, 1984; Bell, 1991) or Bayesian inference (Maher, 1983). Other approaches can be found in literature (Mc Neil, 1983; Nguyen 1983; Ben-Akiva *et al.*, 1985). A systematic review of the statistical formulations of different estimators can be found in Cascetta and Nguyen (1986).

5.1.1 *Problem statement*

To state formally the problem, it is necessary to express the relationship between the vector of observed flows and the unknown O-D demand flows.

The flow f_a, using the link a of the network in the reference period, can be expressed in general as the sum of flows on the paths including that link:

$$f_a = \sum_k \delta_{ak} h_k$$

Path flows, in turn, can be expressed as the product of the O-D demand flow by the percentage (fraction) of users who choose each path:

$$f_a = \sum_k \delta_{ak} h_k = \sum_k \delta_{ak} \sum_i p_{ki} d_i \tag{5.1}$$

where δ_{ak} are the elements of the link-path incidence matrix Δ and p_{ki} is the percentage of the flow d_i between the ith O-D pair using path k. Note that in the previous expression the variables introduced (link flows, O-D flows, path flows and percentages of path choice) indicate the "true" values relative to the system and to the reference period under study.

Equation (5.1) can be expressed differently as

$$f_a = \sum_i d_i \sum_k \delta_{ak} p_{ki} = \sum_i m_{a,od} d_i \tag{5.2}$$

or

$$f_a = m_a^T d$$

where $m_{a,od} = \sum_k \delta_{ak} p_{ki}$ is the assignment fraction, i.e. the fraction of the flow d_i using the link a and m_a is the column vector obtained by ranking these coefficients. Using a matrix notation, (5.2) becomes

$$f = \Delta h = \Delta P d = M d \tag{5.3}$$

where the matrices Δ and $\boldsymbol{P}$ have a number of rows and columns equal to the links where the traffic counts are available and to the paths including them. The matrix $\boldsymbol{M}$, or *assignment matrix,* has dimensions $(n_a \times n_{OD})$ equal to the product of the number of links (or the number of links with counts) by the number of O-D pairs on the network. The relationship (5.3) connecting link flows and O-D demand vector is known as the assignment relationship or map.

When different paths exist between each O-D pair, the elements of this relationship, $m_{a,od}$, are not defined in a unique fashion and must therefore be estimated by the analyst. Path choice and network assignment models usually adopted (Sheffi, 1985; Cascetta, 1998; Ortuzar and Willumsen, 1994)) provide methods for obtaining estimates $\hat{p}_{ki}$ of the fraction p_{ki} and therefore, through the equation (5.2), estimates $\hat{m}_{a,od}$ of the fractions $m_{a,od}$.

In the case of pre-trip, deterministic or probabilistic path choice models, usually adopted for road networks or public transport services with low frequency and high regularity, the percentages $\hat{p}_{ki}$ can be expressed as probabilities of choosing each path k connecting the ith O-D-pair as a function of the path costs vector $\boldsymbol{g}$:

$$\hat{p}_{ki} = p[k/i](g) \tag{5.4}$$

In the case of mixed pre-trip/en-route path choice models, usually used for high-frequency public transport networks (Spiess and Florian, 1989; Nguyen and Pallottino, 1986), the percentages (probability) of path use can be obtained starting from those of hyperpath j use, $p_j[j/i]$, in turn functions of the vector of the hyperpath costs $\boldsymbol{x}$, and of the probabilities q_{kj} of following the path k within each hyperpath j (Cascetta, 1998; Cascetta and Cantarella, 1998):

$$\hat{p}_{ki} = \sum_j q_{kj} p[j/i](x) \tag{5.5}$$

To make explicit the dependence of the assignment matrix elements $\hat{m}_{a,od}$ on the path choice model, and through this on the link costs $\boldsymbol{c}$, the matrix $\hat{M}$ can be formally expressed as:

$$\hat{M} = \Delta \hat{P}(c) \tag{5.6}$$

$$M = \Delta Q \hat{P}(c) \tag{5.7}$$

obtained by substituting the expressions (5.4) and (5.5) in (5.2) for road and public transport networks respectively and leaving as understood the relationship between additive path costs and link cost $C^{ADD} = A^T c$.

If the link and path costs are known, an estimate $\hat{M}$ of the "true" assignment matrix $\boldsymbol{M}$ can be calculated through the path choice models (5.4) and (5.5).

It is to be expected that the estimate $\hat{M}$ should diverge from the true assignment matrix $\boldsymbol{M}$ because of the different approximations implicit in the assignment model (network extraction, cost functions, path choice model, etc.). So by substituting $\hat{M}$ in the equation (5.3), a vector ε^{SIM} of errors deriving from the assignment model, should be added:

$$f = Md = (\hat{M} + E^{SIM})d = \hat{M}d + \varepsilon^{SIM} \tag{5.8}$$

where E^{SIM} is the matrix of deviations between the true assignment matrix and that obtained with the assignment model and ε^{SIM} is the vector of deviations, or assignment errors, between the flow vector resulting from the assignment of "true" demand and "true" flows.

In other words, if the "true" vector of the demand flows $\boldsymbol{d}$ were known, its assignment to the network would produce a flow vector $\boldsymbol{v}$:

$$v = Md = v(d) \tag{5.9}$$

which would diverge from the "true" link flows vector $\boldsymbol{f}$; these deviations are the components of the vector ε^{SIM}:

$$f = v + \varepsilon^{SIM}$$

A further cause of error is related to flow counts. Like all the measures, traffic counts are affected by errors depending, among other things, on the technique used (manual counts, automatic counts, etc.). Furthermore, the counts are usually carried out over several days, sometimes different for different network links, while the "true" demand vector $\boldsymbol{d}$ represents the average O-D flows in periods with similar characteristics (e.g. average weekday peak hour). Basically, if $\hat{f}$ is the measured flows vector, this will diverge from the "true" vector $\boldsymbol{f}$ of a residual vector ε^{OBS} representing measurement errors:

$$\hat{f} = f + \varepsilon^{OBS} \tag{5.10}$$

By combining equation (5.8) and (5.10), it is possible to express the relationship between the count vector $\hat{f}$, the assignment matrix $\hat{M}$ and the "true" O-D demand flows vector as:

$$\hat{f} = \hat{M}d + \varepsilon^{SIM} + \varepsilon^{OBS} = v(d) + \varepsilon \tag{5.11}$$

where the vector ε is the algebraic sum of the vectors ε^{SIM} and ε^{OBS}.

It is usually assumed that the assignment model and the counts are unbiased estimators of the "true" flows, that is that the vector ε is a zero mean random vector $E(\varepsilon) = 0$. The experimental evidence at present available confirms this assumption (Bovy and Jansen, 1983; Cascetta, Nuzzolo, Velardi, 1986).

The information on the O-D demand contained in traffic counts, represented by the stochastic equations system (5.11), is not usually sufficient to estimate the vector $\boldsymbol{d}$. This is evident from the fact that, even assuming the errors vector ε to be zero, the number of independent equations of the linear system expressed by (5.3) is usually significantly lower than the number of unknowns, i.e. the number of elements of the demand vector $\boldsymbol{d}$ to be estimated.

If we add to this the fact that the vector ε is in general different from zero, and this means that the linear equations system $\hat{f} = \hat{M}d$ may not admit a solution, it is evident that the information on the vector $\boldsymbol{d}$ given by the counts must be combined with that from other sources to obtain an estimate of the unknown O-D demand.

The additional information can be of two types, corresponding to two different statistical approaches: sampling or experimental information derived from demand direct surveys, or non-experimental information representing the a priori knowledge of the analyst, such as old estimates to be updated or estimates from an O-D demand model. In the former case reference can be made to the classic theory of statistical interference, while in the latter recourse is made to the Bayesian theory of estimation. The two methods, whose statistical foundations will be described in the following sections, introduce many estimators, which can, however, be represented in a formally analogous way.

In fact, if $\hat{d}_i$ is an initial estimate of the demand vector generic component, and $\hat{d}$ is the whole vector representing the initial information, or the information on demand vector not given by the flows, the problem of estimating the O-D demand using traffic counts can be expressed in a general form as:

$$d^* = \underset{x \geq 0}{\operatorname{argmin}} \lfloor z_1(x, \hat{d}) + z_2(v(x), \hat{f}) \rfloor \tag{5.12}$$

where $\boldsymbol{x}$ is the unknown demand vector. The two functions $z_1(x, \hat{d})$ and $z_2(v(x), \hat{f})$ can be considered as "distance" measures of the unknown demand vector $\boldsymbol{x}$ from the a priori estimate $\hat{d}$ and of the flows vector obtained by assigning $\boldsymbol{x}$ to the network, $\boldsymbol{v}$ $(\boldsymbol{x})$, from the traffic counts vector $\hat{f}$ respectively.

The problem of estimating demand by using traffic counts, expressed in the form of (5.12), can therefore be interpreted intuitively as that of computing the vector $\boldsymbol{d^*}$ closest to the a priori estimate $\hat{d}$, and that, once assigned to the network, produces the flows closest to those measured $\hat{f}$.

In the most general case, the functional form of the two terms z_1 and z_2, as has been said, depends on the type of information available (experimental or non-experimental) and on the probability laws associated with such information. The statistical bases of the various estimators and their resulting functional forms will be described in the following.

5.1.2 Classic estimators

Vector $\boldsymbol{d}$ classic estimators can be deduced through the Maximum Likelihood theory or the Generalized Least Squares *(GLS)* theory, depending on whether explicit assumptions on the probability distribution of random residuals ε^{SIM} and ε^{OBS} are made or not.

Maximum Likelihood Estimators. Maximum Likelihood (*ML*) estimators d^{ML} of the unknown vector $\boldsymbol{d}$ are obtained by maximizing the probability of observing the results of sampling surveys and the flows counted on network links. Under the generally acceptable assumption that these two probabilities are independent, the Maximum Likelihood estimator can be expressed as:

$$d^{ML} = \underset{x \in S}{\operatorname{argmax}} \lfloor lnL(n|x) + lnL(\hat{f}|x) \rfloor . \tag{5.13}$$

where

$\mathbf{x}$ is the "unknown" demand vector of dimensions $(n_{OD} \times 1)$ whose components x_{rs} represent the trip flows between the O-D pair (r, s), indicated from now on with the double index instead of the single index i.

$\mathbf{n}$ is the vector of results of direct demand surveys with dimensions $(n_{OD} \times 1)$. Components of the vector $\boldsymbol{n}$, n_{rs}, represent the number of observed trips in the sample between the O-D pair (r, s) also known as demand counts.

$\hat{\mathbf{f}}$ is the observed flows or traffic counts vector, of dimension $(n_a \times 1)$ with components $\hat{f}_a$.

$lnL(\mathbf{n}|\mathbf{x})$ is the log-likelihood function of the demand counts conditional to the value of the demand vector, or rather the logarithm of the probability of observing the sampling vector $\boldsymbol{n}$ if $\boldsymbol{x}$ is the (true) demand vector.

$lnL(\hat{\mathbf{f}}|\mathbf{x})$ is the log-likelihood function of the traffic counts, or rather the logarithm of the probability of observing the vector of the counts $\hat{f}$ if $\boldsymbol{x}$ is the (true) demand vector.

S is the feasibility set of the (true) demand vector, usually coincident with the non-negative octant, or $S = \{x : x \geq 0\}$.

The *Maximum Likelihood* estimators can therefore be obtained by solving the constrained maximization problem expressed by (5.13). To specify these estimators it is necessary to explicit the log-likelihood functions $lnL(n|x)$ and $lnL(\hat{f}|x)$, or to formulate hypotheses on the probability laws conditional to the demand vector $\boldsymbol{x}$, of demand counts $\boldsymbol{n}$ and of traffic counts $\hat{f}$.

With respect to *traffic counts*, it is usually assumed that they are random variables whose mean is given by the flows $\boldsymbol{v}(\boldsymbol{x})$ obtained by assigning the demand $\boldsymbol{x}$, as expressed by (5.11) and by the further hypothesis of zero mean of the vector ε^{SIM}. The most widely used probability laws are Poisson and Multivariate Normal.

If it is assumed that the traffic counts on each link a are independently distributed as *Poisson* random variables with mean equal to values $v_a(x)$ (Van Zuylen and Branston, 1982), that is:

$$E\lfloor \hat{f}_a \rfloor = v_a(x) = \hat{m}_a^T x$$

the probability of observing the vector $\hat{f}$ is given by the product of the probability of observing the individual components:

$$L(\hat{f}|x) = \prod_{a=1\ldots n_A} \frac{exp(-v_a(x))v_a(x)^{\hat{f}_a}}{\hat{f}_a!} \tag{5.14}$$

and the log-likelihood function becomes:

$$lnL(\hat{f}|x) \cong \sum_{a=1\ldots n_A} (\hat{f}_a ln v_a(x) - v_a(x)) + cost \tag{5.15}$$

where the constant indicates summation terms not depending on the unknown demand vector $\boldsymbol{x}$ and therefore irrelevant to the maximization problem (5.13)(Van Zuylen and Willumsen, 1980; Spiess, 1983a).

If traffic counts are jointly distributed according to a *Normal Multivariate* random variable with a mean vector equal to $\boldsymbol{v}(\boldsymbol{x})$ and a variance-covariance matrix $\boldsymbol{W}$ (Maher, 1983; Cascetta 1984, 1986), or

$$f \leftarrow MVN(v(x), W)$$

the probability function of observing the vector $\hat{f}$ is proportional to:

$$L(\hat{f}|x) \propto exp[-\frac{1}{2}(\hat{f} - v(x))^T W^{-1}(\hat{f} - v(x))] \tag{5.16}$$

and the log-likelihood function becomes:

$$lnL(\hat{f}|x) = -\frac{1}{2}(\hat{f} - v(x))^T W^{-1}(\hat{f} - v(x)) + cost. \tag{5.17}$$

The log-likelihood function of *demand counts* depends on the method used to obtain the sample, or on the type of sampling adopted. The simplest case is that of stratified random sampling by zone of origin. This is the case when a simple random sample of n_r trips is extracted from the d_r trips departing from each origin r (for example, sampling at the toll entrances of a motorway network, or at the cordon sections of the study area). On this hypothesis it can be assumed that the number of trips sampled from each origin to all the destinations is distributed as a multinomial random variable and the probability of observing the whole vector $\boldsymbol{n}$ is the product of these variables extended to all the origins:

$$L(n|x) = \prod_r \lfloor (n_r! / \prod_s n_{rs}) \prod_s (n_{rs}/x_r)^{n_{rs}} \rfloor \tag{5.18}$$

where x_{rs}/x_r is the unknown probability of observing a trip with destination s.

From (5.18) the log-likelihood function is equal to:

$$lnL(n|x) = \sum_{rs} n_{rs} ln x_{rs} + cost \tag{5.19}$$

with the constraint that the trips generated in each zone r are equal to those counted d_{rs}, i.e.,

$$S = \{x : \sum_s x_{rs} = d_r; \qquad x \geq 0\}$$

If the number of trips sampled at each origin is sufficiently high (a few dozen), the multinomial variable can be closely approximated by the product of Poisson variables (one for each O-D pair) with parameter equal to the mean $\alpha_r x_{rs}$, where α_r is the sampling rate associated with origin r:

$$\alpha_r = \frac{n_r}{d_r}.$$

In this case the functions $L(n|x)$ and $lnL(n|x)$ given by (5.18) and (5.19) can be approximated by:

$$L(n|x) = \prod_{rs} \frac{exp(-\alpha_r x_{rs})(\alpha_r x_{rs})^{n_{rs}}}{n_{rs}!} \tag{5.20}$$

and

$$lnL(n|x) = \sum_{rs} (n_{rs} ln(\alpha_r x_{rs}) - \alpha_r x_{rs}) + cost \tag{5.21}$$

For more complex sampling methods, expressions analogous to those described can be deduced; in applications, however, expressions (5.19) and (5.21) are often used as reasonable approximations.

Basically, the Maximum Likelihood estimator d^{ML} is obtained by substituting in the expression (5.13), the expression (5.15) or (5.17) for the log-likelihood function of traffic counts, and the expressions (5.19) or (5.21) for demand counts.

Generalized Least Square Estimators. The other type of estimator derived within the classic theory of statistical interference is that of *Generalized Least Squares (GLS)*. The *GLS* estimator gives the estimate of an unknown parameters vector, in this case the demand flows vector, starting from a system of stochastic equations, linear in the parameters vector. This equations system can be obtained by combining the information on demand contained in traffic counts, expressed by the equation (5.11), and in the direct estimate $\hat{d}$ of the demand vector, obtained by starting from demand counts (Cascetta, 1984):

$$\begin{array}{rclcrclcrcl} \hat{f} & = & \hat{M}x + \varepsilon^{SIM} & \quad & E(\varepsilon) & = & 0 & \quad & Var[\varepsilon] & = & W \\ \hat{d} & = & x + \eta & & E(\eta) & = & 0 & & Var[\eta] & = & Z \end{array} \tag{5.22}$$

where $\hat{d}$ is the O-D demand flows vector, of dimensions $(n_{OD} \times 1)$, whose components $\hat{d}_{rs}$ represent the demand sampling estimates. For example, in the case of rate α simple random sampling, these estimates will be:

$$\hat{d}_{rs} = \frac{n_{rs}}{\alpha}$$

The vector η appearing in (5.22) is the sampling error vector, of dimensions $(n_{OD} \times 1)$, whose components represent the deviations between the true unknown demand vector $\boldsymbol{x}$ and the sampling estimates. On the assumption that the estimators adopted are correct, the vector η has zero means. The variance-covariance matrix $\boldsymbol{Z}$ elements can be estimated by using the expressions relative to variances and covariances of the O-D demand sampling estimates.

The *GLS* estimator of the demand vector can therefore be expressed as:

$$d^{GLS} = \underset{x \in S}{\operatorname{argmin}} \lfloor (\hat{d} - x)^T Z^{-1} (\hat{d} - x) + (\hat{f} - \hat{M}x)^T W^{-1} (\hat{f} - \hat{M}x) \rfloor \tag{5.23}$$

Usually, expression (5.23) is applied in a simplified form assuming that the matrices $\boldsymbol{Z}$ and $\boldsymbol{W}$ are diagonal, i.e. ignoring the covariances between the components of the vector ε^{SIM} and η, both because these are difficult to deduce and to reduce memory space and computing times. In this case (5.23) becomes:

$$d^{GLS} = \underset{x \geq 0}{\operatorname{argmin}} \left[\sum_{rs} \frac{(\hat{d}_{rs} - x_{rs})}{Var[\eta_{rs}]} + \sum_{l} \frac{(\hat{f}_a - \sum_{rs} \hat{m}_{a,rs} x_{rs})^2}{Var[\varepsilon_a]} \right] \tag{5.24}$$

which coincides with the expression (5.13) previously presented. The intuitive interpretation given for (5.13) can be extended to (5.24).

Note also the role of information on the vector $\boldsymbol{d}$ contained in traffic counts. In fact, if this information did not exist, the second term of equations (5.23) and (5.24) would disappear and the estimate of d^{GLS} would coincide with $\hat{d}$ in that this value annuls the quadratic objective function to values always non-negative.

Similar considerations can be made for the Maximum Likelihood estimators previously described, although in this case demonstrating the coincidence of d^{ML} with $\hat{d}$ on the assumption of absence of information on the counts is less immediate.

From the formal point of view, it can be verified that the *GLS* estimator coincides with that of Maximum Likelihood if both the traffic counts and demand estimators $\hat{d}_{rs}$ are assumed to be distributed as Multivariate Normal random variables with mean equal to $\boldsymbol{x}$ and $\hat{M}x$ and dispersion matrices $\boldsymbol{Z}$ and $\boldsymbol{W}$ respectively.

5.1.3 Bayesian estimators

Bayesian methods allow the estimation of unknown parameters by combining experimental, or sampling, information with non-experimental, or "subjective", information. In this particular case, the experimental information on the unknown demand vector is that relative to traffic counts, while non-experimental information may be relative to old O-D matrix estimates to be updated, to estimates obtained with demand models or simply to the analyst "expectations". In each case, $\hat{d}$ will indicate the demand vector made up by non-experimental estimates. Bayesian estimators are obtained from the a posteriori probability function $h(x|\hat{f},\hat{d})$ of the unknown demand vector $\boldsymbol{x}$ conditional to the a priori information $\hat{d}$ and the experimental information $\hat{f}$. According to Bayes theorem, the a posteriori probability is proportional to the product of the a priori probability function $g(x|\hat{d})$, expressing the distribution of subjective probability attributed to the unknown vector (given the a priori estimate $\hat{d}$), by the probability, or Likelihood, $L(\hat{f}|x)$ of experimentally observing the traffic counts $\hat{f}$ conditional to the unknown demand vector $\boldsymbol{x}$:

$$h(x|\hat{f},\hat{d}) \propto L(\hat{f}|x)g(x|\hat{d}) \tag{5.25}$$

A family of Bayesian estimators of demand flows d^B can be obtained by maximizing the a posteriori probability, or its natural logarithm:

$$d^B = \underset{x \in S}{\operatorname{argmax}} \lfloor ln g(x|\hat{d}) + ln L(\hat{f}|x) \rfloor \tag{5.26}$$

The specification of Bayesian estimators therefore depends on the assumptions made for the probability functions $L(\hat{f}|x)$ and $g(x|\hat{d})$. With respect to the function $L(\hat{f}|x)$, equations (5.15) and (5.17), corresponding to the assumptions of independent Poisson and Normal Multivariate random variables respectively, can be used.

The a priori probability function, $g(x|\hat{d})$, can be specified in different ways; the formulations presented in literature are described below.

If it is assumed that the unknown demand vector is a *multinomial random variable* resulting from the distribution of total trips between all possible O-D pairs with probabilities π_{rs} given by the matrix $\hat{d}$:

$$\pi_{rs} = \frac{\hat{d}_{rs}}{\hat{d}_{..}} \qquad \hat{d}_{..} = \sum_{rs} \hat{d}_{rs}$$

the function $g(x|\hat{d})$ can be written as:

$$g(x|\hat{d}) = \frac{(\sum_{rs} x_{rs})!}{\prod_{rs} x_{rs}!} \prod_{rs} (\hat{d}_{rs}/\hat{d}..)^{x_{rs}} \tag{5.27}$$

Using the Stirling approximation ($lnx! \cong xlnx - x$), the logarithm of (5.27) can be expressed as:

$$lng(x|\hat{d}) = \sum_{rs} x_{rs} ln(\sum_{rs} x_{rs}) + \sum_{rs} x_{rs} ln(\hat{d}_{rs}/\hat{d}..x_{rs} \tag{5.28}$$

Furthermore, if the total number of trips ($\sum_{rs} x_{rs} = \hat{d}..$) is assumed to be constant, (5.28) is further simplified in:

$$lng(x|\hat{d}) = -\sum_{rs} x_{rs} ln(x_{rs}/\hat{d}_{rs}) + cost \tag{5.29}$$

The function (5.29), with the sign changed, is known in literature as the *entropy function* of the unknown vector $\boldsymbol{x}$ (Wilson, 1970; Van Zuylen and Willumsen, 1980; Nguyen, 1983b).

Alternatively, it can be assumed that the unknown vector components x_{rs} are independently distributed *Poisson random variables*, of mean (parameter) equal to $\hat{d}_{rs}$. In this case the function $g(x/\hat{d})$ becomes:

$$g(x|\hat{d}) = \prod_{rs} \frac{exp(-\hat{d}_{rs})^{x_{rs}}}{x_{rs}!} \hat{d}_{rs}^{x_{rs}} \tag{5.30}$$

that, using Stirling's approximation, can be expressed as:

$$lng(x|\hat{d}) = -\sum_{rs} x_{rs}[ln(x_{rs}/\hat{d}_{rs}) - 1] + cost \tag{5.31}$$

The function (5.31), with the sign reversed, is also known as the *information function* of the vector $\boldsymbol{x}$.

Finally, it can be assumed that the vector $\boldsymbol{x}$ is distributed according to a *Multivariate Normal random variable* with mean $\hat{d}$ and variance covariance matrix Z_B (Maher, 1983), and in this case the probability function is proportional to:

$$g(x|\hat{d}) \propto exp[-\frac{1}{2}(x - \hat{d})^T Z_B^{-1} (x - \hat{d})]$$

and its logarithm is:

$$lng(x/\hat{d}) = -\frac{1}{2}(x - \hat{d})^T Z_B^{-1} (x - \hat{d}) + cost \tag{5.32}$$

It is easily seen that, if the a priori probability function $g(x|\hat{d})$ and the traffic counts probability function $L(\hat{f}|x)$ are both assumed to be Multivariate Normal variables, i.e. if the expressions (5.17) and (5.32) are substituted in the general expression of the Bayesian estimator (5.26), the resulting Bayesian estimator is formally analogous to the Generalized Least Squares estimator d^{GLS}. However, the correspondence between the two estimators is only formal, in that there are different interpretations of vector $\hat{d}$ and of the variance-covariance matrices $\boldsymbol{Z}$ and Z_B. In the *GLS* estimator, in fact, the vector $\hat{d}$ is deduced as a direct demand estimate from sampling surveys and the matrix $\boldsymbol{Z}$ includes the variances and covariances of this estimate, while in the Bayesian estimator $\hat{d}$ represents the O-D demand vector a priori estimate, and Z_B is made up of variances and covariances summarizing the analyst confidence in the a priori estimate.

The formal analogy of the two estimators should, however, be considered an advantage since it allows the use of the same models and algorithm in very different estimation situations. This generality of *GLS* estimators is one of the reasons that have favored a more and more widespread use of its applications.

5.1.4 Applicative issues

As we have said, different estimators of O-D demand flows combining traffic counts $\hat{f}$ and other information $\hat{d}$ can be expressed in a general form as the vector $\boldsymbol{d^*}$ solving a constrained minimization problem:

$$d^* = \underset{x \in S}{\operatorname{argmin}} \lfloor z_1(x, \hat{d}) + z_2(v(x), \hat{f}) \rfloor \tag{5.33}$$

The application of these methods in practice poses many problems that will be briefly dealt with below.

The *choice of functional form* from among the various possibilities described, obviously depends on the available information about O-D demand flows and therefore on the estimation context (classic or Bayesian). As we saw before, the Generalized Least Squares estimator is robust in so far as it can be adopted in both cases and, as a classic estimator, does not require explicit assumptions on probability, or likelihood, law of traffic and demand counts. Obviously this robustness is paid for in terms of statistical properties which are less satisfactory than those of other estimators in the case that traffic and demand counts probability distributions are known.

In literature there is a number of simulation studies concerning the comparison of statistical performances among the various estimators. Results would seem to confirm the theoretical indications and point to the fact that the *GLS* estimator gives more stable results compared with the various cases considered regarding the quality and quantity of information contained in $\hat{d}$ and $\hat{f}$(Di Gangi, 1988; Cascetta and Nguyen, 1986).

One aspect of *GLS* estimators use relates to the *definition of variance-covariance matrices* $\boldsymbol{Z}$ and $\boldsymbol{W}$. This issue arises only in the case of *GLS* and represents a further level of freedom in that for the other functional forms of $z_1(\cdot)$ and $z_2(\cdot)$ variances and covariances are implicitly defined by the distributions underlying them. Thus, for example, the form (5.15) for the $z_2(v(x), \hat{f})$ implies the assumption that the counts are Poisson independent variables, or that the deviations from the averages $\boldsymbol{v(x)}$ are independent ($Cov(\varepsilon_{l\varepsilon m})=0$) and their variance is equal to the average ($Var[\varepsilon_c]=v_l(x)$).

As has been seen, from the practical point of view, the covariances between the elements of ε and η are usually ignored, or matrices $\boldsymbol{Z}$ and $\boldsymbol{W}$ are assumed to be diagonal. When $\hat{d}$ is a sample estimate of the demand, the sampling error variance η_{rs} can be expressed by starting from the type of sampling adopted. When the Bayesian method is used, instead, variances are a measure of the analyst "confidence" in a priori estimates of demand and therefore cannot be univocally defined. For what concerns the variances of residuals ε^{OBS}, experimental relationship expressing the variation coefficients C_v of the assignment errors for different assignment models as a function of measured flows can be used.

5.1.5 *Solution methods*

The main computational problem relates to *computation of the assignment map* $\boldsymbol{v}(\boldsymbol{x})$, that is calculation of the *assignment matrix* $\hat{M}$, expressed by (5.6) and (5.7). The elements $\hat{m}_{a,rs}$ forming the matrix $\hat{M}$ depend on path choice probabilities, (5.4) and (5.5), which in turn are functions of path (or hyperpath) costs and, basically, of the link costs, as formally expressed by equations (5.6) and (5.7).

In general, given the path or hyperpath choice model, the calculation of the assignment matrix as a function of the link costs vector $\boldsymbol{c}$ can be carried out by means of simple network loading algorithms, either deterministic (all-or-nothing) or probabilistic (stochastic network loading).

In the case of congested networks the link costs depend on the link flows vector $\boldsymbol{f}$. There are therefore two different cases according to whether the link costs vector is known or not.

Uncongested networks (link costs known). When the network is uncongested, or moderately congested, and consequently the link costs can be estimated independently of the flows, it can be assumed that the link costs vector is known, or rather that an estimate $\hat{c}$ is known. Alternatively, the link costs can be assumed to be known for congested networks when flows counts on all the links with significant levels of congestion are available; for these links the costs can be deduced by the cost function in correspondence with the measured flows, $\hat{c}_l = \hat{c}_l(\hat{f}_l)$. A further possibility is to deduce links costs, and in particular travel times, by means of direct network surveys, for example with "pilot vehicles" or surveillance systems based on vehicle location and remote transmission through TLC technologies.

Basically, when the link costs are known, it is possible to obtain an estimate of the assignment matrix $\hat{M}(\hat{c})$ by leaving out of consideration the demand vector, and $\boldsymbol{d}^*$ can be estimated by applying the model (5.33):

$$d^* = \underset{x \in S}{\operatorname{argmin}}[z_1(x.\hat{d}) + z_2(\hat{M}(\hat{c})x, \hat{f})] \tag{5.34}$$

The model (5.34) is a constrained minimization problem that can be solved with various algorithms in relation to the constraints defining the set S. In almost all cases, the feasibility set S is defined by non-negativity constraints of demand vector components ($x_i \geq 0\, \forall i$) and the projected gradient algorithm can be used. It is usually easy to obtain explicit formulae for the objective function gradient. In the case of the *GLS* estimator, for example, and on the assumption of diagonal matrices $\boldsymbol{Z}$ and $\boldsymbol{W}$, the ith generic component of the gradient has the value:

$$
\begin{aligned}
Gr_i &= \frac{\partial}{\partial x_i}\left[\sum_i \frac{(x_i - \hat{d}_i)^2}{Var[\eta_i]} + \sum_a \frac{(\hat{f}_a - \sum_i \hat{m}_{ai} x_i)^2}{Var[\varepsilon_a]}\right] \\
&= 2\left[\frac{(x_i - \hat{d}_i)}{Var[\eta_i]} + \sum_l \frac{\hat{m}_{ai}(\sum_j \hat{m}_{aj} x_j - \hat{f}_a)}{var[\varepsilon_a]}\right]
\end{aligned}
$$

Congested networks (link costs unknown). The link costs vector $\boldsymbol{c}$ might not be known a priori, at least for some of its components. This is the case with congested networks, when the information described above is not available. In these circumstances there is a problem of circular dependence in so far as it is possible to obtain link flows $\boldsymbol{v}(\boldsymbol{d^*})$, and therefore link costs $\boldsymbol{c}(\boldsymbol{v}(\boldsymbol{d^*}))$ estimates, by assigning, with an equilibrium assignment model, the demand vector $\boldsymbol{d^*}$, solution to the problem (5.34), which in its turn to be solved requires link flows and costs.

The estimation problem can be formalized as a bi-level optimization problem or as a fixed-point problem (Cascetta and Postorino, 1998).

In the first case, the problem is formulated as a bi-level mathematical programming problem for DUE (Deterministic User Equilibrium) assignment. Many estimation methods derived from this approach have been proposed in the literature. Among the others, Fisk (1988) extends the entropy model proposed by Van Zuylen and Willumsen (1980) to the congested case by using a bi-level formulation with a DUE approach. In this case counted flows are included in the model as equality deterministic constraints and the existence of O-D demand flows exactly reproducing the observed link flows cannot be guaranteed. Yang (1995) uses a GLS approach to resolve the problem for congested network, by using a bi-level formulation and some influence factor that take into account the results obtained at both the first and second level. Chen and Florian (1994) develop optimal conditions for solving the problem for congested network with a DUE assignment model and propose a coordinate crescent method that guarantees the observance of the optimal conditions, even though convergence is not guaranteed.

With respect to the second case, in order to formalize the fixed-point problem, let $d = \delta(\hat{M})$ be the solution of the estimation problem (5.34) for a known assignment matrix $\hat{M}$:

$$
d = \delta(\hat{M}) = \underset{x \in S}{\operatorname{argmin}}[z_1(x, \hat{d}) + z_2(\hat{M}x, \hat{f})]
$$

When the above problem allows only one solution, the relationship $d = \delta(\hat{M})$ can be considered as a function that associates to each assignment matrix $\hat{M}$ a "traffic counts corrected" estimate of the demand vector $\boldsymbol{d}$.

Furthermore, the assignment matrix $\hat{M}$ can be expressed as a function of the demand flow vector. In fact, by combining the relationship linking the assignment matrix to the link costs, $\hat{M} = \hat{M}(c)$, expressed by (5.6) (or by 1.7), with the cost functions $\boldsymbol{c}=\boldsymbol{c}(\boldsymbol{f})$, and introducing the relationship between link and demand flows through the assignment model, $\boldsymbol{f}=\boldsymbol{v}(\boldsymbol{d})$, we get:

$$
\hat{M} = \hat{M}(c(v(d)))
$$

Estimation of demand flows with traffic counts can be defined through a fixed-point model obtained by combining the two functions $d = \delta(\hat{M})$ and $\hat{M} = \hat{M}(c(v(d)))$:

$$
d^* = \delta(\hat{M}(d^*))
$$

that is

$$d^* = \underset{x \in S}{\text{argmin}}[z_1(x, \hat{d}) + z_2(\hat{M}(c(v(d))))x, \hat{f}] \tag{5.35}$$

The problem (5.35) is computationally more complex than the problem (5.34) in that it is necessary to simultaneously solve the constrained optimization problem (5.34) providing the demand estimate, and the equilibrium assignment problem giving the link flows and costs necessary for calculating the assignment matrix. This problem can be solved by using fixed-point iterative algorithms, which alternately solve the problem of estimation and that of assignment, which leads to the calculation of the matrix $\hat{M} = \hat{M}(c(v(d)))$, by averaging out the results obtained at the point of convergence. For more details on convergence and numerical issues related to the fixed-point formulation, see Cascetta and Postorino (1998).

5.2 DYNAMIC ESTIMATORS OF O-D MATRICES USING TRAFFIC COUNTS

In this section the statistical framework proposed for the static problem is generalized and extended to the dynamic O-D estimation case. This allows to obtain a formal analysis of the statistical performances for different estimators and an efficient use of all available information (as previously seen, in fact, other information sources, such as surveys, historic matrices, or the analyst expectation, can complement traffic counts and significantly improve the quality of final estimates).

In literature, many authors have proposed different methods to estimate time-varying O-D flows using traffic counts. Among the others, Cremer and Keller (1987) propose sequential estimators in the case of a simple network and making use only of traffic counts; Cascetta, *et al.* (1993) propose dynamic estimators obtained by optimizing a two-part objective function; Okutani and Stephanades (1984) use a standard Kalman Filtering approach in order to obtain sequential estimators, while Ashok and Ben-Akiva (1993) prove that the same approach can be used to obtain GLS simultaneous estimators.

In this framework, two types of estimators are discussed. Simultaneous estimators, producing joint estimates of the whole set of O-D matrices, one for each time slice, and sequential estimators producing a sequence of O-D estimates for successive time slices.

For an interesting discussion about dynamic estimation with an unknown $\boldsymbol{M}$ matrix see also Ashok and Ben-Akiva (1993).

5.2.1 Relationship between demand and counts

Relationships between link flows and O-D demand when both are allowed to vary across intervals of the reference period (discrete time within-day dynamics) are far more complicated than in the time-uniform (within-day static) case.

Let the total study period J be divided into n_j intervals $j = 1, \ldots, , n_j$, assumed to have equal length T, so that $J = n_j T$. Let $d_{rs,t}$ represent the number of users moving between O-D pair r,s and leaving the origin during the interval t, $\boldsymbol{d}_t$ the column vector obtained by arranging O-D flows. Let $\hat{d}_{rs,t}$ denote a priori information, or an initial estimate of the true demand $d_{rs,t}$and $\hat{d}_t$ the corresponding vector.

For each interval j a link flow f_{aj} can be associated to each link a of the network, or more precisely to each section of each link, as the number of users crossing the section in that interval. Once defined, the link flows can be "counted". In general, link counts over an interval are affected by measurement errors ε^{OBS}. The measured flow $\hat{f}_{aj}$ is therefore only an estimate of the actual flow f_{aj}. In vector form:

$$\hat{f}_j = f_j + \varepsilon^{OBS} \tag{5.36}$$

however defined, f_{aj} is made up of O-D flows leaving during the same interval or in previous ones and reaching link a in interval j. This can be formally expressed by defining the quantity $m_{aj}^{rs,t} \in [0,1]$ as the fraction of O-D flow $d_{rs,t}$ contributing to the flow on link a in interval j, resulting in:

$$f_{aj} = \sum_{t=1}^{j} \sum_{rs} m_{aj}^{rs,t} d_{rs,t} \tag{5.37}$$

Equation (5.37) can be expressed in matrix form by introducing the $(n_a \times n_{rs})$ assignment fraction matrices $\boldsymbol{M}_{jt,}$analogous to the within-day static counterpart defined in equation (5.3):

$$f_j = \sum_{t=1}^{j} M_{jt} d_t$$

This equation assumes that demands and counts before the first interval are negligible. This assumption introduces a positive bias in O-D estimates for the first interval; it can be easily relaxed if an estimate of O-D demand leaving before the study period is available.

Let $\boldsymbol{h}_{kt}$ be the path flow, i.e. the average number of travelers per time unit following path k between O-D pair r,s and leaving during period t. Path flows can also be expressed as the product of the O-D demand $d_{rs,t}$ and the probability (average fraction) $p(k|t)$ of choosing path k given the departing interval t:

$$h_{kt} = d_{rs,t} \cdot p(k|t) \tag{5.38}$$

In order to express assignment fractions $m_{aj}^{rs,t}$ in terms of path choice probabilities, the formal dependence of link flows on path flows must be introduced:

$$f_{aj} = \sum_{k \in K_{sr}} \sum_{t=1}^{j} b_{aj}^{kt} h_{kt} \tag{5.39}$$

where the summation is extended to all paths belonging to the set K_{rs} of paths connecting O-D pair r,s.

In the above expression the b_{aj}^{kt} is the crossing fraction, or the fraction of path flow h_{kt} crossing link a at interval j; the above fractions depend on how link flows are defined, when each path flow reaches link a, and how it moves on it. It is also worth noting that incidence fractions α are the dynamic counterparts of the zero-one elements of the static link-path incidence matrix. By combining equations (5.38) and (5.39), and comparing with equation (5.37), it results:

$$m_{aj}^{rs,t} = \sum_{k \in K_{rs}} p_{aj}^{kt} p(k|t) \tag{5.40}$$

and then

$$f_{ai} = \sum_{t=1}^{j} \sum_{k \in K_{rs}} b_{aj}^{kt} F_{kt}$$

this equation, expressed in matrix form becomes:

$$f_j = \sum_{t=1}^{j} B_{tj} h_t$$

where $\boldsymbol{B}_{tj}$ is the crossing fraction matrix $B_{tj} = \{b_{aj}^{kt}\}$.

In practice, only estimates $\hat{p}(k|t)$ and $\hat{b}_{aj}^{kt}$ can be obtained through path choice and Dynamic Network Loading (DNL) models. Estimates of assignment fractions can thus be formally expressed as:

$$\hat{m}_{aj}^{rs,t} = \sum_{k \in K_{rs}} \hat{b}_{aj}^{kt} p(k|t) \tag{5.41}$$

The estimated values of crossing fraction $\hat{b}$, path choice probabilities $\hat{p}$ and, consequently, assignment fraction $\hat{m}$ are expected to be different from the true ones. This implies, as already seen in the static context, that even if the true demand vector $\boldsymbol{d}_t$ were known and assigned to the network substituting $\hat{m}$ instead of m in equation (5.37), the resulting link flows would differ from the actual ones by a random error term ε^{SIM} (modeling error):

$$f_{aj} = \sum_{t=1}^{j} \sum_{rs} \hat{m}_{aj}^{rs,t} d_{rs,t} + \varepsilon_j^{SIM} \tag{5.42}$$

or in matrix form

$$f_j = \sum_{t=1}^{j} \hat{M}_{jt} d_t + \varepsilon^{SIM} \tag{5.43}$$

Equation (5.36) and (5.43) can be combined into:

$$\hat{f}_j = \sum_{t=1}^{j} \hat{M}_{jt} d_t + \varepsilon \tag{5.44}$$

where the random vector ε is the sum of the two (independent) vectors ε^{OBS} and ε^{SIM}.

5.2.2 *Dynamic estimation of O-D demand*

In the first section, it has been demonstrated that most O-D demand static estimators can be obtained by solving a constrained optimization problem of the form:

$$d^* = \underset{x \in S}{\operatorname{argmin}}[z_1(x, \hat{d}) + z_2(v(x), \hat{f})] \tag{5.45}$$

In this section the estimators previously proposed for the static context will be extended to the two dynamic estimation cases.

Now the problem is to estimate O-D demand for each interval, $\boldsymbol{d}_t$, by using counts $\hat{f}_j$. Two alternative approaches are possible. The first, referred to as *simultaneous*, looks for an estimator giving, in a single step, the whole O-D demand pattern $(d_1, \ldots, d_n)$ by using counts simultaneously over all intervals. The second approach, referred to as *sequential*, produces at each step the O-D demand vector for one period, using counts relative to that period and to the previous one and, possibly, O-D estimates relative to previous periods. In the following sub-sections the two estimators will be described separately.

Simultaneous Estimators. Static estimators can be extended in a straightforward manner to the simultaneous estimation framework. In this case, however, the single unknown demand vector has to be replaced by the n_j vectors $(\boldsymbol{x}_1, \ldots, \boldsymbol{x}_j, \ldots, \boldsymbol{x}_n)$. Likewise, the counted flow vector is replaced by $(\hat{f}_1, \ldots \hat{f}_j, \ldots \hat{f}_{n_j})$. The general form of the estimators then becomes:

$$(d_1^* \ldots d_{n_j}^*) = \underset{x_1 \geq 0 \ldots x_{n_j} \geq 0}{\operatorname{argmin}} \; [z_1(x_1 \ldots x_{n_j}; \hat{d}_1 \ldots \hat{d}_{n_j}) + z_2(x_1 \ldots x_{n_j}; \hat{f}_1 \ldots \hat{f}_{n_j})] \quad (5.46)$$

All reported specifications of objective functions $z_1(\cdot)$ and $z_2(\cdot)$ can be extended and substituted in equation (5.46) thus obtaining ML, GLS, or Bayes estimators depending on the distribution assumptions made on the vectors $\hat{d}_j$. For example, in the case of GLS estimator, they become (Cascetta, Inaudi and Marquis, 1993):

$$z_1 = \sum_{j=1}^{n} (x_j - d_j)^T Z_j^{-1} (x_j - d_j)$$

$$z_2 = \sum_{j=1}^{n} \left(\sum_{t=1}^{j} \hat{M}_{tj} x_t - f_j\right)^T W^{-1} \left(\sum_{t=1}^{j} \hat{M}_{tj} x_t - f_j\right)$$

Nguyen, Morello and Pallottino (1988) have proposed different simultaneous estimators on a general transit network.

Sequential Estimators. In this context, an O-D demand vector is estimated for a single interval j at each time interval. There are two advantages to this approach. The first is the reduction of computational complexity by breaking down a large optimization problem into a number of smaller and more manageable ones; the second is that estimates obtained for an interval can, but not necessarily have to, be used as initial estimates in subsequent estimations.

The main idea in this approach is that of expressing the counts of a period j as a linear (stochastic) function of the unknown demand of the same period only. This is achieved by equating the demand relative to previous periods to the already computed estimates d_t^*:

$$\hat{f}_j = \sum_{t=1}^{j-1} \hat{P}_{tj} d_t^* + \hat{P}_{jj} x_j + \varepsilon_j \quad (5.47)$$

The general formulation of the static estimation problem can be adapted to this context, giving rise to:

$$d_n^+ = \underset{x_j \geq 0}{\operatorname{argmin}} [z_1(x_j, d) + z_2(x_j | d_1^*, \ldots, d_{j-1}^*; f_j)] \quad (5.48)$$

where $f_j = \sum_{t=1}^{j-1} \hat{M}_{tj} d_t^* + \hat{M}_{jj} x_j - \varepsilon_j$.

In the case of GLS estimator, the objective functions z_1 and z_2 become (Cascetta and Cantarella, 1991):

$$z_1 = (x_j - d_j)^T Z_j^{-1} (x_j - d_j)$$

$$z_2 = (\sum_{t=1}^{j-1} \hat{M}_{tj} d_t + \hat{M}_{jj} x_j - f_j)^T W^{-1} (\sum_{t=1}^{j-1} \hat{M}_{tj} d_t + \hat{M}_{jj} x_j - f_j)$$

Finally, it is worth quoting Ashok and Ben-Akiva (1993). The authors have proposed a Kalman Filter based approach for real-time estimation of O-D flows, introducing the concept of *deviation* of O-D flows from historical estimates. In their formulation the set of decision variables is defined in terms of O-D deviations conforming to an autoregressive process, and the Kalman Filter procedure is derived from state-space modeling theory. It has also been proved that there is a strong connection between the Kalman Filter and the Generalize Least Square estimators and that it is possible to obtain a formulation that can be viewed as a simultaneous estimator.

5.3 ESTIMATION OF DEMAND MODELS PARAMETERS FROM TRAFFIC COUNTS

In this section a general statistical framework for the problem of efficiently combining traffic counts with other data sources for estimating parameters of pre-specified aggregated travel demand models is proposed.

The aggregated information on transportation demand contained in counts can also be used to "improve" the estimation of demand models parameters. In the Authors' opinion the potential of these methods, both from the theoretical and applicative point of views, have been so far largely underlooked. In fact, this problem has been less researched with respect to direct O-D flow estimation using traffic counts. Nevertheless, in literature there are several works proposing estimators of model parameters from traffic counts. For example, Fricker (1996), deals with the joint estimation of path choice model parameters and O-D flows on congested networks, Cascetta and Russo (1997), propose a general class of estimation methods for evaluating parameters of demand models considering traffic counts with other available information. Finally, Cascetta (1998) proposes a method combining estimation of path choice model parameters and O-D demand flows. Further comments on this problem can also be found in Hogberg (1976), Cascetta (1986), Willumsen and Tamin (1989), Di Gangi (1989).

5.3.1 *Problem statement*

Once specified, demand models can be seen as functions relating the demand vector to the activity system variables $\boldsymbol{SE}$ and to the transport system variables $\boldsymbol{T}$ through the unknown parameters vector β.

$$\boldsymbol{d}=\boldsymbol{d}(\boldsymbol{SE},\ \boldsymbol{T},\ \beta) \tag{5.49}$$

For a given configuration of the two systems, i.e. for given values of $\boldsymbol{SE}$ and $\boldsymbol{T}$, (5.49) can be considered as a relationship between the demand and the unknown vector β. An estimate $\hat{\beta}$ of the unknown parameters vector can be obtained by using disaggregated results of sampling surveys, through well established econometric methods (see for instance Ben-Akiva Lerman, 1985). Since it is an estimate, the vector $\hat{\beta}$ is the determination of a random variable. In other words, the estimate of the generic component $\hat{\beta}_i$ diverges from the "real" parameter value by an unknown quantity σ_i:

$$\hat{\beta}_i = \beta_i + \sigma_i \tag{5.50}$$

In the case that β_i have been estimated through the Maximum Likelihood method, the variance of the error σ_i can be deduced by means of the inverse of the Hessian matrix of the Log-Likelihood function.

Alternatively, in a Bayesian approach the vector $\hat{\beta}$ can indicate parameters values obtained in a study area of similar characteristics, or simply the analyst a priori expectations. In this case, the expression (5.50) can be seen as a relationship between the "true" parameter and an initial value, and the variance of σ_i is a measure of the analyst "confidence" in the initial estimate. The two approaches coincide if the a priori estimates are obtained from sampling surveys.

To use traffic counts on some network links, it is necessary to express the relationship linking these to the unknown parameters of a demand model. To calibrate overall demand models, e.g. the sequence generation/ distribution/ modal split, it is necessary to have counts on the different modal network links. Let $\hat{f}_m$ be the vector of the measured flows (counts) on the mode m network and $\hat{f}$ be the vector obtained by sequentially ranking all the vectors $\hat{f}_m$.

The relationship between the traffic counts vector $\hat{f}_m$ and the "true" demand vector relative to mode m, $\boldsymbol{d}_m$, is basically analogous to (5.11) which now becomes:

$$\hat{f}_m = \hat{M}_m d_m + \varepsilon_m^{SIM} + \varepsilon_m^{OBS} \tag{5.51}$$

The demand vector relative to mode m obtained through the model can be expressed as $\boldsymbol{d}_m(\beta)$, where, for simplicity sake, the vectors $\boldsymbol{SE}$ and $\boldsymbol{T}$ are understood. Even if the "true" parameters vector β were known, the demand obtained from the model would diverge from the "true" one of a vector ε_m^{MOD}:

$$d_m = d_m(\beta) + \varepsilon_m^{MOD} \tag{5.52}$$

and by substituting (5.52) in (5.51) we get:

$$\hat{f}_m = \hat{M}_m d_m(\beta) + \varepsilon_m \tag{5.53}$$

where the deviations vector ε_m is the sum of the various error components:

$$\varepsilon_m = \varepsilon_m^{SIM} + \varepsilon_m^{OBS} + \hat{M}_m \varepsilon_m^{MOD}$$

and has zero mean if the vectors $\varepsilon_m^{SIM}, \varepsilon_m^{OBS}, \varepsilon_m^{MOD}$ have zero mean.

The relationship (5.53) can be extended to the set of links belonging to different modal networks:

$$\hat{f} = \hat{M} d(\beta) + \varepsilon \tag{5.54}$$

where the vectors $\hat{f}, d(\beta)$ and ε are obtained by sequentially ordering the vectors relative to different modes for which traffic counts are available, and similarly the assignment matrix $\hat{M}$ is given by the sequential ordering of the individual modal assignment matrices.

The relationships (5.50) and (5.54) represent in synthetic form the relationships between the unknown vector β and the available information, that is the estimate $\hat{\beta}$ and the counts $\hat{f}$. This information can be combined in different ways, obtaining different estimators of β, in relation to the classic or Bayesian interpretation of the initial estimate $\hat{\beta}$ and to the assumptions on the probability distribution of vectors σ and ε. Also for this estimation problem, it is in fact possible to obtain Maximum Likelihood, Generalized Least Squares and Bayesian estimators completely analogous to those described in section 1.3 and 1.4, which can be expressed in the general form:

$$\beta^* = \underset{b \in S_B}{\operatorname{argmin}}[z_1(b, \hat{\beta}) + z_2(\hat{M}d(b), \hat{f})] \tag{5.55}$$

Note that in (5.55) the unknown parameters vector of $\boldsymbol{b}$ has replaced the unknown O-D demand vector $\boldsymbol{x}$. This means that a vector of much smaller size (dozens of components instead of hundreds or thousand) is being estimated, note also that the optimization problem is "more non-linear" than in the direct demand estimation case, since the non-linearity of demand models as function of unknown parameters is added to the non-linearity of functions $z_1(\cdot)$ and $z_2(\cdot)$. The parameters feasibility set S_B may be coincident with the entire Euclidean space, or constraints can be imposed on the "expected" signs of the coefficients (e.g. negative cost coefficients etc.)

The model (5.55) can also be specified in the case that only the aggregated information on the demand, traffic counts or other sources are used. On this assumption, we get an exclusively aggregated estimator in which the distance term $z_2(\cdot)$ between the values of observed traffic counts and the estimates obtained by assigning demand flows as generated by demand models, is minimized with respect to model parameters:

$$\beta^* = \underset{b \in S_B}{\operatorname{argmin}}\, z_2(\hat{M}d(b), \hat{f})$$

In this case it is possible to use only traffic counts, since the number of independent counts is generally larger than the number of unknown demand parameters. Also for the estimation of parameters from counts the most robust specification with respect to the modeling assumptions and the most widely used in the so-far limited applications is that of Non-Linear Generalized Least Squares ($NLGS$), which in its simplified form becomes:

$$\beta^* = \underset{b \in S_b}{\operatorname{argmin}} \left[\sum_i \frac{(b_i - \hat{\beta}_i)^2}{Var[\sigma_i]} + \sum_a \frac{(\hat{f}_a - \sum_i \hat{m}_{ai} d_i(b))^2}{Var[\varepsilon_a]} \right] \tag{5.56}$$

where $\boldsymbol{b}$ is the current value of the parameters vector and S_B its feasibility set, for example non-positivity constraints for parameters of "cost" attributes (i.e. travel time, monetary cost, addition times, etc).

In order to solve problem (5.56), a gradient or projected gradient algorithm can be used according to whether constraints on the components of the unknown vector $\boldsymbol{b}$ have been encountered. In this case, however, the whole procedure and the gradient of (5.56) must be relative to the unknown parameters b_i (Cascetta and Russo, 1997). The kth component of the (5.56) objective function gradient can therefore be expressed as:

$$
\begin{aligned}
GR_k &= \frac{\partial}{\partial b_k}\left[\sum_i \frac{(b_i-\hat{\beta}_i)^2}{Var[\sigma_i]} + \sum_a \frac{\left(\hat{f}_a - \sum_i \hat{m}_{ai} d_i(\mathbf{b})\right)^2}{Var[\varepsilon_a]}\right] \\
&= \frac{2(b_k-\hat{\beta}_k)}{Var[\sigma_k]} + 2\sum_a \frac{\left(\sum_i \hat{m}_{ai} d_i(\mathbf{b}) - \hat{f}\right)}{Var[\varepsilon_a]} \cdot \sum_i \hat{m}_{ai} \frac{\partial d_i(\mathbf{b})}{\partial b_k}
\end{aligned}
\tag{5.57}
$$

The calculation of the partial derivative of demand function on the ith O-D pair with respect to the generic component k of the unknown parameters vector obviously depends on the specification adopted for the model sequence being calibrated. In more complex cases, the analytical calculation of these derivatives can be very cumbersome, or even impossible (e.g. for Probit models). In these cases recourse is made to numerical computation methods of the partial derivatives.

The applications of the described method on rather limited mode have dealt mainly with parameters calibration for rather simple aggregated demand models (Tamin and Willumsen, 1989), such as, gravity models of traditional four stage models (Cascetta and Russo, 1997). The results obtained are generally satisfactory.

On the assumption that the vector $\hat{\beta}$ is obtained by calibrating the demand model parameters on the basis of a sample of observations, the model (5.55) can be considered as a two-stage mixed disaggregated/aggregated estimator. In other words, at the first stage the vector $\hat{\beta}$ is estimated from the disaggregated data and at the second this estimate is "corrected" using the aggregated information contained in traffic count.

It is also possible to formulate a "simultaneous" mixed estimator, such as a Maximum Likelihood estimator, which maximizes the probability $p[j(i)](h)$ of observing the choices $j(i)$ of a users sample and the measurement of traffic counts $\hat{f}_a$. In this case, assuming that the information is independent and that the $j(i)$ are obtained with a simple random sample, the estimate β^* can be obtained by combining the log-likelihood function relative to the parameters vector β with one of the functions $z_2(\cdot)$ previously described expressing the logarithm of the probability of observing the counts as a function of the assignment matrix and of the demand model parameters vector β:

$$
\beta^* = \underset{b \in S_b}{\text{argmax}}[\sum_i lnp[j(i)](b) + z_2(\hat{M}d(b), \hat{f})]
$$

The literature does not describe experiments with the simultaneous estimator or comparisons with the sequential estimator.

One final consideration relates to path choice model parameters. In all the previous analysis, in fact, it has been assumed that the path choice model providing the elements $\hat{p}_{ki}$ of the matrix $\hat{P}$, and therefore the matrix $\hat{M}$, has been specified and calibrated, that is that the parameters β^{PATH} appearing in the systematic utility (equal to the generalized cost with the sign changed) and those relative to the random residuals distribution are known.

These parameters are consequently not included in the vector β to be estimated, the less so in the demand flow vector. However, it is possible to specify the estimation problem in order to use traffic counts to improve the initial estimate β^{PATH} of these parameters. In this case, the general expression of the model (5.55) becomes:

$$\beta^* = \operatorname*{argmin}_{b \in S_b}[z_1(b, \hat{\beta}) + z_2(\hat{M}(b))d(b), \hat{f}] \tag{5.58}$$

where the vector β^{PATH} has been included in the general parameters vector β and in the variables vector so as to optimize $\boldsymbol{b}$. It can be seen from a comparison of the expressions (5.55) and (5.58) that in the latter case the non-linearity of the model further increases because also the assignment matrix elements are a function of the unknown parameters vector, or rather of some of its components.

A similar argument can be extended to the case of O-D demand vector and path choice model parameters combined estimate. In this case we get a formulation of the following type:

$$\beta^*, d^* = \operatorname*{argmin}_{\substack{b \in S_b \\ x \in S_d}}[z_1(x, d) + z_2(b, \hat{\beta}) + z_3(\hat{M}(b)x, \hat{f})] \tag{5.59}$$

where the vector β coincides with β^{PATH}.

In this case the solution to the problem (5.59) simultaneously gives the estimate of path choice model parameters and the demand vector minimizing the "divergences" from their respective initial estimates and from the traffic counts observed.

As it was noted previously, these latter applications have not been pursued beyond the stage of formal specification and the theoretical and numerical analysis of their properties requires further research.

Conclusions

Use of traffic counts for the estimation of travel demand is a well-established area of research over the world since a few decades. Its relevance is confirmed both by the amount of research and the countless number of applications in real-life contexts. This area has branched out in many sectors related to static-dynamic O-D estimation and calibration of demand parameters. In particular, dynamic O-D estimation and model calibration received relatively less attention. While the former is justified by the recent developments of Dynamic Traffic Assignment models and computational powers able to make the problem valuable in real cases, delay in the latter sector is less justifiable.

Further research and applications should cover these two areas, also trying to integrate in the process other information sources (e.g. different surveys and/or counting information), and taking into account that many of the arguments proposed for the estimation of demand models parameters can be relatively easily extended to the case of dynamic estimation.

References

Ashok K. and Ben-Akiva M. (1993) "Dynamic Origin-destination Matrix Estimation and Prediction for Real Time-Traffic Management Systems" In C. F. Daganzo editor, International Symposium on Transportation and Traffic Theory, pp. 465-484, Elsevier Science Publ. Co.

Bell M., (1991) "The estimation of origin-destination matrices by constrained generalized least squares", *Transp. Res.*, vol. 25B (2), pp. 13-22.

Ben Akiva M. E., Macke P. P., Hsu P.S. (1985) "Alternative methods to estimate route-level trip tables and expand on-board surveys" *TRB* 1037, pp. 1-11.

Ben Akiva M., Lerman S. (1985) "Discrete Choice Analysis: Theory and Application to Travel Demand" MIT Press, Cambridge, Mass.

Bovy P.H.L., Jansen G.R.M. (1983) "Network aggregation effect upon equilibrium assignment outcomes: an empirical investigation" *Transp. Sci.*, vol. 17 (3), pp. 240-262.

Cascetta E. (1984) "Estimation of trip matrices from traffic counts and survey data: a generalized least squares estimator" *Transp. Res.*, vol. 18B (4/5), pp. 289-299.

Cascetta E. (1986) "A class of travel demand estimators using traffic flows" *CRT Publication No. 375*, Universit de Montréal, Montréal, Canada.

Cascetta E., Nuzzolo A., Velardi V. (1986) "Un'analisi sperimentale dei modelli di assegnazione alle reti urbane di trasporto privato" *Atti del IV convegno nazionale del PFT-CNR*, Torino.

Cascetta E., Nguyen S. (1986) "A unified framework for estimating or updating Origin-Destination matrices from traffic counts" *Transp. Res.*, vol. 22B, pp. 437-455.

Cascetta E., Inaudi D., Marquis G. (1993) "Dynamic estimators of Origin-Destination matrices using traffic counts" *Transp. Science* vol. 27-4.

Cascettta E., Cantarella G. E. (1998) "Stochastic assignment to transportation networks: models and algorithms" in Equilibrium and Advanced Transportation Modeling, P. Marcotte, S. Nguyen (Eds.) (*Proceedings of the international Colloquium*, Montréal, Canada).

Cascetta E., Russo F. (1997) "Calibrating aggregate travel demand models with traffic counts: estimators and statistical performances" *Transportation*, vol. 24 (3), pp. 271-293.

Cascetta E., Postorino M.N. (1997) "Fixed point models for the estimation of O-D matrices using traffic counts on congested networks" submitted to *Transp. Sci.*,

Cascetta E. (1998) " Teoria e Metodi dell'Ingegneria dei Trasporti", UTET, Torino.

Cremer M. and Keller H. (1987) "A new class of dynamic methods for the identification of Origin-Destination flows" *Transp. Res.* vol. 21B(2), pp. 117-132.

Di Gangi M. (1988) "Una valutazione delle prestazioni statistiche degli estimatori della matrice O/D che combinano i risultati di indagini e/o modelli con i conteggi di flussi di traffico" *Ricerca Operativa*, No. 51, pp. 23-59.

Fisk C. S. (1988) "On combining maximum entropy trip matrix estimation with user optimal assignment. Transp. Res. vol. 22B(1), pp. 241-246.

Florian M., Chen Y. (1995) "A coordinate Descent Method for the Bi-level O-D Matrix Adjustment Problem" *Int. Trans. Op. Res.*, vol. 2, pp. 165-179.

Hogberg P. (1976) "Estimation of parameters in models for traffic prediction: a non-linear regression approach" *Transp. Res.* vol. 10B, pp. 263-265.

Liu S. and Fricker J. D. (1996) "Estimation of trip table and the θ parameter in a stochastic network" *Transp. Res.* vol. 17B.

Maher M. J. (1983) "Inferences on trip matrices from observation on link volumes: a Bayesian statistical approach" *Transp. Res.* vol. 17B, pp. 435.447

Mc Neil S. (1983) "Quadratic matrix entry estimation methods". *Ph.D. thesis*, Dept. of Civil Engineering, Carnegie-Mellon University.

Nguyen S. (1983a) "Inferring origin-destination demands from network data". Proceedings of the 1983 AIRO conference, Naples, pp. 53-102.

Nguyen S., Pallottino S. (1986) "Assegnazione dei passeggeri ad un sistema di linee urbane: determinazione degli ipercammini minimi" *Ricerca Operativa*, No. 39, pp. 207-230.

Okutani I., Stephanades Y. (1984) "Dynamic prediction of traffic volume through Kalman Filtering theory" *Transp. Res.*vol. 18B, pp. 1-11.

Ortuzar J. de D., Willumsen L.G. (1994) "Modelling Transport" John Wiley and Sons, II ed.

Sheffi Y. (1985) "Urban transportation networks" Prentice Hall, Englewood Cliff, NJ.

Spiess H. (1983) "A maximum Likelihood model for estimating Origin-Destination Matrices" *CRT Publication No* 293, Centre de Recherche sur les transports, Université de Montréal.

Spiess H., Florian M., (1989) "Optimal strategies: a new assignment model for transit networks" *Transp. Res.*, vol. 23B (2), pp. 82-102.

Van Zuylen J.H., Willumsen L.G. (1980) "The most likely trip matrix estimated from traffic counts" *Transp. Res.*, vol.14B (2), pp. 281-293.

Van Zuylen J. H. and Branston D. M. (1982) "Consistent link flows estimation from counts". *Transp. Res.* vol. 13A, pp. 473-476.

Willumsen L. G. (1984) "Estimating time-dependent trip matrices from traffic counts" *Proceedings of 9th International Symposium on Transportation and Traffic Theory.*VNU Science Press, Utrecht.

Willumsen L. G. and Tamin O. Z. (1989) "Transport demand model estimation from traffic counts". *Transportation* vol. 16, pp. 3-26.

Wilson A. G. (1970) "Entropy in Urban and Regional Modeling" Methuen, Inc. New York.

Yang H., (1995) "Heuristic algorithms for the bi-level origin/destination matrix estimation problem" *Transp. Res.*vol. 29B (2).

6 PATH RECOVERY/RECONSTRUCTION AND APPLICATIONS IN NONLINEAR MULTIMODAL MULTICOMMODITY NETWORKS

Teodor Gabriel Crainic

Gina Dufour

Michael Florian

Diane Larin

Abstract: Two issues are addressed in this paper: (1) How to reconstruct the product-specific path information based on the link and transfer flows that result from a non-linear, multimodal, multiproduct network optimization formulation and (2) how to use this information efficiently for various *path analyses* for transportation planning purposes. The paper also touches upon the challenge of introducing a major development into a mature and complex software such as the STAN decision support system for the strategic analysis and planning of multimodal multiproduct freight transportation. Results from actual applications are used to illustrate the path analysis capabilities of STAN.

6.1 INTRODUCTION

Network optimization models and methods are at the core of most transportation planning systems for passengers as for freight (Crainic, 1999; Crainic and Laporte, 1997; Florian and Hearn, 1995). Due principally to the size of the respective networks, most such systems yield results in terms of link flows only, and path information is not readily available.

Path information is required for many analyses, however, particularly so for the study of multiproduct freight movements through multimodal transportation systems

M. Gendreau and P. Marcotte (eds.), Transportation and Network Analysis: Current Trends, 95–108.

which form the object of our interest. Indeed, freight is hauled through the transportation services and facilities following multimodal routes or *paths*. Moreover, paths are specific to each *market*: a product has to be transported between a given pair of origin and destination zones (an *OD pair*). Thus, while link-based flow distribution is sufficient for network-wide productivity measures and analyses, it is not appropriate for more in-depth studies of particular markets.

Two issues are therefore of importance. First, how to reconstruct the product-specific path information based on the link flows that result from a nonlinear, multimodal, multiproduct network optimization formulation and, second, how to use this information efficiently for various *path analyses*. Concerning the first issue, it is well-known from network flow theory that the problem admits many equivalent solutions, numerous path patterns yielding the same link flows. Our study further underscores the point. The second issue is strongly related to the computer implementation of the methodology and is particularly sensitive when the planning software is already in use in various firms and agencies. One then has to conform the new development to the general design concept of the software and perform a seamless integration. This emphasizes again the strong interactions between operations research models and methodologies and the corresponding computer implementations when a practical utilization of research results is contemplated.

Our developments relate to *STAN* (*Strategic Transportation ANalysis*), a multimode, multiproduct method and interactive-graphic decision support system for strategic analysis and planning of freight transportation (Guélat, Florian, and Crainic, 1990; Crainic, Florian, and Léal, 1990; Crainic *et al.*, 1990; Crainic, Florian, and Larin, 1994). The focus is on the proper representation of the network and its several transportation modes and intermodal transfer operations, the various criteria used to determine the movement of freight, and the associated estimation of the traffic distribution over the transportation system. The simultaneous demand for transportation services by several product groups as well as their interactions and competition for limited resources are explicitly integrated. Although used by many organizations in more than 15 countries around the world, STAN is a system in continuous evolution. The aim is to enhance the functionalities and capabilities it offers to the user and it is as often prompted by user-perceived needs as by our own vision of the development of the field. A major recent development concerns the capability to perform *path analyses*. This goes beyond the simple access to the multimodal routes used for each market. Path analysis offers the possibility to interactively select and examine paths according to generalized, user-defined criteria, to perform *select element* (link or intermodal transfer) studies, and to compute demand or performance matrices for selected products and OD pairs.

The goal of this paper is to present the path retrieving problem, to review possible methodological approaches, and to describe our implementation. We also aim to document the process of introducing a major development into a mature and complex software, a significant software-development challenge. Two applications illustrate the path analysis capabilities of STAN: a select element analysis and the adjustment of demand matrices according to observed flows.

The paper is organized as follows. Section 6.2 describes the path analysis issue and examines various methodological approaches. Section 6.3 briefly recalls the STAN modelling and algorithmic framework and the main characteristics of the decision support system. It also introduces the path reconstruction strategy we adopted. Section 6.4 addresses a number of implementation issues and illustrates the capabilities of the new development. We conclude in Section 6.5.

6.2 PATH RECONSTRUCTION

Let $\mathcal{G} = \{\mathcal{N}, \mathcal{M}, \mathcal{A}, \mathcal{T}\}$ be a multimodal network, defined by the sets of nodes $\mathcal{N}$, transportation modes $\mathcal{M}$, links $\mathcal{A}$, and intermodal transfers $\mathcal{T}$. Since several transportation modes are usually available between two adjacent nodes, one defines links as uni-directional and uni-modal: $\mathcal{A} = \{a \mid (i,j,m),\ i,j \in \mathcal{N},\ m \in \mathcal{M}\}$. Consequently, a transfer that may take place at a given node between two, not necessarily different, modes is defined as $t = (j, a_1, a_2) = (i, j, m_1, k, m_2),\ t \in \mathcal{T}$.

Let $\mathcal{P}$ stand for the set of products that have to move through $\mathcal{G}$. Each product $p \in \mathcal{P}$ is shipped from some origins $o \in \mathcal{N}$ to some destinations $d \in \mathcal{N}$ of the network using a given subset of modes $m(p) \subseteq \mathcal{M}$. A demand matrix $g^{m(p)}$, associated with product p, indicates the amount of cargo that has to be shipped. Let $\mathcal{K}^{m(p)} = \bigcup_{o,d} \mathcal{K}_{od}^{m(p)}$ represent the set of all paths in $\mathcal{G}$ available for the traffic of this product (i.e., using modes in $m(p)$), for all OD pairs.

Product flows on the links and transfers of the multimodal network constitute the decision variables of the model: $v = (v_a^p,\ v_t^p)$, for $p \in \mathcal{P},\ a \in \mathcal{A},\ t \in \mathcal{T}$. Functions $f_a^p(v)$ and $f_t^p(v)$ represent the "cost" of moving product p on the links and through the transfers of $\mathcal{G}$. The normative, minimum cost, multimode, multiproduct assignment model may then be formulated as follows:

$$\text{Minimize } F(v) = \sum_{p \in \mathcal{P}} \left(\sum_{a \in \mathcal{A}} f_a^p(v) + \sum_{t \in \mathcal{T}} f_t^p(v) \right), \tag{6.1}$$

$$\text{s.t.} \sum_{l \in \mathcal{K}_{od}^{m(p)}} h_l^p = g_{od}^{m(p)} \qquad o,\ d \in \mathcal{N},\ p \in \mathcal{P},\ m(p) \subseteq \mathcal{M}, \tag{6.2}$$

$$h_l^p \geq 0 \qquad l \in \mathcal{K}_{od}^{m(p)},\ o,\ d \in \mathcal{N},\ p \in \mathcal{P},\ m(p) \subseteq \mathcal{M}, \tag{6.3}$$

with the definitional relationships between link (transfer) and path flows

$$v_a^p = \sum_{l \in \mathcal{K}^{m(p)}} \delta_{al} h_l^p, \qquad a \in \mathcal{A},\ p \in \mathcal{P}, \tag{6.4}$$

$$v_t^p = \sum_{l \in \mathcal{K}^{m(p)}} \delta_{tl} h_l^p, \qquad t \in \mathcal{T},\ p \in \mathcal{P}, \tag{6.5}$$

where h_l^p is the flow on path $l \in \mathcal{K}_{od}^{m(p)}$, while $\delta_{al} = 1$ if $a \in l$ (and 0, otherwise) and $\delta_{tl} = 1$ if $t \in l$ (and 0, otherwise) are the indicator functions which identify the arcs and transfers, respectively, of a particular path. $F(v)$ is assumed convex and we let $\Omega = \bigcup_p \Omega^p$ represent the feasible solution space defined by constraints (6.2) to (6.5).

Many algorithms generate multimodal paths in order to identify optimal solutions to this formulation. It is the case, for example, for the Frank-Wolfe algorithm which forms the basis for the solution method embedded in STAN: a path is generated for each OD pair at each iteration. Generally, however, there is a huge number of paths. To illustrate, consider that for a very small application with 67 zones (a zone is both an origin and a destination) and 9 demand matrices, 20453 OD pairs had positive demand, which meant 20453 paths generated at each iteration. Consequently, the actual paths are usually discarded and results are available only in terms of corresponding link and transfer flows $\bar{v}_a^p$ and $\bar{v}_t^p$, respectively. Thus, prior to any path analysis, one must (1) reconstruct the paths that correspond to the optimal solution and (2) determine the amount of product flow on each path.

To reconstruct the paths, one may take advantage of the optimality conditions of the formulation (6.1) - (6.5) which state that for each product p and origin-destination pair (o, d), paths with positive flows have equal marginal costs, while marginal costs are larger for paths with zero flow:

$$
\begin{aligned}
\frac{\partial F(v)}{\partial h_l^p} &= C_{od}^p \quad \text{if} \quad h_l^p > 0, \\
\frac{\partial F(v)}{\partial h_l^p} &> C_{od}^p \quad \text{if} \quad h_l^p = 0,\ l \in \mathcal{K}_{od}^{m(p)}.
\end{aligned} \tag{6.6}
$$

An adaptation of a *k-shortest-path* method may then be used to determine the paths that satisfy the optimality conditions of the nonlinear assignment model.

Several approaches are then possible to determine the flow distribution over these paths. We have experimented with greedy-like heuristics, which have proved successful in some previous studies (Drissi-Kaïtouni, 1988). For the multimodal case, the results were truly disappointing, however. In many cases, the heuristics failed even to identify feasible solutions. Considerable more success was achieved by using optimization-based methods which are described next.

Note that once the paths are known, path flows have to satisfy the following system of equations:

$$
\sum_{l \in \mathcal{K}_{od}^{m(p)}} h_l^p = g_{od}^{m(p)} \qquad o,\ d \in \mathcal{N},\ p \in \mathcal{P},\ m(p) \subseteq \mathcal{M}, \tag{6.7}
$$

$$
\bar{v}_a^p = \sum_{l \in \mathcal{K}^{m(p)}} \delta_{al} h_l^p \quad a \in \mathcal{A},\ p \in \mathcal{P}, \tag{6.8}
$$

$$
\bar{v}_t^p = \sum_{l \in \mathcal{K}^{m(p)}} \delta_{tl} h_l^p \quad t \in \mathcal{T},\ p, \in \mathcal{P}, \tag{6.9}
$$

where $\bar{v}_a^p$ and $\bar{v}_t^p$ are the known link and transfer product flows, and $h_l^p \geq 0,\ l \in \mathcal{K}^{m(p)}$ (to simplify the presentation, we do not change the notation for the path sets and assume that sets $\mathcal{K}^{m(p)}$ include only paths with equal - and minimum - marginal cost). These relations correspond to the feasible region (constraints (6.2) - (6.5)) of the assignment model. To solve the system of equations might become tedious, however, especially for realistically sized problem instances. Better to let an optimization formulation guide the computation.

Assume that one is ready to tolerate a given deviation from the true link and transfer flows:

$$-\epsilon_a \leq \sum_{l\in\mathcal{K}^{m(p)}} \delta_{al} h_l^p - \bar{v}_a^p \leq \epsilon_a, \qquad a \in \mathcal{A},\ p \in \mathcal{P}, \tag{6.10}$$

$$-\epsilon_t \leq \sum_{l\in\mathcal{K}^{m(p)}} \delta_{tl} h_l^p - \bar{v}_t^p \leq \epsilon_t, \qquad t \in \mathcal{T},\ p \in \mathcal{P}. \tag{6.11}$$

By using auxiliary variables z_a (z_t) and x_a (x_t) to indicate positive and negative deviations in the link (transfer) flows, one may formulate an optimization problem to minimize the total deviation:

$$\text{Minimize} \quad \sum_{a\in\mathcal{A}} (z_a + x_a) + \sum_{t\in\mathcal{T}} (z_t + x_t) \tag{6.12}$$

$$\text{s.t.} \quad \sum_{l\in\mathcal{K}_{od}^{m(p)}} h_l^p = g_{od}^{m(p)} \quad o,\ d \in \mathcal{N},\ p \in \mathcal{P},\ m(p) \subseteq \mathcal{M}, \tag{6.13}$$

$$\sum_{l\in\mathcal{K}^p} \delta_{al} h_l - \bar{v}_a^p + z_a - x_a = 0 \qquad a \in \mathcal{A},\ p \in \mathcal{P}, \tag{6.14}$$

$$\sum_{l\in\mathcal{K}^p} \delta_{tl} h_l - \bar{v}_t^p + z_t - x_t = 0 \qquad t \in \mathcal{T},\ p \in \mathcal{P}, \tag{6.15}$$

$$0 \leq z_a,\ x_a \leq \epsilon_a \qquad a \in \mathcal{A}, \tag{6.16}$$

$$0 \leq z_t,\ x_t \leq \epsilon_t \qquad t \in \mathcal{T}, \tag{6.17}$$

$$h_l^p \geq 0 \qquad l \in \mathcal{K}_{od}^{m(p)},\ o,\ d \in \mathcal{N},\ p \in \mathcal{P},\ m(p) \subseteq \mathcal{M}. \tag{6.18}$$

A tighter formulation may be obtained by introducing in the previous model the deviation in the total cost for each product. Denote

$$F^p(v) = \sum_{a\in\mathcal{A}} f_a^p(v) + \sum_{t\in\mathcal{T}} f_t^p(v) \tag{6.19}$$

the total cost for product p resulting from the assignment and

$$C_l^p(h) = \sum_{a\in\mathcal{A}} \delta_{al} f_a^p(h_l^p) + \sum_{t\in\mathcal{T}} \delta_{tl} f_t^p(h_l^p) \tag{6.20}$$

the cost of path l for product p. The difference between the total transportation cost for each product p may then be written as

$$\sum_{l\in\mathcal{K}^{m(p)}} C_l^p(h) - F^p(v) + z_p - x_p = 0 \quad p \in \mathcal{P}, \tag{6.21}$$

$$0 \leq z_p,\ x_p \leq \epsilon_p \quad p \in \mathcal{P}, \tag{6.22}$$

that yields a second optimization formulation

$$\text{Minimize} \quad \sum_{a\in\mathcal{A}} (z_a + x_a) + \sum_{t\in\mathcal{T}} (z_t + x_t) + \sum_{p\in\mathcal{P}} (z_p + x_p) \tag{6.23}$$

subject to constraints (6.13) to (6.18), (6.21), and (6.22).

A third optimization formulation may be obtained by directly minimizing the total product cost and flow deviation:

$$\text{Minimize} \sum_{p \in \mathcal{P}} \sum_{l \in \mathcal{K}^{m(p)}} C_l^p(h) + \sum_{a \in \mathcal{A}} (z_a + x_a) + \sum_{t \in \mathcal{T}} (z_t + x_t) \tag{6.24}$$

subject to constraints (6.13) to (6.18).

Experimental results on relatively small networks (with dimensions similar to those of the STAN demonstration application) yielded interesting results (Charland, 1996): almost no difference could be observed between the total assignment cost and that of the generated paths. These observations point to interesting avenues for issues related to construction of path flows and extraction of OD demand matrices when only link (and transfer) flows are available.

In the particular case of the STAN system, however, the approach did not appear adequate. On the one hand, the experimental results indicate that computations might become heavy for large applications. Furthermore, because the multimodal formulations have multiple solutions, there is no guarantee that the paths that are generated are those used by the assignment procedure. This implies that disaggregated selections and analyses based on paths become difficult to perform. Moreover, the methods presented in this section require that all the paths, for all products and demand matrices, be extracted simultaneously. This, again, is not efficient for actual applications. On the other hand, these same approaches are based on the premise that link and transfer flows correspond to the optimal solution of the assignment model. This hypothesis might not be verified in many applications where the assignment procedure is stopped as soon as reasonably good solutions (measured in terms of relative gap to the optimal solution) are obtained. We therefore adopted a solution that requires the storage of additional information (other than the link and transfer product flows), but permits to efficiently generate the paths actually used during assignment and to compute the flows on these paths when the assignment procedure was stopped. This approach is presented in the next section together with a brief presentation of the STAN transportation planning system.

6.3 THE STAN MODELLING FRAMEWORK AND SYSTEM

In STAN the movements of several commodities through the transportation networks and services of several carriers are considered simultaneously, in order to assess the evolution of a given transportation system and its response to various modifications in its environment. The main goal is to evaluate and compare scenarios based on flow distribution and the performances of the system. Scenarios reflect planned or forecast modifications to the infrastructure and services offered, evolution of goods production, consumption, and transportation demand in terms of volume, composition, and spatial distribution, changes in the economic, legal, and regulatory environment, and so on. These questions are often part of cost-benefit analyses and comparative studies of investment alternatives, especially when the available monetary resources are limited. Thus, planning and regulatory agencies are particularly interested in such issues, as are international agencies involved in financing projects in developing countries. Private firms are also interested when involved, for example, in the financing of transportation infrastructure or services, or in the planning, management, and operations of goods distribution services using several transportation modes.

STAN is both a general modelling and analysis framework and an interactive graphic computer system offering a comprehensive toolkit aimed at performing strategic analyses of multimodal multicommodity transportation systems. As a modelling framework, STAN offers capabilities to represent the main functional elements of such systems – i.e., transportation infrastructure and services, supplies and demands, costs and other performance evaluation criteria, etc. –, as well as the means to simulate the associated traffic behavior (Guélat, Florian, and Crainic, 1990; Crainic, Florian, and Léal, 1990). On the other hand, STAN, the computer system, is composed of a series of modules used to input, process, display, and output the pertinent information (Crainic *et al.*, 1990; Crainic, Florian, and Larin, 1994; Crainic *et al.*, 1998; Larin *et al.*, 2000).

The STAN modelling framework is that of a multimodal network, made up of modes, nodes, links, and intermodal transfers, on which multiple products are to be moved by specific vehicles and convoys between given origin and destination points. Here, a mode is a means of transportation having its own characteristics, such as operating and depreciation costs, energy consumption, vehicle type and capacity. Depending on the scope and level of detail of the contemplated strategic study, a mode may represent a carrier or part of its network corresponding to a particular transportation service, a specific vehicle or ship type or size, an aggregation of several carrier networks, or specific transportation infrastructures such as highway networks or ports.

The network representation follows the concept and notation of Section 6.2. Note that parallel links are used to represent situations where more than one mode is available between two adjacent nodes. This network representation is compact and enables easy identification of the flow of goods by mode, as well as various cost functions (e.g., operating cost, time delay, energy consumption, levels of emissions and noise, transportation or storage risk, etc.) by product and mode. Furthermore, the network model resembles the physical network, since, for example, the rail and road infrastructures are physically different. Also, when different services are operated on the same infrastructure, such as diesel and electric trains on rail lines, a separate link may be assigned to each service to capture the different behaviour, characteristics, and performance levels. A path in this network therefore consists of a sequence of directed links of a mode, a possible transfer to another mode, a sequence of directed links of the second mode, etc. This representation also allows for the restriction of flows of certain commodities and OD pairs to given subsets of modes (e.g., iron ore may be shipped only by rail and ship) to capture *a priori* mode choices and restrictions that occur in the operation of freight networks and transshipment facilities.

A product is any commodity (collection of similar products) – goods or passengers – that generates a link flow. The demand for each product is exogenous and is specified by a *set* of OD matrices. The mode choice for each product is also exogenous and is indicated by defining for each of these OD matrices a subset of modes allowed for transporting the corresponding demand. By using this modelling capability, one may, for example, indicate that the traffic out of certain regions has to use rail, while in other regions there is a choice between rail and barges. In order to avoid complicating the presentation, however, in the following we continue to use the $g^{m(p)}$ notation of Section 6.2.

Product flows on the multimode network constitute the decision variables of the model and STAN makes use of formulation (6.1) - (6.5) to simulate the distribution of the multicommodity flows over the multimodal network. In order to simplify somewhat the implementation, however, the cost functions are written with respect to average cost functions $s_a^p(v)$, on links, and $s_t^p(v)$, on transfers:

$$f_a^p(v) = s_a^p(v)v_a^p \quad a \in \mathcal{A},\ p \in \mathcal{P}, \tag{6.25}$$
$$f_t^p(v) = s_t^p(v)v_t^p \quad t \in \mathcal{T},\ p \in \mathcal{P}. \tag{6.26}$$

This network model allows for a detailed representation of the transportation infrastructure, facilities and services as well as the simultaneous assignment of multiple products on multiple modes. It captures the competition of products for the service capacity available, a feature of particular relevance when alternative scenarios of network capacity expansion are considered. It allows for the specification and combination of a wide variety of performance measures and assignment criteria, including user-optimum type of functions when the nature of a particular product requires it. Furthermore, the model is sufficiently flexible to represent the transport infrastructure of one carrier only.

The algorithm implemented in the STAN system exploits the natural decomposition by product of the problem and results in a Gauss-Seidel-like procedure which allows the solution of large size problems in reasonable computational times. Figure 6.1 gives a schematic description of the algorithm (see Guélat *et al.*, 1990, for a detailed analysis).

Initialization. $k = 0$, v^0 is the initial solution.

Major cycle. $k = k + 1$; For each product $p \in \mathcal{P}$ do

Descent Direction. Compute the *descent direction* $d^k = y^k - v^k$, where y^k is the "all-or-nothing" solution to the linearized problem

$$\text{Min } F(v^k) + \nabla F(v^k)(v - v^k) \text{ subject to } v \in \Omega^p. \tag{6.27}$$

This involves solving a shortest marginal cost path problem for each OD pair.

Line Search. Find the *step length* λ_k^p solution of $\text{Min}_{0 \le \lambda \le 1} F(v^k + \lambda d^k)$.

Flow Update. $v^{k+1} = v^k + \lambda_k^p d^k$.

Stopping Criteria. The algorithm stops either when the relative gap is below a given threshold or when a predefined number of major cycles have been performed.

Figure 6.1 The STAN Assignment Algorithm

The model and algorithm are embedded in the STAN interactive-graphic system where they are complemented by a large number of tools to define and compare scenarios, to input, display, analyze, modify, and output data, and to specify the network and assignment models. Matrix-based computing tools may be used to implement a wide gamut of mode choice and demand models. Similarly, a network calculator allows the combination of network data and results to implement various traffic-related performance and analysis models. The data required by the STAN system is organized into a dedicated and strictly structured data bank. See Larin *et al.* (2000) for a detailed description of the STAN system, components, interfaces, and tools.

Modifications and additions to the models, methods, or system functionalities have to fit into this framework and function seamlessly with the other system components. Moreover, the fact that the system is already in use, requires that the compatibility with previous versions of the software be guaranteed. The introduction of path analysis capabilities has to abide by these rules.

We already indicated several reasons that justify the decision not to implement the approaches described in Section 6.2. This led to the re-appraisal of the possibility to store the information relative to the paths generated during the assignment. The number of paths and the corresponding storage requirements are huge, however. Moreover, one cannot decide *a priori* on the maximum number of iterations, since it depends on the actual application and required precision. Consequently, it is not possible to evaluate *a priori* the total space required to store the paths. Thus, one cannot store the path information in a fixed-size data structure.

Two conclusions emerged from this analysis. First, it is not possible to include the path information in the present structure of the STAN data bank (its size is fixed and determined by the user when it is created). To modify this structure would represent a huge investment as well as raise significant compatibility issues. Therefore, path information should be stored outside the STAN data bank. Second, even by today's computer storage standards, to store the full path information is impractical. Therefore, one should store only the information needed to reconstruct the paths and determine the traffic distribution.

To reconstruct the paths generated at each iteration of the assignment procedure, one needs to store only the product marginal costs on the links and transfers of the network at that iteration. Applying the multimodal shortest path routine used by the assignment procedure (at step *Descent Direction*) will then yield the correct paths for each product and origin-destination pair. Assuming N iterations have been performed, the total storage requirements are of $O((|\mathcal{A}| + |\mathcal{T}|) * |\mathcal{P}| * N)$, which is manageable.

To compute the flows present on the paths when the procedure was stopped, only the step sizes λ_k^p used at each iteration (*Line Search* step) are required. Weights w_k^p, $k = 0, 1, ..., N$, give the distribution of the OD flows of product p among the $N+1$ paths that correspond to the $N+1$ iterations that were performed (including iteration 0 - the initialization). Given that $\lambda_0^p = 1$ (the "all-or-nothing" solution: all flow goes to the initial path), weights are computed as follows:

$$w_N^p = \lambda_N^p \qquad k = 0 \tag{6.28}$$

$$w_{N-1}^p = \lambda_{N-1}^p(1 - \lambda_N^p) \qquad k = 1 \tag{6.29}$$

$$w_{N-k}^p = \lambda_{N-k}^p(1 - \lambda_{N-k-1}^p)(1 - \lambda_{N-k-2}^p)\dots(1 - \lambda_N^p) \quad k = 2, ..., N \tag{6.30}$$

6.4 PATH ANALYSIS IMPLEMENTATION AND APPLICATIONS

The information required to reconstruct exactly the paths and flows, the product link and transfer marginal costs as well as the step lengths, are stored during the assignment procedure in a file external to the STAN data bank. The information is stored for all products and all OD demand matrices. In this way, the information is available for any number and variety of path analyses without the need to perform additional assignment procedures.

For efficiency reasons, however, only one product and one demand matrix may be analyzed at a time. Constraints may be specified on the selected demand matrix to restrict the analysis to well defined OD pairs. The number of paths to be analyzed may be reduced further by imposing restrictions on some of their basic attributes (e.g., length, costs, etc.), as well as on any attribute that may be obtained by a combination (e.g., a sum) of attributes of the links and transfers that make them up. The paths are generated, selected, and analyzed, iteration by iteration, and OD pair by OD pair, according to the following procedure:

- For a product p and demand matrix $g^{m(p)}$ and for each iteration $k = 0, ..., N$:
  - For each origin o:
    - Generate the set (tree) of *shortest marginal cost paths* to all destinations.
    - For each destination d:
      - ◇ Compute the path partial volumes: $w^p_{N-k} g^{m(p)}_{od}$;
      - ◇ Compute path attributes and perform selection operations;
      - ◇ Perform path computations;
      - ◇ Report results.

Many types of analyses may be performed by using this procedure. The simplest path analysis would only output the selected paths. Paths may be output as reports, text files, or plots. Figure 6.2 illustrates two multimodal paths for grains moved from origins in the Canadian Prairie Provinces (nodes 45 and 53) to a port on the Canadian Pacific coast (node 62).

More complex analyses are based on a *select element analysis* approach. Here, ad-hoc, analysis-specific, path attributes are created by combining attributes of the links and transfers that make them up. Total path travel time, waiting time in terminals, transshipment costs, emissions, etc. are a number of such ad-hoc attributes, that may then be used to select paths of particular interest. The same mechanism may also be used to select paths that fulfill certain conditions, such as using some or all of a number of particular facilities or services: ports, intermodal platforms, border crossings, service lines, and so on. One may thus compute and save global results with respect to selected sets of links and transfers. The global results that may be saved for further network or matrix analyses are:

1. The volumes of the given product moved by using the selected paths saved as link and transfer attributes.

2. The values of the ad-hoc attribute aggregated for each OD pair over all the selected paths and saved in a matrix.

3. The active demand, that is, the demand actually assigned to the selected paths, saved in a matrix. A typical application of this feature is to identify the OD pairs that make use of selected facilities or services and to compute the associated demand.

To illustrate this process, assume that one is interested in the analysis of the role of a strategic rail line (it runs through a mountain pass) in the distribution of grain

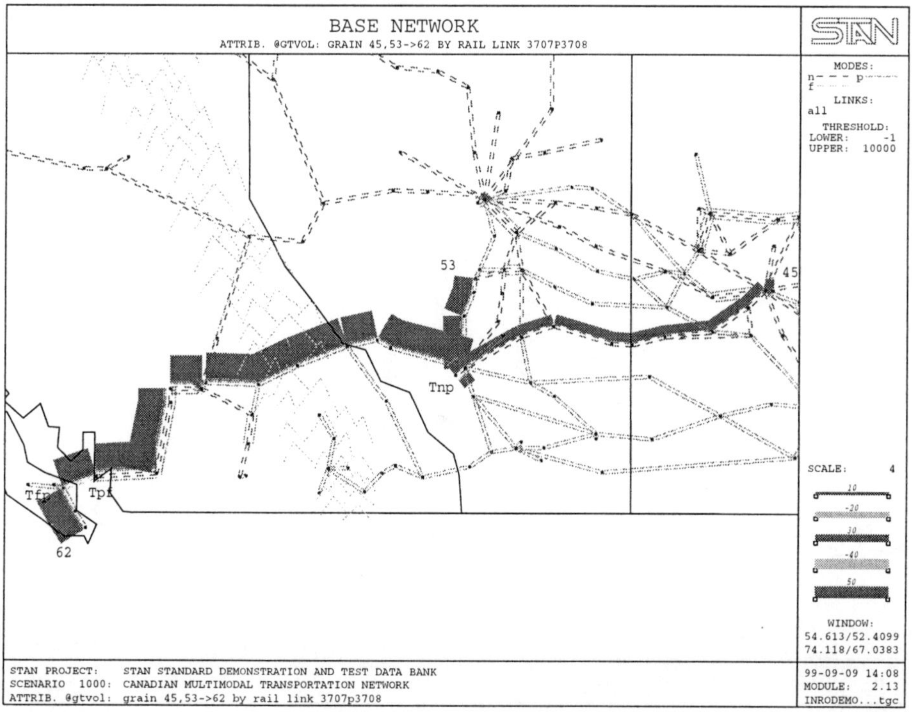

Figure 6.2 Two Multimodal Paths

in Western Canada (Figure 6.2). First, the network calculator is used to mark the line (the attribute of the appropriate link is set to 1). The path analysis module is then used to select the paths that use that line (the link attributes are summed up and only paths with a combined attribute equal to 1 are kept) to move grain from origins in the main producing provinces to the Canadian ports on the Pacific. Flows are saved in link attributes and the corresponding active demand in a matrix.

Figure 6.2 illustrates two such paths. The first, starting at node 53, makes use of a rail and a ferry mode, with transfers between train and ferry to load and unload the train on the ferry taking place at nodes Tpf and Tfp, respectively. The other path, originating at node 45, starts on a different rail mode (representing a different company), but transfers at node Tnp to take advantage of the other company's facilities to cross the mountains; the rest of the path is identical to the first one. Figure 6.3 displays a plot of the corresponding active demand: the height of the chimneys is proportional to the flow sent (zones East of the mountains) or received (ports on the coast, West of the mountains).

More complex analyses may be performed based on paths. An interesting application is the simultaneous adjustment of several demand matrices to observed flows. The issue of adjusting demand matrices to data obtained from surveys and counting posts is documented in the passenger traffic literature (see, for example, Florian and Chen, 1995, and references therein). Significantly less has been achieved with respect to freight transportation. Simply stated, the issue is to slightly modify the values of

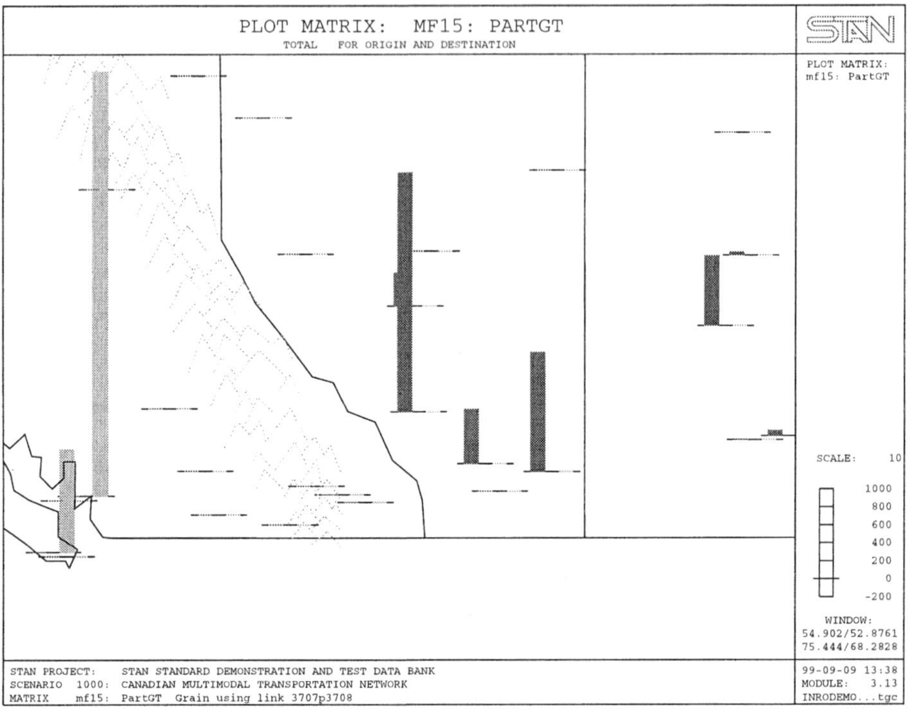

Figure 6.3 Partial Demand

an OD matrix such that the flows resulting from its assignment be as close as possible to some observed measures. The difficulty of the problem is increased when freight transportation is contemplated due to the multiproduct (and multimode) nature of the transportation system. Several cases may be defined according to the number of (product) matrices to be adjusted and the number of different observations: product specific or aggregated. The first case is quite common when observations come from rail or port facilities. The later is also quite frequent and corresponds, for example, to truck counting without inquiry into the products transported.

An in-depth treatment of this issue is well beyond the scope of this paper. Crainic *et al.* (2000) formulate the problem as a bi-level optimization model, where the master problem minimizes the sum of the (squared) differences between the link and transfer flows corresponding to the observed and adjusted matrices, while the lower level sub-problem performs the multimodal multiproduct assignment for the modified demand matrices. Under certain conditions, this bi-level matrix adjustment optimization problem can be solved by a *steepest descent*-type of algorithm (Spiess, 1990; Florian and Chen, 1995) that may be implemented very efficiently by using the tools offered in STAN: the path analysis, and the network and matrix calculators. See Crainic *et al.* (2000) for details and results.

6.5 CONCLUSIONS

Path information is required for many transportation planning analyses, in particular for the study of multiproduct freight movements through multimodal transportation systems. Unfortunately, due principally to the size of the networks and the huge number of the corresponding paths, most planning systems yield results in terms of link flows only, and path information is not readily available. We have described the problem and presented a number of mathematical programming approaches that open interesting perspectives regarding the construction of path flows and extraction of OD demand matrices from link (and transfer) flows.

We have also examined issues related to the integration of such a major development into a complex and mature software system. Our developments relate principally to STAN, a multimode, multiproduct method and interactive-graphic decision support system for strategic analysis and planning of freight transportation.

The path recovery approach that we selected allows a seamless integration of the new development into the STAN planning system. It offers the important advantage of an exact reconstruction of the paths generated during the assignment procedure and the computation of the flows on these paths when the assignment was terminated. The method requires that a limited amount of data be saved during the assignment. In return, any number of path analyses may be realized, without the need of additional assignments. Of the many types of analyses that are possible, two important applications have been described: select element (link or transfer) analyses, and the adjustment of product demand matrices according to observed data. These applications illustrate the wide range and flexibility of our *path analysis* methodology as well as of the STAN transportation planning system.

Acknowledgments

Funding for this project has been provided by the Natural Sciences and Engineering Research Council of Canada and by the Fonds F.C.A.R. of the Province of Quebec.

References

Charland, Y. (1996). Reconstitution d'itinéraires optimaux dans un réseau multimodes-multiproduits. M.Sc. Dissertation, Département d'informatique et recherche opérationnelle, Université de Montréal, Montréal, QC, Canada.

Crainic, T.G. (1999). Long Haul Freight Transportation. In Hall, R.W., editor, *Handbook of Transportation Science*, pages 433–491. Kluwer, Norwell, MA.

Crainic, T.G., Dufour, G., Florian, M., and Larin, D. (1999). Path Analysis in STAN. In C. Zopounidis and D. Despotis, editors, *Proceedings 5th International Conference of the Decision Sciences Institute*, pages 2060–2064. New Technologies Publications, Athens, Greece.

Crainic, T.G., Dufour, G., Florian, M., and Larin, D., and Leve, Z. (2000). Demand Matrix Adjustment for Multimodal Freight Networks. Publication CRT-2000-34, Centre de recherche sur les transports, Université de Montréal, Montréal, QC, Canada.

Crainic, T.G., Florian, M., Guélat, J., and Spiess, H. (1990). Strategic Planning of Freight Transportation: STAN, An Interactive-Graphic System. *Transportation Research Record*, 1283:97–124.

Crainic, T.G., Florian, M., and Larin, D. (1994). STAN: New Developments. In Al S. Khade and R. Brown, editors, *Proceedings of the 23rd Annual Meeting of the Western Decision Sciences Institute*, pages 493–498. School of Business Administration, California State University, Stanislaus CA.

Crainic, T.G., Florian, M., and Léal, J.-E. (1990). A Model for the Strategic Planning of National Freight Transportation by Rail. *Transportation Science*, 24(1):1–24.

Crainic, T.G. and Laporte, G. (1997). Planning Models for Freight Transportation. *European Journal of Operational Research*, 97(3):409–438.

Drissi-Kaïtouni, O. (1988). An Algorithm for the Decomposition of Arc Flows into Path Flows for the GSPEP. Publication CRT 569, Centre de recherche sur les transports, Université de Montréal, Montréal, QC, Canada.

Florian, M. and Chen, Y. (1995). A Coordinate Descent Method for the Bi-level O-D Matrix Adjustment Problem. *International Transactions in Operations Research*, 2(2):165–179.

Florian, M. and Hearn, D. (1995). Networks Equilibrium Models and Algorithms. In Ball, M., Magnanti, T.L., Monma, C.L., and Nemhauser, G.L., editors, *Network Routing*, volume 8 of *Handbooks in Operations Research and Management Science*, pages 485–550. North-Holland, Amsterdam.

Frank, M. and Wolfe, P. (1956). An Algorithm for Quadratic Programming. *Naval Research Logistics Quaterly*, 3:95–110.

Guélat, J., Florian, M., and Crainic, T.G. (1990). A Multimode Multiproduct Network Assignment Model for Strategic Planning of Freight Flows. *Transportation Science*, 24(1):25–39.

Larin, D., Crainic, T.G., Simonka, G., James-Lefebvre, L., Dufour, G., and Florian, M. (2000). *STAN User's Manual, Release 6*. INRO Consultants, Inc., Montréal, QC, Canada.

Spiess, H. (1990). A Gradient Approach for the O-D Matrix Adjustment Problem. Publication 693, Centre de recherche sur les transports, Université de Montréal, Montréal, QC, Canada.

7 LINEAR-IN-PARAMETERS LOGIT MODEL DERIVED FROM THE EFFICIENCY PRINCIPLE

Sven Erlander

Abstract: The linear-in-parameters logit model is a discrete choice model which can be derived in many ways, e.g. by the additive random utility maximizing approach. It can also be derived from the efficiency principle. The efficiency approach offers a new way of testing the model against observations. This paper derives the linear-in-parameters logit model from the efficiency principle and shows how the basic efficiency assumption, and hence the linear-in-parameters logit model, can be tested against observations.

7.1 INTRODUCTION

The linear-in-parameters logit model is a discrete choice model, where the choice probabilities are given by the multinomial logit formula

$$p_k = \frac{\exp(\sum_{s=1}^{S} \beta_s v_{ks})}{\sum_{k=1}^{K} \exp(\sum_{s=1}^{S} \beta_s v_{ks})}, \tag{7.1}$$

where the attributes are measured by the quantities $v_{ks}, s = 1, \ldots, S$, and where the parameters β_s indicate the relative importance of the attributes (see e.g. Ben-Akiva and Lerman, 1985, p. 108). A higher value of v_{ks} will give a higher probability for alternative k being chosen, provided that the parameter β_s is positive. The converse holds if β_s is negative.

Early contributions to the use of the simple logit model, $S = 1$, (Luce, 1959; McFadden, 1974) in transportation studies were given by M. Florian (Florian, 1974; Florian and Fox, 1976 and Florian, 1977).

The standard way of deriving the linear-in-parameters logit model is to use the additive random utility maximizing approach with (generalized) extreme value dis-

M. Gendreau and P. Marcotte (eds.), Transportation and Network Analysis: Current Trends, 109–117.

tribution for the non observable random part of the utility (see e.g. Ben-Akiva and Lerman, 1985).

However, the linear-in-parameters logit model can equally well be derived from the Efficiency Principle. This way of formulating the underlying basic assumptions is instructive. It offers necessary and sufficient theoretical conditions for the model to hold. Furthermore, the conditions can be investigated empirically, thus offering a new approach to testing the model.

The (Cost-)Efficiency Principle was introduced by Smith, 1978 as a way of deriving the gravity model in transportation planning. It has since then been extended in many directions (Smith, 1983; Smith, 1988; Erlander, 1985; Lundgren, 1989; Erlander and Smith, 1990; Erlander, 1998). This paper is based on Erlander, 2000.

7.2 LINEAR-IN-PARAMETERS LOGIT MODEL

We shall now discuss how (7.1) can be derived from the efficiency principle.

Consider a group of decision makers, each one choosing exactly one alternative k out of a set of alternatives, $k, k = 1, \ldots, K$, belonging to a common choice set. To each alternative in the choice set there are S attributes which are measured by the observable quantities v_{ks}, $s = 1, \ldots, S$.

We shall introduce the ideas by an informal argument. The basic idea is that the likelihood of choosing a specified set of alternatives would be higher if the corresponding values of the observed attribute values $[v_{ks}]$ are higher.

The probability of choosing alternative k would be an increasing function of the values of the attributes. Let z_k be the number of times that alternative k is chosen. The relative frequencies $(z_k / \sum_k z_k)$ can be used as an estimate of the probability of choosing alternative k. Hence, on the average, we would expect the relative frequencies $(z_k / \sum_k z_k)$ to increase with increasing values of v_{ks}. There are many increasing functions that would be possible candidates for the function we are looking for. A slightly more sophisticated approach would be to consider the sums $\sum_k z_k v_{ks}$ or, equivalently, the averages $\bar{v}_s = (1/N) \sum_k z_k v_{ks}$, where $N = \sum_k z_k$. The likelihood of the decisions leading to the frequencies $z_k, k = 1, \ldots, K$ would be an increasing function of the sums $\sum_k z_k v_{ks}$ or the average utilities $\bar{v}_s$. This simple idea can be formulated as the Efficiency Principle. In this way the functional form of the choice probabilities will be determined . The same idea is at the root of the testing procedure that we are going to discuss later on. We shall now make the argument precise.

Consider a sample of N independent decision makers. Let the random variable X_n be the decision taken by decision maker $n, n = 1, \ldots, N$,

$$X_n = k \;\text{ if decision maker } n \text{ chooses alternative } k.$$

Furthermore, let z_k be the number of decision makers choosing alternative k, $\sum_{k=1}^{K} z_k = N$. Let us now assume that the decision makers act independently of each other, each following the same probability distribution $[p_k]$,

$$p_k = \Pr(\text{decision maker number } n \text{ chooses alternative } k) = \Pr(X_n = k).$$

The efficiency principle is a way to determine the form of the so far unknown probability distribution $[p_k]$.

Let k_n denote the alternative chosen by decision maker number $n, n = 1, \ldots, N$. Then the probability of the event $[X_1 = k_1, \ldots, X_N = k_N]$ is given by the likelihood

$$\Pr(X_1 = k_1, \ldots, X_N = k_N) = p_{k_1} \cdots\ p_{k_N} = \prod_{k=1}^{K} p_k^{z_k}.$$

The total values of the attributes of all the N decision makers is given, for $s = 1, \ldots, S$, by

$$\sum_{n=1}^{N} v_{k_n s} = \sum_{k=1}^{K} z_k v_{ks}. \tag{7.2}$$

We shall now compare the likelihood of two independent samples. Let $z^1 = [z_k^1]$ and $z^2 = [z_k^2]$ be the outcome of the decisions of N decision makers in two independent samples.

The efficiency principle (Smith, 1978), formula (7.3) below, assumes that samples with greater total attribute value in all S components have greater likelihood. Consider first one specific attribute s. The likelihood of the sample would be higher if the sum $\sum_{k=1}^{K} z_k v_{ks}$ is higher. Thus $\sum_{k=1}^{K} z_k^1 v_{ks} \geq \sum_{k=1}^{K} z_k^2 v_{ks}$ would imply that the likelihood of the first sample is higher than the likelihood of the second sample. This should hold for each attribute $s, s = 1, \ldots, S$, provided that the influence on the choices as measured by v_{ks} is positive. If the influence is negative, e.g. if the attribute is a cost, the sign of v_{ks} is reversed. The argument can be used for each s leading to the following formulation.

Proposition 1 *LINEAR-IN-PARAMETERS LOGIT MODEL derived from the EFFICIENCY PRINCIPLE. For* $k = 1, \ldots, K$, *let the* S *attributes of alternative* k *be measured by the observable quantities* $v_{ks}, s = 1, \ldots, S$. *Assume that for every size* N *of independent samples* z^1 *and* z^2, *the EFFICIENCY ASSUMPTION*

$$[\sum_{k=1}^{K} z_k^1 v_{ks} \geq \sum_{k=1}^{K} z_k^2 v_{ks}, \quad s = 1, \ldots, S] \implies [\prod_{k=1}^{K} p_k^{z_k^1} \geq \prod_{k=1}^{K} p_k^{z_k^2}] \tag{7.3}$$

holds.

Then the probability distribution $[p_k]$ *is EFFICIENT with respect to* $[v_{ks}]$ *and is given by,*

$$p_k = \frac{\exp(\sum_{s=1}^{S} \beta_s v_{ks})}{\sum_{k=1}^{K} \exp(\sum_{s=1}^{S} \beta_s v_{ks})}, k = 1, \ldots, K, \tag{7.4}$$

for some $\beta_s \geq 0, \quad s = 1, \ldots, S$.

This is a particular case of Theorem 2.1 in Erlander and Smith, 1990.

Thus, the Efficiency Assumption, formula (7.3), implies the form of the probabilities given by formula (7.4). The opposite implication is also true; formula (7.4) implies formula (7.3), which easily follows by substituting (7.4) into (7.3). Note, that Proposition 1 is a theoretical result: For those samples that satisfy the S conditions

$$\sum_{k=1}^{K} z_k^1 v_{ks} \geq \sum_{k=1}^{K} z_k^2 v_{ks}, \quad s = 1, \ldots, S,$$

it is assumed that the likelihood of the first sample is larger than the likelihood of the second sample. Nothing is said about all other possible samples.

The Efficiency Assumption seems to be a natural assumption in many applications. In fact, it is hard to imagine a situation such that the Efficiency Assumption would not hold at least as a first order formulation, if the choices are positively influenced by the $[v_{ks}]$. It is surprising that the specific form of the probabilities, formula (7.4), follows from this rather weak assumption.

We have thus seen how the linear-in-parameters logit model can be derived from the efficiency principle. Note that our derivation utilizes nothing but observable quantities. The standard additive random utility maximizing approach requires the introduction of non observable random utilities with special stochastic properties. Hence, the efficiency principle approach leads to the same choice probabilities, formula (7.4), with less restrictive assumptions. Also, the choice probability formula (7.4) satisfies the Efficiency Assumption, formula (7.3), irrespective of way of derivation (for a discussion of the modelling aspects, see Erlander, 1998).

MAXIMUM LIKELIHOOD ESTIMATION OF THE PARAMETERS

Let there be $r = 1, \dots, R$ samples $[z_k^r]$ of size N with utilities/attributes v_{ks}, $k = 1, \dots, K$, and $s = 1, \dots, S$, and let

$$v_s = \sum_{k=1}^{K} p_k v_{ks}, \quad \bar{v}_s^r = \frac{1}{N} \sum_{k=1}^{K} z_k^r v_{ks}, \quad \bar{v}_s = \frac{1}{R} \sum_{r=1}^{R} \bar{v}_s^r = \frac{1}{RN} \sum_{r=1}^{R} \sum_{k=1}^{K} z_k^r v_{ks}. \tag{7.5}$$

Maximum likelihood estimation of the parameters means maximizing the likelihood $\prod_{r=1}^{R} \prod_{k=1}^{K} p_k^{z_k^r}$ when

$$p_k = \frac{\exp(\sum_{s=1}^{S} \beta_s v_{ks})}{\sum_{k=1}^{K} \exp(\sum_{s=1}^{S} \beta_s v_{ks})}.$$

This is equivalent to solving the likelihood equations

$$\bar{v}_s = v_s, \quad s = 1, \dots, S, \tag{7.6}$$

for $\hat{\beta} = [\hat{\beta}_s, \ s = 1, \dots, S]$.

Under regularity conditions, as $RN \to \infty$, $\sqrt{RN}(\hat{\beta} - \beta)$ is asymptotically normal with mean zero and covariance matrix $[I(\beta)]^{-1}$, where $I(\beta)$ is the information matrix

$$I_{st}(\beta) = \sum_{k=1}^{K} p_k (v_{ks} - \sum_{k=1}^{K} p_k v_{ks})(v_{kt} - \sum_{k=1}^{K} p_k v_{kt}).$$

(See e.g. Lehman, 1983, Theorem 6.4.1 p. 429, Ex. 6.5.3 p. 438.)

7.3 TESTING THE EFFICIENCY ASSUMPTION

The basic efficiency assumption (7.3) can be tested by empirical observations. We shall now discuss how this can be done.

The basic idea is to compare two or more independent samples. Let there be R independent samples $[z_k^r, k = 1, \dots, K]$, $r = 1, \dots, R$ of size N. If the efficiency assumption is true, then (7.3) would hold for each pairwise comparison. Hence, if we order the samples according to the observed values of $\sum_{k=1}^{K} z_k v_{ks}$ for a specific s, the efficiency assumption becomes

$$[\sum_{k=1}^{K} z_k^1 v_{ks} \geq \sum_{k=1}^{K} z_k^2 v_{ks} \geq \ldots \geq \sum_{k=1}^{K} z_k^R v_k] \implies$$

$$[\prod_{k=1}^{K} p_k^{z_k^1} \geq \prod_{k=1}^{K} p_k^{z_k^2} \geq \ldots \geq \prod_{k=1}^{K} p_k^{z_k^R}],$$

or equivalently,

$$[\sum_{k=1}^{K} (z_k^1/N) v_{ks} \geq \sum_{k=1}^{K} (z_k^2/N) v_{ks} \geq \ldots \geq \sum_{k=1}^{K} (z_k^R/N) v_{ks}] \implies$$

$$[\sum_{k=1}^{K} (z_k^1/N) \log p_k \geq \sum_{k=1}^{K} (z_k^2/N) \log p_k \geq \ldots \geq \sum_{k=1}^{K} (z_k^R/N) \log p_k].$$

By plotting $\sum_{k=1}^{K} (z_k^r/N) \log p_k$ against $\sum_{k=1}^{K} (z_k^r/N) v_{ks}$ for $r = 1, \ldots, R$, we would have a graphical test of the efficiency assumption (7.3) for each $s, s = 1, \ldots, S$. However, the probability distribution p is unknown, and has to be estimated. A natural estimate of the probability p_k in sample r is the relative frequency z_k^r/N; $\bar{p}^r = [\bar{p}_k^r]$, $\bar{p}_k^r = z_k^r/N, k = 1, \ldots, K$. (We use the notation $\bar{p}$ to indicate that the expression is an average based on the relative frequencies of the sample.) $\bar{p}^r$ maximizes the likelihood function, or equivalently, the logarithm of the likelihood function divided by N,

$$\sum_{k=1}^{K} (z_k^r/N) \log \bar{p}_k^r = \max_{[p_k]} \sum_{k=1}^{K} (z_k^r/N) \log p_k, \text{ subject to } \sum_{k=1}^{K} p_k = 1.$$

Let the entropy of the probability distribution $p = [p_k]$ be defined by

$$H(p) = -\sum_{k=1}^{K} p_k \log p_k,$$

and correspondingly for the observed distribution $\bar{p}^r$,

$$H(\bar{p}^r) = -\sum_{k=1}^{K} \bar{p}_k^r \log \bar{p}_k^r = -\sum_{k=1}^{K} (z_k^r/N) \log(z_k^r/N).$$

Hence, the maximum value of the logarithm of the likelihood function divided by N is equal to the observed negative entropy,

$$\max_{[p_k]} \frac{1}{N} \log \prod_{k=1}^{K} p_k^{z_k^r} = \max_{[p_k]} \sum_{k=1}^{K} (z_k^r/N) \log p_k = \sum_{k=1}^{K} (z_k^r/N) \log(z_k^r/N) = -H(\bar{p}^r). \quad (7.7)$$

Let

$$v_s = \sum_{k=1}^{K} p_k v_{ks}, \text{ and } \bar{v}_s^r = \sum_{k=1}^{K} (z_k^r/N) v_{ks} = \sum_{k=1}^{K} \bar{p}_k^r v_{ks}.$$

Replacing the log-likelihood expressions with the equivalent observed entropy values the test statement can be written

$$[\bar{v}_s^1 \geq \bar{v}_s^2 \geq \ldots \geq \bar{v}_s^R] \implies [-H(\bar{p}^1) \geq -H(\bar{p}^2) \geq \ldots \geq -H(\bar{p}^R)]. \tag{7.8}$$

Hence, we have arrived at a graphical test for the efficiency assumption, and thus also a graphical test of the linear-in-parameters logit model:

> If the efficiency assumption holds, then a graphical plot of the observed sample entropies $H(\bar{p}^r)$ against the observed average sample attribute values $\bar{v}_s^r = \sum_{k=1}^{K}(z_k^r/N)v_{ks}$ would show a pattern of decreased entropy for increased sample attribute value for each $s = 1, \ldots, S$.

Note that the efficiency assumption should hold for *any* values of the components $\bar{v}_s^r$ of (7.8). Therefore, in the following the values $\bar{v}_s^r, r = 1, \ldots, R$, will be viewed as given constants, in spite of the fact that they have been generated by R samples. However, the statement above will not hold always even if the efficiency assumption is true, since we have introduced some randomness by replacing the true values p by the observed values $\bar{p}^r$.

Furthermore, in real world applications it would be unrealistic to expect the efficiency assumption to hold exactly. If the number of observations tends to infinity, $N \to \infty$, we will in most cases discover deviations from the formula. What kind of random deviations are to be expected?

A lower bound for the observed negative entropy is given by

$$-H(\bar{p}) \geq -H(p) + \sum_{s=1}^{S} \beta_s(\bar{v}_s - v_s), \text{ where} \tag{7.9}$$

$$p_k = \frac{\exp(\sum_{s=1}^{S} \beta_s v_{ks})}{\sum_{k=1}^{K} \exp(\sum_{s=1}^{S} \beta_s v_{ks})}, k = 1, \ldots, K.$$

The next proposition gives the approximate stochastic behavior of the observed negative entropy.

Proposition 2 *ASYMPTOTIC PROPERTIES. Assume that the efficiency assumption (7.3) is true. Consider* $\bar{v}_s$, $s = 1, \ldots, S$ *as given. Then, for large* N, *and for given* $\bar{v}_s$, $s = 1, \ldots, S$, *the stochastic variable* $2N[-H(\bar{p}) + H(p) - \sum_{s=1}^{S} \beta_s(\bar{v}_s - v_s)]$ *has a (conditional)* χ^2*-distribution with* $(K - 1 - S)$ *degrees of freedom. The conditional expected value and the conditional variance are approximately given by*

$$E[-H(\bar{p}) \mid \bar{v}_s, s = 1, \ldots, S] \approx -H(p) + \sum_{s=1}^{S} \beta_s(\bar{v}_s - v_s) + \frac{K - 1 - S}{2N}, \textit{ and}$$

$$Var[-H(\bar{p}) \mid \bar{v}_s, s = 1, \ldots, S] \approx \frac{K - 1 - S}{2N^2}.$$

Proof. We have

$$-H(\bar{p}) = \sum_{k=1}^{K}(z_k/N)\log p_k + \frac{1}{N}\sum_{k=1}^{K} z_k \log(z_k/Np_k). \tag{7.10}$$

By using

$$\begin{aligned}\sum_{k=1}^{K}(z_k/N)\log p_k &= \sum_{s=1}^{S}\beta_s\sum_{k=1}^{K}(z_k/N)v_{ks} - \sum_{k=1}^{K}(z_k/N)\log\sum_{k=1}^{K}\exp(\sum_{s=1}^{S}\beta v_{ks})\\ &= -H(p) + \sum_{s=1}^{S}\beta_s(\bar{v}_s - v_s),\end{aligned}$$

and

$$\sum_{k=1}^{K} z_k \log(z_k/Np_k) \approx \frac{1}{2}\sum_{k=1}^{K}\frac{(z_k - Np_k)^2}{Np_k}, \tag{7.11}$$

(See e.g. Kullback, 1959, p. 113-114.) we obtain

$$-H(\bar{p}) \approx -H(p) + \sum_{s=1}^{S}\beta_s(\bar{v}_s - v_s) + \frac{1}{2N}\sum_{k=1}^{K}\frac{(z_k - Np_k)^2}{Np_k}. \tag{7.12}$$

The observed frequencies $[z_k]$ have a multinomial probability distribution, and hence, for large N, z_k is approximately normal with expected value $= Np_k$ and variance $= Np_k(1-p_k) \approx Np_k$. Hence, for large N, for given $\bar{v}_s, s = 1,\dots,S$, the sum

$$\sum_{k=1}^{K}\frac{(z_k - Np_k)^2}{Np_k} \tag{7.13}$$

has a χ^2-distribution with $(K-1-S)$ degrees of freedom since there are the $(1+S)$ linear relationships $\sum_{k=1}^{K} z_k = N$ and $\sum_{k=1}^{K} z_k v_{ks} = \bar{v}_s$ (See e.g. Stuart and Ord, 1987, p. 1166). But this is the conditional distribution that we are looking for.

Recall that the expected value and variance of a χ^2-distributed variable with K degrees of freedom are, respectively, $E[\chi^2(K)] = K$ and $Var[\chi^2(K)] = 2K$. □

We can now construct a test by using the lower bound for the observed negative entropy given by formula (7.9) and an upper bound obtained from the asymptotic results of Proposition 2, as we shall now show.

Let $\bar{P}_k = \sum_{r=1}^{R} z_k^r/(RN)$ be the full sample estimate of p_k, and let $\bar{v}_s = \sum_{k=1}^{K}\bar{P}_k v_{ks}$. Let $\hat{\beta} = [\beta_s, s = 1,\dots,S]$ be the full sample maximum likelihood estimate of β, obtained by solving the likelihood equations

$$\bar{v}_s = v_s,\ s = 1,\dots,S,\ \text{when}\ p_k = \frac{\exp(\sum_{s=1}^{S}\beta_s v_{ks})}{\sum_{k=1}^{K}\exp(\sum_{s=1}^{S}\beta_s v_{ks})}. \tag{7.14}$$

EFFICIENCY TEST. If the efficiency assumption (7.3) holds, then for large N, the expected number of points falling between the bounds defined by

$$\begin{aligned} \Pr(-H(\bar{P}) + \textstyle\sum_{s=1}^{S} \hat{\beta}_s(\bar{v}_s^r - \bar{v}_s) &\leq -H(\bar{p}^r) \\ \leq -H(\bar{P}) + \textstyle\sum_{s=1}^{S} \hat{\beta}_s(\bar{v}_s^r - \bar{v}_s) &+ \chi_\alpha^2(K-1-S)/2N) \\ &\approx (1-\alpha), \quad r = 1, \ldots, R, \end{aligned} \tag{7.15}$$

is approximately $(1-\alpha)R$. Here $\chi_\alpha^2(K-1-S)$ is the upper α-tail of a χ^2-distribution with $(K-1-S)$ degrees of freedom.

7.4 CONCLUSIONS

The efficiency assumption (7.3) is a straightforward behavioral assumption applicable in many situations. We have seen that the linear-in-parameters logit model (7.1) follows from this behavioral assumption. The reverse implication is also true; the linear-in-parameters logit model (7.1) satisfies the behavioral assumption (7.3). This means that we can investigate whether the linear-in-parameters logit model holds by investigating if the efficiency assumption (7.3) is true. The efficiency test (7.15) offers a way of testing the behavioral assumption (7.3) and hence can be used as a test for the linear-in-parameters logit model (7.1) itself. If more than $(1-\alpha)R$ points fall outside of the intervals given by (7.15), then this is an indication that the linear-in-parameters logit model has to be refuted.

The efficiency assumption (7.3) can be interpreted as a definition of utility maximizing behavior (Erlander and Lundgren, 2000). By using this approach the linear-in-parameters logit model can be derived in a new way from utility maximizing behavior.

Acknowledgements The author has benefited from discussions with Jan T. Lundgren.

References

Ben-Akiva, M. and Lerman, S.R. (1985). *Discrete Choice Analysis: Theory and Application to Travel Demand.* Cambridge, Massachusetts: The MIT Press.

Erlander, S. (1985). On the principle of monotone likelihood and loglinear models, *Mathematical Programming*, 21:137–151.

Erlander, S. (1998). Efficiency and the logit model, *Annals of Operations Research*, 82:203–218.

Erlander, S. (2000). A graphical test for utility maximizing behavior, Technical report LiTH-MAT-R-2000-11, Department of Mathematics, Linköping University, Linköping, Sweden.

Erlander, S. and Smith, T.E. (1990). General representation theorems for efficient population behavior, *Applied Mathematics and Computation*, 36:173–217.

Erlander, S. and Lundgren, J.T. (2000). Cost minimizing behavior in random discrete choice modeling, Technical Report LiTH-MAT-R-2000-08, Department of Mathematics, Linköping University, Linköping, Sweden.

Florian, M. (1974). On modelling congestion in dial's probabilistic assignment model, *Transportation Research*, 8:85–86.

Florian, M. and Fox, B. (1976). On the probabilistic origin of dial's multipath traffic assignment model, *Transportation Research*, 10:339–341.

Florian, M. (1977). A traffic equilibrium model of travel by car and public transit modes, *Transportation Science*, 8:166–179.

Kullback, S. (1959). *Information Theory and Statistics*: New York: Wiley.

Lehmann, E.L. (1983). *Theory of Point Estimation.* New York: Wiley.

Luce, R.D. (1959). *Individual Choice Behavior: A Theoretical Analysis.* New York: Wiley.

Lundgren, J.T. (1989). *Optimization Approaches to Travel Demand Modelling*, PhD thesis, Department of Mathematics, Linköping Institute of Technology, Linköping, Sweden.

McFadden, D. (1974). Conditional logit analysis of qualitative choice behavior, in: Zarembka, P. (ed.), *Frontiers of Econometrics*, pages 105–142. New York: Academic Press.

Smith, T.E. (1978). A cost-efficiency principle of spatial interaction behavior, *Regional Science and Urban Economics*, 8:313–337.

Smith, T.E. (1983). A cost-efficiency approach to the analysis of congested spatial-interaction behavior, *Environment and Planning A*, 15:435–464.

Smith, T.E. (1988). A cost-efficiency theory of dispersed network equilibria, *Environment and Planning A*, 20:231–266.

Stuart, A. and Ord, J.K. (1987). *Kendall's Advanced Theory of Statistics. Volume 1. Distribution Theory.* London: Charles Griffin.

8 A MULTI-CLASS MULTI-MODE VARIABLE DEMAND NETWORK EQUILIBRIUM MODEL WITH HIERARCHICAL LOGIT STRUCTURES

Michael Florian

Jia Hao Wu

Shuguang He

Abstract: We consider a multi-class multi-mode variable demand network equilibrium model where the mode choice model is given by aggregate hierarchical logit structures and the destination choice is specified as a multi-proportional entropy type trip distribution model. The travel time of transit vehicles depends on the travel time of other vehicles using the road network. A variational inequality formulation captures all the model components in an integrated form. A solution algorithm, based on a Block Gauss-Seidel decomposition approach coupled with the method of successive averages results in an efficient algorithm which successively solves network equilibrium models with fixed demands and multi-dimensional trip distribution models. Numerical results obtained with an implementation of the model with the EMME/2 software package are presented based on data originating from the city of Santiago, Chile.

8.1 INTRODUCTION

Over the past twenty years, the models used in transportation planning for congested urban networks received a lot of attention from researchers and practitioners. One of the principal issues studied and debated is that of the consistency of the models used for predicting the demand for travel and the network models used to determine the levels of service. One seeks to ensure that the levels of service which are used to compute the demand would be reproduced if that demand were used again to

M. Gendreau and P. Marcotte (eds.), Transportation and Network Analysis: Current Trends, 119–133.

compute the levels of service. The keywords that we find in the literature regarding these models are "integrated", "combined", "simultaneous", "feedback" and "variable demand". The recent San Francisco Bay Area lawsuit (see Garrett and Wachs (1996), p.199) shows that the importance of this issue has social and political impacts.

The model presented in this paper is inspired from the work carried out for the ESTRAUS project in Santiago, Chile (ESTRAUS, 1985-1998) and is a complex multi-class multi-mode variable demand network equilibrium model with hierarchical structures which, until now, was not formulated in an integrated way as a variational inequality. This formulation permits the rigorous mathematical analysis of the model structure and the development of an efficient solution algorithm.

The remainder of the paper is organized as follows. In the next section, we provide a literature review of the related problems. In Section 8.3, we introduce the notations and develop the variational inequality formulation. Section 8.4 provides the analysis of the mathematical structure of the model by using the corresponding KKT conditions, and the solution algorithm is given in Section 8.5. In Section 8.6, we provide computational results, and some conclusions are presented in the last section.

8.2 A BRIEF LITERATURE REVIEW

The development of network equilibrium models that may be formulated as optimization problems originates with the seminal contribution of Beckmann, McGuire and Winsten (1956). The introduction of variable demand network equilibrium models for one mode, where the destination choice is determined by entropy type distribution models, can be found in the contributions of Florian, Ferland and Nguyen (1975), and Evans (1976).

This model, referred to as combined trip distribution and assignment models, was extended to two modes and mode choice by Florian and Nguyen (1978) when the travel times by two modes, say auto and transit, are not related. These models are convex cost multi-commodity network optimization models that may be solved by the linear approximation method of Frank and Wolfe (1956) or by the partial linear approximation method first suggested by Evans (1976). Florian (1977) formulated a two-mode network equilibrium model where the transit travel times depend on the auto travel times, and the auto travel times account for the pressure of the transit vehicles which share the capacity of the roads.

This model was recognized to be a special instance of a variational inequality formulation following the work of Smith (1979) and Dafermos (1980) for network equilibrium models with asymmetric link cost functions. It was further analyzed by Fisk and Nguyen (1982) and Florian and Spiess (1982) and solved by an adaptation of the nonlinear Jacobi method. Sufficient conditions for the convergence of the algorithm were developed by Dafermos (1982) and Pang and Chan (1982). Friesz (1981) developed an optimization formulation of the multi-class combined trip distribution, mode choice and assignment model where the travel costs depend on the mode chosen, and Fisk and Boyce (1983) developed a variational inequality formulation for this model where the travel costs are not mode interrelated and the mode choice is given by a multinomial logit function. In the attempt to introduce the trip generation model in an integrated model, Safwat and Magnanti (1988) developed an integrated model which has an equivalent convex cost optimization formulation, and Safwat, Nabil and Walton (1988) provide a calibration and computational experience with this model.

Lam and Hai-Jun (1992) present a multi-class combined trip distribution mode choice and assignment model formulated as well as an equivalent convex cost optimization problem and solve it with the partial linearization method. The first formulation of a network equilibrium model with hierarchical logit demand functions is that of Fernandez et al. (1994) where the choice of stations for the "park-n-ride" mode is given by a lower nest of the demand function. The model is formulated as an equivalent convex cost optimization model and the solution method suggested is an adaptation of the partial linear approximation algorithm. Oppenheim (1995) presents several versions of combined trip distribution mode choice and assignment models. The importance of properly stating and solving integrated network equilibrium models is stressed by Boyce (1998). A recent contribution is that of Abrahamsson and Lundqvist (1999) who developed a convex cost optimization model for the combined trip distribution, mode choice and assignment models with hierarchical choices, where distribution may precede mode choice or vice versa. In their model, the transit travel times are independent of the auto travel times.

The model developed in this paper is motivated by the ESTRAUS model developed for the city of Santiago, Chile and described in various internal technical reports. In this model, the trip productions are given by class and purpose of travel. The mode choice model is an aggregated hierarchical logit function (Ortuzar and Willumsen 1996, p.219). There are thirteen classes of travelers, three travel purposes and multiple modes of travel, which include walk, auto, multiple transit modes and combined modes ("park-n-ride"). The model components were calibrated by using survey data, and a computational procedure was used for its numerical solution without the statement of an integrated model. In this paper, we formulate the model as a variational inequality and show that the Karush-Kuhn-Tucker (KKT) conditions replicate all the model components. Then a solution algorithm based on a Block Gauss-Seidel decomposition of the model, which is akin to the partial linear approximation method of Evans (1976), is provided to develop a solution algorithm.

8.3 MODEL FORMULATION

In order to present the model components and the variational inequality formulation, the following notation is used. The arcs of the road network are designated by $a, a \in A$, where A is the set of arcs. The segments of the transit lines which compose the transit networks are designates $b, b \in B$, where B is the set of segments of the transit network. $a(b)$ designates that segment b of a transit line segment uses link a of the road network. The demand for travel by user class $m, m \in M$, for purpose $p, p \in P$, class $n, n \in N$ and mode $m, m \in M$ for the origin-destination pair (i, j), is denoted T_{ij}^{pnm} where P, N, M are the sets of purposes, classes and modes. This demand may use paths $R_{ij}^{pnm} \subset R$ where R is the set of all routes $R = \bigcup_{(i,j),p,n,m} R_{ij}^{pnm}$. It is also useful to identify M_1 as the subset of modes which use private vehicles (auto, taxi), M_2 as the subset of transit related modes and M_3 as the walk mode.

A hierarchical logit structure for each purpose has three levels: the root, the groups and the modes. G^p denotes a set of groups or nests of nodes for purpose p, and g denotes a set of modes in a group or nest. These are illustrated in Figure 8.1, where $G^p = \{g_1, g_2, g_3\}, g_1 = \{m_1\}, g_2 = \{m_2\}$ and $g_3 = \{m_3, m_4, m_5\}$.

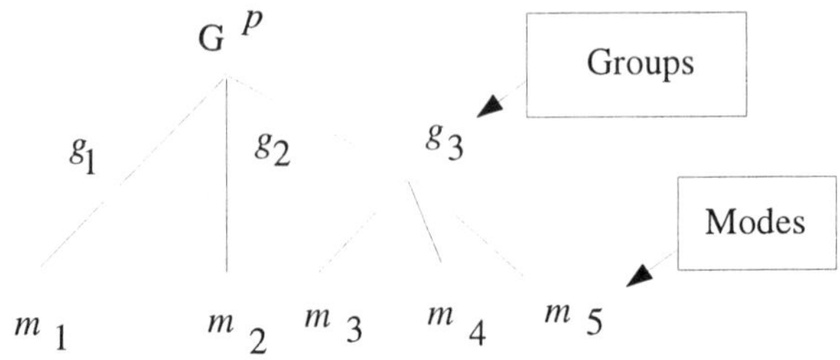

Figure 8.1 Nested logit structure for purpose p

The inclusion coefficients of each group are given as

$$\phi_g^p = \begin{cases} 1 & \text{if } |g| = 1 \\ \phi_g^p & \text{otherwise} \end{cases} \quad \text{all } g, p.$$

The production of trips are specified by origin i, purpose p and class n as O_i^{pn} while the attractions of trips are specified by destination j and purpose p, D_j^p. The travel demands T_{ij}^{pnm} give rise to path flows $h_r^{pnm}, r \in R_{ij}^{pnm}$. The feasible Ω region of the demands for travel and the path flows is stated next, where dual variables, which will be used later in the analysis are indicated in brackets.

$$\sum_g \sum_j T_{ij}^{png} = O_i^{pn}, \quad \forall i, p, n \qquad (\alpha_i^{pn}) \tag{8.1}$$

$$\sum_g \sum_n \sum_i T_{ij}^{png} = D_j^p, \quad \forall j, p \qquad (\xi_i^p) \tag{8.2}$$

$$\sum_{m \in g} T_{ij}^{pnm} - T_{ij}^{png} = 0, \quad \forall j, p, n, g \qquad (L_{ij}^{png}) \tag{8.3}$$

$$\phi_g^p \left(\sum_{r \in R^m} h_r^{pnm} - T_{ij}^{pnm} \right) = 0, \quad \forall j, p, n, m \in g \qquad (\mu_{ij}^{pnm}) \tag{8.4}$$

$$h_r^{pnm} \geq 0, \quad \forall r, p, n, m \qquad (\gamma_r^{pnm}) \tag{8.5}$$

$$T_{ij}^{pnm} > 0, \quad \forall (i,j), p, n, m \tag{8.6}$$

$$T_{ij}^{png} > 0, \quad \forall (i,j), p, n, g \tag{8.7}$$

Equations (8.1) and (8.2) are the balance equations for productions and attractions. Equation (8.3) relates mode totals to group totals. Equations (8.4) are the conservation of flow equations for path flows. Note that the multiplication by ϕ_g^p is required for the analysis to follow. Equations (8.5), (8.6) and (8.7) are the usual positive and nonnegativity requirements on path flows and demands respectively.

The presentation method that we have chosen is to state the variational inequality formulation first and then to show that the KKT conditions recover the model components. The variational inequality formulation supposes that β^{pn} is a calibrated parameter for all (p, n) and $\phi_g^p = (0, 1)$ is the inclusion coefficient.

Find $(h^*, T^*) \in \Omega$ which satisfies

$$\sum_{pn} \sum_{(ij)} \Bigg[\sum_{g \subseteq G^p} \Bigg(\sum_{m \in g} \Big(\sum_{r} \phi_g^p C_r^{pnm}(h^*, T^*)(h_r - h_r^*) \Big) + \sum_{m \in g} \phi_g^p ln(T_{ij}^{pnm*}/T_{ij}^{png*})(T_{ij}^{pnm} - T_{ij}^{pnm*}) \Bigg) + \frac{1}{\beta^{pn}} ln T_{ij}^{png*}(T_{ij}^{png} - T_{ij}^{png*}) \Bigg] \geq 0, \quad \forall (h,T) \in \Omega. \quad (8.8)$$

It is worthwhile to note that (8.8) may be rewritten in the equivalent form:

$$\sum_{pn} \sum_{(ij)} \Bigg[\sum_{g \in G^p} \Bigg(\sum_{m \in g} \Big(\sum_{r} \phi_g^p C_r^{pnm}(h^*, T^*)(h_r - h_r^*) \Big) + \sum_{m \in g} \phi_g^p ln(T_{ij}^{pnm*})(T_{ij}^{pnm} - T_{ij}^{pnm*}) \Bigg) + \left(\frac{1}{\beta^{pn}} - \phi_g^p \right) ln T_{ij}^{png*}(T_{ij}^{png} - T_{ij}^{png*}) \Bigg] \geq 0, \quad \forall (h,T) \in \Omega. \quad (8.9)$$

This formulation looks similar to some mathematical structures of combined trip distribution mode choice and assignment models which have equivalent convex cost optimization formulations. (See Fernandez et al. (1994) and Boyce (1998), p.81.) Although the model presented here does not have an equivalent convex cost optimization formulation, the variational inequality approach provides a powerful way to analyze the model structure.

8.4 THE ANALYSIS OF THE MATHEMATICAL STRUCTURE OF THE MODEL

In this section, we derive and analyse the KKT conditions of the variational inequality formulation (8.35), which are given below.

$$T_{ij}^{png} \quad : \quad \frac{1}{\beta^{pn}} ln T_{ij}^{png} - \alpha_i^{pn} - \xi_i^p + L_{ij}^{png} = 0 \quad (8.10)$$

$$T_{ij}^{pnm} \quad : \quad \phi_g^p ln(T_{ij}^{pnm}/T_{ij}^{png}) - L_{ij}^{png} + \phi_g^p \mu_{ij}^{pnm} = 0 \quad (8.11)$$

$$h_r^{pnm} \quad : \quad \phi_g^p C_r^{pnm} - \phi_g^p \mu_{ij}^{pnm} - \gamma_r^{pnm} = 0 \quad (8.12)$$

Complementarity:

$$h_r^{pnm} \gamma_r^{pnm} = 0 \quad (8.13)$$

$$\gamma_r^{pnm} \geq \text{ and (8.1)–(8.7)}$$

From (8.10), we have

$$T_{ij}^{png} = e^{\beta^{pn}(\alpha_i^{pn} + \xi_j^p)} e^{-\beta^{pn} L_{ij}^{png}}. \quad (8.14)$$

From (8.11), we have

$$T_{ij}^{pnm}/T_{ij}^{png} = e^{(1/\phi_g^p)L_{ij}^{png}} e^{-\mu_{ij}^{pnm}}, m \in g. \tag{8.15}$$

Using (8.15) and (8.3), we have

$$e^{-(1/\phi_g^p)L_{ij}^{png}} = \sum_{m' \in g} e^{-\mu_{ij}^{pnm'}} \tag{8.16}$$

which implies

$$L_{ij}^{png} = -\phi_g^p ln \left(\sum_{m' \in g} e^{-\mu_{ij}^{pnm'}} \right) \tag{8.17}$$

From (8.14) and (8.15), it follows that

$$T_{ij}^{pnm} = e^{\beta^{pn}(\alpha_i^{pn}+\xi_j^p)} e^{(1/\phi_g^p - \beta^{pn})L_{ij}^{png}} e^{-\mu_{ij}^{pnm}}. \tag{8.18}$$

The analysis of (8.10)–(8.18) leads to the following propositions which emphasize the fact that the variational inequality formulation (8.8) implies both group and mode proportions ($P_{ij}^{png}, P_{ij}^{pnm}$) of demands computed by the hierarchical logit structure as well as the network equilibrium path choice conditions.

Proposition 1 *(Group and Mode Proportions)*

$$P_{ij}^{png} = \frac{T_{ij}^{png}}{T_{ij}^{pn}} = \frac{e^{-\beta^{pn} L_{ij}^{png}}}{\sum_{g' \subseteq G^p} e^{-\beta^{pn} L_{ij}^{png'}}} = \frac{e^{-\beta^{pn} L_{ij}^{png}}}{e^{-\beta^{pn} L_{ij}^{pn}}}, \quad g \in G^p \tag{8.19}$$

and

$$P_{ij}^{pnm} = \frac{T_{ij}^{pnm}}{T_{ij}^{png}} = \frac{e^{-u_{ij}^{pnm}}}{\sum_{m' \in g} e^{-u_{ij}^{pnm'}}} = \frac{e^{-u_{ij}^{pnm}}}{e^{-(1/\phi_g^p)L_{ij}^{png}}}, \quad m \in M. \tag{8.20}$$

Proposition 2 *(Multinomial Logit Model)*

If $\phi_g^p = 1$ $(\forall g \in G^p)$, then we obtain the traditional multinomial logit structure for purpose p, i.e.,

$$P_{ij}^{pnm} = \frac{T_{ij}^{pnm}}{T_{ij}^{pn}} = \frac{e^{-u_{ij}^{pnm}}}{\sum_{m' \in M} e^{-u_{ij}^{pnm'}}}, \quad m \in M.$$

Proposition 3 *(Multi-Class Multi-Mode Equilibrium Conditions)*

$$C_r^{pnm} \begin{cases} = \mu_{ij}^{pnm} & \text{if} \quad h_r^{pnm} > 0 \\ \geq \mu_{ij}^{pnm} & \text{if} \quad h_r^{pnm} = 0 \end{cases}, \quad \forall r, p, n, m \in M.$$

Proof. Conditions (8.12) and (8.13) imply the equilibrium conditions.

The KKT conditions are also used for the algorithm development. It is important to recall the way that costs are propagated upward in the hierarchical logit demand structure. We define $L_{ij}^{png} = -\phi_g^p ln(\sum_{m' \in g} e^{-\mu_{ij}^{pnm'}}), \forall(i,j), p, n, g$ as group logsum for purpose p, class n and group g (this definition is particular to the ESTRAUS model which inspired this model), where ϕ_g^p is the inclusion coefficient of group g (see Ortuzar and Willumsen 1996, p.219).

We define $L_{ij}^{pn} = -\frac{1}{\beta^{pn}} ln \sum_{g \subseteq G^p} e^{-\beta^{pn} L_{ij}^{png}}, \forall(i,j), p, n$ as top logsum for purpose p and class n. The total demands for each p, n may be obtained as the solution of the following multi-proportional matrix balancing problem (combined trip distribution/mode choice model), by noting that $T_{ij}^{pn} = \sum_{g \subseteq G^p} G_{ij}^{png}$,

$$T_{ij}^{pn} = e^{\beta^{pn}(\alpha_i^{pn} + \xi_j^p)} e^{-\beta^{pn} L_{ij}^{pn}} \tag{8.21}$$

$$\sum_j T_{ij}^{pn} = O_i^{pn}, \quad \forall i, p, n \tag{8.22}$$

$$\sum_n \sum_i T_{ij}^{pn} = D_j^p, \quad \forall j, p \tag{8.23}$$

$$T_{ij}^{pn} > 0, \quad \forall ij, p, n \tag{8.24}$$

which may be easily shown to be equivalent to the following entropy maximization problem:

$$\begin{aligned}
&\max - \sum_{pn} \sum_{w} \frac{1}{\beta^{pn}} T_w^{pn} (ln T_w^{pn} - 1) \\
\text{s.t.} \quad &\sum_j T_{ij}^{pn} = O_i^{pn}, \quad \forall i, p, n \\
&\sum_n \sum_i T_{ij}^{pn} = D_j^p, \quad \forall j, p \qquad\qquad (8.25) \\
&\sum_i \sum_j T_{ij}^{pn} L_{ij}^{pn} = \overline{L}^{pn}, \quad \forall p, n \\
&T_{ij}^{np} > 0, \quad \forall (i,j), p, n
\end{aligned}$$

where $\overline{L}^{pn}$ is a constant (unknown) such as total generalized cost, and the dual variable associated with (8.25) is set to be 1. This is exactly the trip distribution component of the ESTRAUS model.

8.5 A SOLUTION ALGORITHM

In this section, we develop the solution algorithm used to obtain the solution of the variational inequality (8.9). We employ a Block Gauss-Seidel method (Equilibration Algorithm) where the network equilibrium formulation for given demands are computed in one block and the trip distribution/mode choice computations are computed

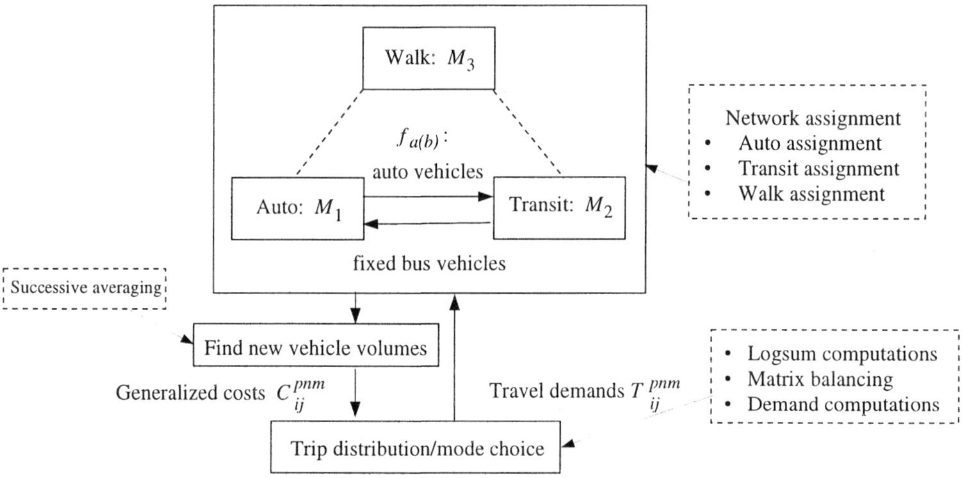

Figure 8.2 Block diagram of Equilibration Algorithm

in the other. The demand matrices for trips that use the vehicle network are computed at every successive iteration by using the method of successive averages over the vehicle flows. The solution algorithm flow chart is shown in the block diagram of Figure 8.2.

The network block requires the solution of the variational inequality formulation subject to constraints (8.27) to (8.34) as seen below. Since there are three distinct network modes M_1, M_2, M_3, one obtains three corresponding variational inequality problems which have equivalent optimization problems via a diagonalization. The network block is hence given by

$$\sum_{pn} \sum_{m \in m} \sum_{ij} \sum_{r \in R_{ij}^{pnm}} C_r^{pnm}(h^*, T)(h_r - h_r^*) \geq 0, \ \forall h \in \Omega(T). \tag{8.26}$$

$$\text{s.t.} \quad \sum_r h_r^{pmn} = T_{ij}^{pnm}, \ \forall (ij), p, n, m \tag{8.27}$$

$$h_r^{pnm} \geq 0, \ \forall p, n, m \in M, r \in R_{ij}^{pnm}, (ij) \in W \tag{8.28}$$

where

$$f_a^1 = \sum_{pn} \sum_{m \in M_1} \sum_{r \in R_1} h_r^{pnm} \delta_{ar}, \quad \forall a \in A \tag{8.29}$$

$$f_b^2 = \sum_{pn} \sum_{m \in M_2} \sum_{r \in R_2} h_r^{pnm} \delta_{br}, \quad \forall b \in B \tag{8.30}$$

$$f_a^3 = \sum_{p} \sum_{m \in M_3} \sum_{r \in R_3} h_r^{pnm} \delta_{ar}, \quad \forall a \in A \tag{8.31}$$

$$C_r^{pnm} = \sum_{a \in A} c_a^{pnm}(f_a^1) \delta_{ar}, \quad \forall p, n, m \in M_1, r \in R_{ij}^{pnm} \tag{8.32}$$

$$C_r^{pnm} = \sum_{b \in B} c_b^{pnm}(f_{a(b)}^1)\delta_{br}, \quad \forall p,n,m \in M_2, r \in R_{ij}^{pnm} \tag{8.33}$$

$$C_r^{pnm} = \sum_{a \in A} c_a^{pnm}\delta_{ar}, \quad \forall p,n,m \in M_3, r \in R_{ij}^{pnm} \tag{8.34}$$

The use of the nonlinear Jacobi method permits to solve the three distinct network modes in sequence. First, the auto network problem (8.35)–(8.39) is solved as a fixed demand network equilibrium model for the network M_1.

$$\sum_{pn}\sum_{ij}\sum_{m \in M_1}\sum_{r} C_r^{pnm}(h^*,T)(h_r - h_r^*) \geq 0, \quad \forall h \in \Omega(T). \tag{8.35}$$

$$\text{s.t.} \quad \sum_{r} h_r^{pmn} = T_{ij}^{pnm}, \quad \forall (ij), p, n, m \in M_1 \tag{8.36}$$

$$h_r^{pnm} \geq 0, \quad \forall p,n,m \in M_1, r \in R_{ij}^{pnm} \tag{8.37}$$

where

$$f_a^1 = \sum_{pn}\sum_{m \in M_1}\sum_{r \in R_1} h_r^{pnm}\delta_{ar}, \quad \forall a \in A \tag{8.38}$$

$$C_r^{pnm} = \sum_{a \in A} c_a^{pnm}(f_a^1)\delta_{ar}, \quad \forall p,n,m \in M_1, r \in R_{ij}^{pnm} \tag{8.39}$$

Thus, the transit link costs can be determined to compute the transit flows (8.40)–(8.44) with link costs $c_b^{pnm}(f_{a(b)}^1)$ to compute the corresponding origin-destination costs.

$$\sum_{pn}\sum_{ij}\sum_{m \in M_2}\sum_{r} C_r^{pnm}(h^*,T)(h_r - h_r^*) \geq 0, \quad \forall h \in \Omega(T). \tag{8.40}$$

$$\text{s.t.} \quad \sum_{r} h_r^{pmn} = T_{ij}^{pnm}, \quad \forall (ij), p, n, m \in M_2 \tag{8.41}$$

$$h_r^{pnm} \geq 0, \quad \forall p,n,m \in M_2, r \tag{8.42}$$

where

$$f_b^2 = \sum_{pn}\sum_{m \in M_2}\sum_{r \in R_2} h_r^{pnm}\delta_{br}, \quad \forall b \in B \tag{8.43}$$

$$C_r^{pnm} = \sum_{b \in B} c_b^{pnm}(f_a^1)\delta_{br}, \quad \forall p,n,m \in M_2, r \in R_{ij}^{pnm} \tag{8.44}$$

Finally, the walk network model may be solved by an "all-or-nothing" assignment as the solution of (8.45) to (8.49).

$$\sum_{pn}\sum_{ij}\sum_{m \in M_3}\sum_{r} C_r^{pnm}(h^*,T)(h_r - h_r^*) \geq 0, \quad \forall h \in \Omega(T). \tag{8.45}$$

$$\text{s.t.} \quad \sum_{r} h_r^{pmn} = T_{ij}^{pnm}, \quad \forall (ij), p, n, m \in M_3 \tag{8.46}$$

$$h_r^{pnm} \geq 0, \quad \forall p,n,m \in M_3, r \in R_{ij}^{pnm} \tag{8.47}$$

where

$$f_a^3 = \sum_p \sum_{m \in M_3} \sum_{r \in R_3} h_r^{pnm} \delta_{ar}, \quad \forall a \in A \tag{8.48}$$

$$C_r^{pnm} = \sum_{a \in A} c_a^{pnm} \delta_{ar}, \quad \forall p, n, m \in M_3, r \in R_{ij}^{pnm} \tag{8.49}$$

The demand block requires the computation of the multi-proportional trip distribution model described in the previous sections, which is solved by an adaptation of the two-dimensional balancing method.

If $\phi_g^p = 1(\forall g \in G_p)$ and there are no constraints (8.2), then (8.18) can be expressed as

$$T_{ij}^{pnm} = e^{\beta^{pn} \alpha_i^{pn}} e^{-u_{ij}^{pnm}}. \tag{8.50}$$

Using (8.1), (8.3) and (8.45), we obtain

$$e^{\beta^{pn} \alpha_i^{pn}} = O_i^{pn} / \sum_g \sum_m \sum_j e^{-\mu_{ij}^{pnm}} \tag{8.51}$$

With (8.27) and (8.51), we have

$$T_{ij}^{pnm} = O_i^{pn} e^{-u_{ij}^{pnm}} / \sum_g \sum_{m' \in g} \sum_j e^{-u_{ij}^{pnm'}}. \tag{8.52}$$

We call this case a "single constraint". The logsum costs L_{ij}^{pn} are computed by straightforward matrix computations according to the formulae stated in the previous section. The Equilibration Algorithm which employs the nonlinear Jacobi method for the network blocks and the method of successive averages to dampen the demand is summarized below.

Equilibration Algorithm

Step 0 (Initialization): $l = 0$

Choose an initial demand $\overline{T}^0$ (which equals the sum of auto and taxi trips) and compute $f^{1,0}$

Step 1 (Computation of generalized costs): $l = l + 1$,

Compute generalized costs $C^{pnm,l}, \forall(p, n, m)$ based on the current $f_a^{1,l}, c_a(f_a^{1,l})$ and compute transit costs $c_b(f_{a(b)}^{1,l})$.

Compute logsum matrices L_{ij}^{png} and $L_{ij}^{pn}, \forall(i, j), (p, n, g)$.

Step 2 (Combined trip distribution/mode choice model):

Carry out multi-proportional balancing (8.21)–(8.24) to obtain T_{ij}^{pn}, $\forall(p, n)$ and compute the O-D matrix of auto drivers $T^{p,n,m,l} \forall(p, n, m)$ (using the dual variables from the trip distrbution model).

Step 3 (Vehicle network equilibrium assignment):

Compute (8.35)–(8.39) for $\overline{f}^{1,l}$ with $T_{ij}^{pnm} \forall(i, j)(p, n, m \in M_1)$

Step 4 (Convergence test):
If $||f^{1,l} - \overline{f}^{1,l}|| \leq \varepsilon \quad (\varepsilon > 0)$, to to Step 6.

Step 5 (Method of Successive Averages):
Compute $f^{1,l+1} = f^{1,l}(x) + \overline{f}^{1,l}(1-x), (x = 1/(l+1))$ and return to Step 1.

Step 6 (Computation of final results): compute $T_{ij}^{pnm}, \forall(i,j), p, n, m$
Compute final f^1, f^2, f^3.
Compute final $C_{ij}^{pnm}, \forall(i,j), p, n, m$.

8.6 COMPUTATIONAL RESULTS (STGO MODEL)

This model is implemented with EMME/2 and is applied in Santiago, Chile. The model is called STGO. And while there are many details, such as parameters and equations (required for computation), it would serve no purpose to discuss them here. However, we provide the following basic information to specify the STGO model. See Table 8.1 for the STGO base network specification, Figure 8.3 for the network and Table 8.2 for the travel modes.

Table 8.1 STGO base network specifications

Supply Modes		12	Regular nodes	1091
Travel Modes		11	Directional links	5606
Centroids	Demands zones	264	Transit lines	1273
	Parking Zones	47	Transit line segments	45753
Classes		13	Trip purposes	3

Table 8.2 Travel modes

Travel mode	Description	Travel mode	Description
1 (wk)	Walk	7 (mt)	Metro
2 (ad)	Auto-driver	8 (bm)	Bus-metro
3 (ap)	Auto-passenger	9 (dm)	Auto-driver-metro
4 (tx)	Taxi	10 (pm)	Auto-passenger-metro
5 (tc)	Taxi-col	11 (tm)	Taxi-col-metro
6 (bs)	Bus		

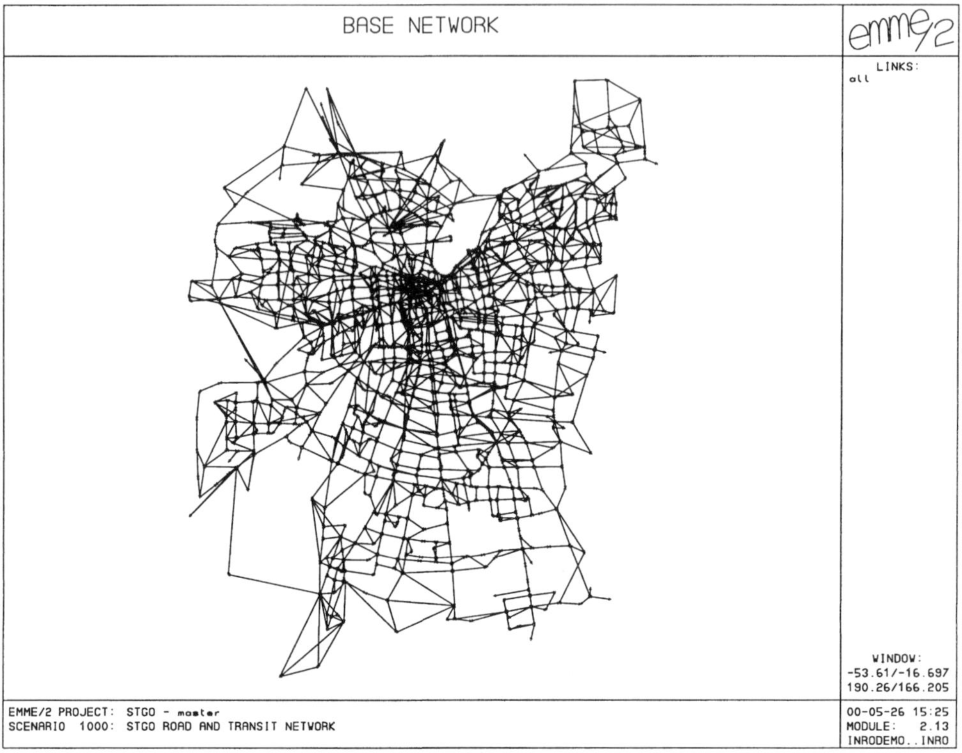

Figure 8.3 Santiago Network

We note that different groups (nests) of travel modes are established according to trip purposes, with different parameters. Here the EMME/2 implementation on a PC can effectively deal with the combined modes, such as park and ride (see modes 9 and 10).

The implementation has the following features. Transit in-vehicle times depend on the vehicle travel time in the vehicle network via transit functions. The generalized costs for transit related travel modes depend on the fares, mode specific walk times, in-vehicle times, waiting times and boarding times. The generalized costs for vehicle related travel modes depend on vehicle time and distances. The interactions between M_1, M_2, M_3 are given in Figure 8.4 where the dash lines imply no interactions implemented and the solid lines mean that there are interactions. In Step 1, the transit related generalized costs are computed as done in the ESTRAUS model where five transit assignments are performed with each for one transit mode. The corresponding computed components (such as in-vehicle times and fares) are used to compute the generalized costs. In Step 3, the vehicle network equilibrium assignment (8.35)–(8.39) is performed as done in the ESTRAUS model with a set of calibrated volume-delay functions. On a PC (PII 450 MHz, 96 Mb RAM), it took 20 min per iteration. About ten iterations are needed to obtain very similar results to that of ESTRAUS, if the initial demand is zero. If we want to perform a full run for a new scenario, we may only need three to five iterations.

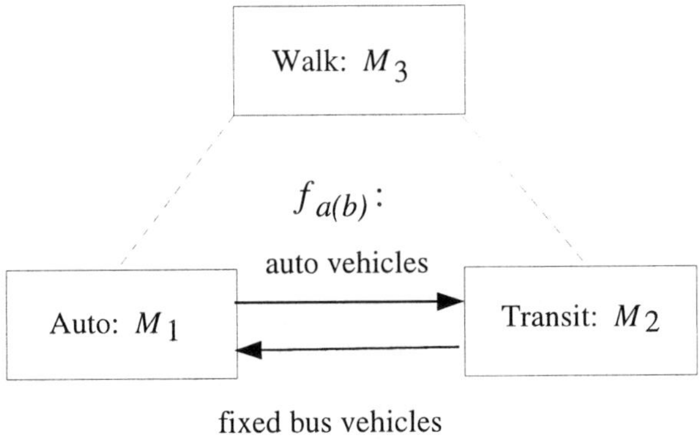

Figure 8.4 Interactions among the three assignment components

Table 8.3 Convergence of the procedure (30 loops)

	DEMAND		LINK VOLUME RELATIVE DIFFERENCE		
LOOPS	Demand	Average demand relative difference (percentage)	Total	Max.	Average
1	96.720695	0	3616	1	1
2	329579.5	94.80645	3962.34545	1	0.999583
3	321322	49.46157	1853.61218	55.236835	0.46832
...	...	...	...	...	...
9	294608.687	1.113139	874.864257	99.95684	0.222555
10	294363.687	0.832489	830.680847	89.338211	0.211154
11	294199.062	0.627432	782.620056	95.591217	0.198988
12	294104	0.53044	644.677001	77.937889	0.163957
...	...	...	...	...	...
30	293717.406	0.078193	333.711791	27.862077	0.084849
31	293736.562	0.05948	307.420928	28.45512	0.078164

The convergence of the Equilibration Algorithm for the STGO model is reported in Table 8.3. We see that after about ten iterations, the total demand stabilizes. Figure 8.5 shows very similar demand totals by mode computed by both the ESTRAUS and STGO models.

The advantages of using the variational inequality formulation are that it is more flexible and general by comparison with the limited optimization formulation. Based on this formulation, we derived a Block Gauss-Seidel method. We have not seen any optimization formulation developed so far which can result in the corresponding equations being used in practice.

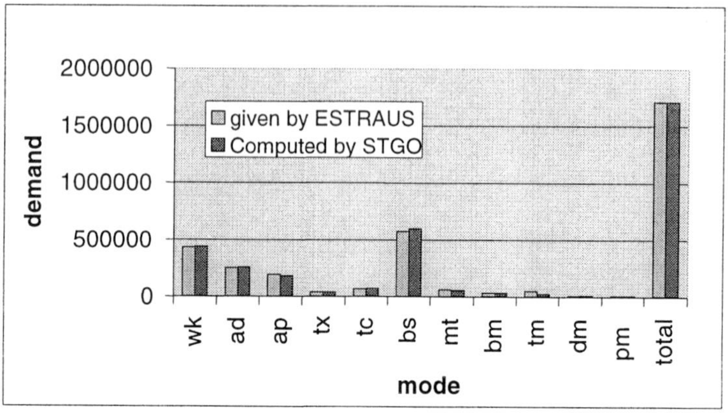

Figure 8.5 Demand totals for each mode between STGO and ESTRAUS

Acknowledgements The authors would thank Prof. Maximo Bosch, Mr. Manuel Diaz, Mr. Christian Lopez and Mr. Fernando Bravo for all the discussions during the project and Dr. Vladimir Livshits for a part of EMME/2 macro development.

References

Abrahamsson, T. and Lundqvist, L. (1999). "Formulation and Estimation of Combined Network Equilibrium Models with Applications to Stockholm," *Transportation Science*, 33, 1, 80–100.

Beckmann, M., McGuire, C.B. and Winsten, C.B. (1956). *Studies in the Economics of Transportation.* New Haven, CT: Yale University Press.

Boyce, D. (1998). "Long-Term Advances in the State of the Art of Travel Forecasting Methods", *Equilibrium and Advanced Transportation Modeling*, Edited by Marcotte, P. and Nguyen, S., Kluwer Acacemic Publishers.

Dafermos, S. (1980). "Traffic Equilibria and Variational Inequalities," *Transportation Science*, 14, 42–54.

Dafermos, S. (1982). "Relaxation Algorithms for the General Asymmetric Traffic Equilibrium Problem," *Transportation Science*, 16(2), 231–240.

ESTRAUS, (1998). Internal report.

Evans, S.P. (1976). "Derivation and Analysis of Some Models for Combining Trip Distribution and Assignment," *Transportation Research*, 10, 37–57.

Fernandez, E., De Cea, J., Florian, M. and Cabrera, E. (1994). "Network Equilibrium Models with Combined Modes," *Transportation Science*, 3, 182–192.

Fisk, C.S. and Nguyen, S. (1982). "Solution Algorithms for Network Equilibrium Models with Asymmetric User Costs," *Transportation Science*, 3, 361–381.

Fisk, C.S. and Boyce, D.E. (1983). "Alternative Variational Inequality Formulations of the Network Equilibrium-Travel Choice Problem," *Transportation Science*, 4, 454–463.

Florian, M. (1977). "A Traffic Equilibrium Model of Travel by Car and Public Transit Modes," *Transportation Science*, 2, 166–179.

Florian, M., Ferland, J. and Nguyen, S. (1975). "On the Combined Distributed-assignment of Traffic," *Transportation Science*, 9, 45–53.

Florian, M. and Nguyen, S. (1978). "A Combined Trip Distribution Modal Split and Trip Assignment Model," *Transportation Research*, 12, 241–246.

Florian, M. and Spiess, H. (1982). "The Convergence of Diagonalization Algorithms for Asymmetric Network Equilibrium Problems," *Transportation Research*, 16B, 447–483.

Frank, M. and Wolfe, P. (1956). "An Algorithm for Quadratic Programming," *Naval Research Logistics Quarterly*, 3, 95–110.

Friesz, T.L. (1981). "An Equivalent Optimization Problem for Combined Multi-class Distribution, Assignment and Modal Split which Obviates Symmetry Restrictions." *Transportation Research*, 15B, 5, 361–369.

Garrett, M. and Wachs, M. (1996). *Transportation Planning on Trail: The Clean Air Act and Travel Forecasting.* Thousand Oaks, CA: Sage Publications.

Lam, W.H.K. and Hai-Jun, H. (1992). "A Combined Trip Distribution and Assignment Model for Multiple User Classes," *Transportation Research*, 26B, 4, 275–287.

Oppenheim, N. (1995). *Urban Travel Demand Modeling.* A Wiley-Interscience Publication, John Wiley & Sons, Inc.

Ortuzar, J. de D. and Willumsen, L.G. (1996). *Modeling Transport.* John Wiley & Sons, Second Edition.

Pang, J.S. and Chan, D. (1982). "Iterative Methods for Variational and Complementarity Problems," *Mathematical Programming*, 24, 284–313.

Safwat, K. and Magnanti, T. (1988). "A Combined Trip Generation, Trip Distribution, Modal Split, and Trip Assignment Model," *Transportation Science*, 1, 14–30.

Safwat, K., Nabil, A. and Walton, C.M. (1988). "Computational Experience with an Application of a Simultaneous Transportation Equilibrium Model to Urban Travel in Austin, Texas," *Transportation Research*, 22B, 6, 457–467.

Smith, M.J. (1979). "The Existence, Uniqueness and Stability of Traffic Equilibria," *Transportation Research*, 13B, 289–294.

9 A TOLL PRICING FRAMEWORK FOR TRAFFIC ASSIGNMENT PROBLEMS WITH ELASTIC DEMAND

Donald W. Hearn
Mehmet B. Yildirim

Abstract: This paper extends the notion of toll pricing and the toll pricing framework previously developed for fixed demand traffic assignment (Bergendorff, Hearn and Ramana, 1997; Hearn and Ramana, 1998) to the problem with elastic demand. The system problem maximizes net benefit to the network users (Gartner, 1980; Yang and Huang, 1998) and the user problem is the usual one of finding equilibrium with elastic demand. We define and characterize $\mathcal{T}$, the set of all tolls for the user problem that achieve the system optimal solution. When solutions to the two problems are unique, $\mathcal{T}$ is a polyhedron defined by the optimal solution of the system problem, similar to the case in (Bergendorff, Hearn and Ramana, 1997; Hearn and Ramana, 1998). The Toll Pricing Framework in (Hearn and Ramana, 1998) is also extended to allow optimization of secondary criteria over $\mathcal{T}$. Examples include minimizing the number of toll booths and minimizing the maximum toll on any link. A numerical example illustrates the results.

9.1 INTRODUCTION

Motivation for congestion toll pricing stems from the fact that traffic jams and congestion on the roads are major problems that have direct economic impact on metropolitan areas in all countries. A recent study estimates that Bangkok loses one third of its potential output due to congestion (Commuting, 1998). In 1994, Arnott and Small estimated that the total cost of delay in 39 metropolitan areas of the United States is \$48 billion per year, an amount that translates to \$640 per driver.

Economists (Arnott and Small, 1994; Beckmann, McGuire and Winston, 1956) believe that roads are underpriced resources because users only experience their own

M. Gendreau and P. Marcotte (eds.), Transportation and Network Analysis: Current Trends, 135–145.

traffic delays as their cost. To improve system utilization, the standard proposal is to charge (toll) individual users for the delays (externalities) they impose on other users. When developed in the mathematical context of traffic assignment (TA) models, it leads to a specific formula for what are known as *marginal social cost pricing* (MSCP) tolls. A simple derivation shows that when MSCP tolls are imposed, the equilibrium model for user behavior has a solution which agrees precisely with the solution of another (untolled) TA model that measures total impact on the system. The first of these models is known as the user optimal/equilibrium model and the second as the system optimal model. References on TA models include Sheffi (1985), Patriksson (1994) and Florian and Hearn (1995).

It is now recognized that MSCP tolls are simply one possibility for defining tolls to achieve a system optimal solution. Bergendorff, Hearn and Ramana (1997) introduce a notion of alternative tolls and formally characterize the *toll set* for fixed demand TA models. A primary result is that under typical uniqueness conditions for the fixed demand TA problem, the toll set is a polyhedron. When the user problem is perturbed by any toll vector in the toll set, the user optimal solution is the system optimal solution. Hearn and Ramana (1998) further define a Toll Pricing Framework where the determination of specific tolls can be carried out by optimization methods such as linear or integer programming. The approach of defining alternative tolls to obtain system optimal flows is closely related to that of constraining user optimal flows to satisfy bound constraints. That approach appears in (Hearn, 1980; Hearn and Ribera, 1980; Bergendorff, 1995) and more recently in (Larsson and Patriksson, 1998).

This paper extends results in Bergendorff, Hearn and Ramana, 1997; Hearn and Ramana, 1998 to TA problems with elastic demand. The system problem is taken to be that of maximizing net benefit of the network users (Gartner, 1980; Yang and Huang, 1998) and the user equilibrium problem is the standard elastic demand TA problem. As before, when solutions to the two problems are unique, the toll set is a polyhedron defined by the optimal solution of the system problem. A new property, unique to the elastic demand case, is that all valid tolls generate the same total revenue. Thus MSCP tolls, in the aggregate, are no more expensive than other tolling schemes (Larsson and Patriksson, 1998). (This is very much not the situation in the fixed demand case, where the differences can be significant as shown in Bergendorff, Hearn and Ramana, 1997; Hearn and Ramana, 1998.) However, MSCP tolls have other disadvantages, and we extend the Toll Pricing Framework (Hearn and Ramana, 1998) to allow optimization of secondary criteria over the toll set. Examples include minimizing the number of toll booths and minimizing the maximum toll on any link. A previously employed numerical example (Hearn and Ramana, 1998) is modified with linear demand functions to illustrate the results.

9.2 TRAFFIC ASSIGNMENT MODELS WITH ELASTIC DEMAND

This section introduces notation and then defines the system and user equilibrium models. The system model assumes that the goal of transportation planners is to maximize net user benefit (Gartner, 1980; Yang and Huang, 1998), while the equilibrium problem models the usual Wardrop principle (Wardrop, 1952) of users choosing routes from which there is no unilateral improvement available.

9.2.1 Notation

Employing notation similar to (Bergendorff, Hearn and Ramana, 1997; Hearn and Ramana, 1998), let $\mathcal{G} = (\mathcal{N}, \mathcal{A})$ be a directed network with $\mathcal{N}$ being the node set and $\mathcal{A}$ being the set of links, and denote the incidence matrix of $\mathcal{G}$ by A. Let $\mathcal{K}$ be a set indexing the origin-destination pairs (o, d) where both $o \in \mathcal{N}$ and $d \in \mathcal{N}$. Then we may write $k = (o, d)$ for $k \in \mathcal{K}$ and refer to k as a "commodity." The demand for travel from some origin o to destination d is expressed as $t_k(c_k)$ where t_k is a nonnegative invertible demand function of c_k, the generalized cost of travel for commodity k. For brevity we write t_k rather than $t_k(c_k)$, and denote the inverse of t_k by $w_k(t_k)$. The t_k are components of the vector t and the inverse demand functions will be the components of the vector function $w(t)$. The kth *commodity flow* vector is denoted by the variable x^k, and the sum of all the commodity flow vectors is the *aggregate flow* vector v. We assume that a continuously differentiable cost map $s : \mathcal{A} \to \mathcal{A}$ is given. When the aggregate flow in the network is v, the cost incurred by a user on link a is given by $s_a(v)$. The Jacobian of s is denoted by ∇s. Note that $\nabla_{x^k} s = \nabla_v s$ since $v = \sum_k x^k$.

The system defining the feasible flows and demands is given by:

$$\begin{array}{rcll} v & = & \sum_{k \in \mathcal{K}} x^k & \\ Ax^k & = & t_k E_k & \forall k \in \mathcal{K} \\ x^k & \geq & 0 & \forall k \in \mathcal{K} \\ t_k & \geq & 0 & \forall k \in \mathcal{K}. \end{array}$$

where $E_k = e_o - e_d$, and e_o and e_d are unit vectors. E_k is thus a column incidence vector for commodity $k = (o, d)$ with $+1$ in position o and -1 in position d. We define the set of aggregate feasible flows and demands as

$$V = \{(v, t) |\text{ there exists } x^k \text{ satisfying the system above }\}.$$

This is a *node-link formulation* of the constraints for the elastic demand TA models.

9.2.2 The System Problem

In the system problem, the traffic planner maximizes net user benefit (Florian, 1984; Gartner, 1980; Yang and Huang, 1998), the difference between total user benefit and the system cost. From basic economic principles, the total network user benefit from travel is $\sum_{k \in \mathcal{K}} \int_0^{t_k} w_k(z)dz$ and the system cost is defined by $s(v)^T v$. Thus, using minimization in the objective, the elastic demand system problem (SOPT-ED) is

$$\begin{array}{ll} \min & s(v)^T v - \sum_{k \in \mathcal{K}} \int_0^{t_k} w_k(z)dz \\ \text{subject to} & (v, t) \in V. \end{array} \qquad \text{(SOPT-ED)}$$

The following lemma characterizes the Karush-Kuhn-Tucker (KKT) points of SOPT-ED:

Lemma 1 *A feasible point $(\bar{v}, \bar{t})$ is a KKT point for SOPT-ED if and only if there exists ρ^k $\forall k \in \mathcal{K}$ such that the following holds:*

$$\begin{array}{rcll} (s(\bar{v}) + \nabla s(\bar{v})\bar{v}) & \geq & A^T \rho^k & \forall k \in \mathcal{K} \\ w_k(\bar{t}_k) & \leq & E_k^T \rho^k & \forall k \in \mathcal{K} \\ (s(\bar{v}) + \nabla s(\bar{v})\bar{v})^T \bar{v} & = & w(\bar{t})^T \bar{t}. & \end{array}$$

Proof: The first two conditions follow from stationarity of the Lagrangian function for SOPT-ED, and the equality constraint is the aggregated complementarity condition. To prove this, we use the disaggregate complementarity conditions:

$$(\bar{x}^k)^T \gamma^k = 0 = (\bar{x}^k)^T(-A^T \rho^k + \nabla s(\bar{v})\bar{v} + s(\bar{v}))$$

and

$$\bar{t}_k \delta_k = 0 = \bar{t}_k(-w_k(\bar{t}_k) + E_k^T \rho^k)$$

where the vector γ^k and scalar δ_k are KKT multipliers for the constraints $-x^k \leq 0$ and $-t_k \leq 0$, respectively. Summing these gives

$$\begin{aligned} 0 &= \sum_k (\bar{x}^k)^T(-A^T \rho^k + \nabla s(\bar{v})\bar{v} + s(\bar{v})) + \sum_k \bar{t}_k(-w_k(\bar{t}_k) + E_k^T \rho^k) \\ &= \sum_k (\bar{x}^k)^T(-A^T \rho^k) + \sum_k (\bar{x}^k)^T(\nabla s(\bar{v})\bar{v} + s(\bar{v})) + \sum_k \bar{t}_k E_k^T \rho^k - (\bar{t})^T w(\bar{t}) \\ &= \bar{v}^T(\nabla s(\bar{v})\bar{v} + s(\bar{v})) - (\bar{t})^T w(\bar{t}) + \sum_k (\bar{x}^k)^T(-A^T \rho^k) + \sum_k (A\bar{x}^k)^T \rho^k \end{aligned}$$

Therefore $(s(\bar{v}) + \nabla s(\bar{v})\bar{v})^T \bar{v} = w(\bar{t})^T \bar{t}$. □

9.2.3 *The User Problem*

Wardrop's first principle states that the cost of travel (which is usually measured by travel time) on all routes actually used are equal, and they are less than or equal to the cost which would be experienced by a single vehicle on any unused route. In other words, if a feasible flow pattern has the property that there is no incentive for any user to deviate from the currently chosen route, then it is a *user equilibrium flow*. Mathematically, the user problem is stated as a variational inequality (Florian and Hearn, 1995):

Find $(\bar{v}, \bar{t}) \in V$ such that

$$s(\bar{v})^T(v - \bar{v}) - w(\bar{t})^T(t - \bar{t}) \geq 0 \quad \forall (v,t) \in V. \qquad \text{(UOPT-ED)}$$

The UOPT-ED solutions are characterized via the following lemma.

Lemma 2 $(\bar{v}, \bar{t}) \in V$ *solves UOPT-ED if and only if* $s(\bar{v})^T \bar{v} = w(\bar{t})^T \bar{t}$ *and there exists* ρ^k $\forall k \in \mathcal{K}$ *such that*

$$\begin{aligned} s(\bar{v}) &\geq A^T \rho^k \ \ \forall k \in \mathcal{K} \\ w_k(\bar{t}_k) &\leq E_k^T \rho^k \ \ \forall k \in \mathcal{K}. \end{aligned}$$

Proof: The vector $(\bar{v}, \bar{t})$ is user optimal if and only if (UOPT-ED) holds. Equivalently,

$$s(\bar{v})^T v - w(\bar{t})^T t \geq s(\bar{v})^T \bar{v} - w(\bar{t})^T \bar{t} \quad \forall (v,t) \in V$$

Thus $(\bar{v}, \bar{t})$ is user optimal if and only if it solves the linear program

$$\min\{s(\bar{v})^T v - w(\bar{t})^T t | v = \sum_k x^k \text{ and } Ax^k = t_k E_k, x^k \geq 0, t_k \geq 0 \ \ \forall k \in \mathcal{K}\}$$

whose dual is

$$\max\{\sum_k 0^T\rho^k | A^T\rho^k \leq s(\bar{v}), w_k(\bar{t}_k) \leq E_k^T\rho^k \quad \forall k \in \mathcal{K}\}.$$

Therefore, the optimality conditions ensure that there exist $\rho^k \ \forall\, k \in \mathcal{K}$ such that

$$\begin{aligned} s(\bar{v}) &\geq A^T\rho^k \quad \forall k \in \mathcal{K} \\ w_k(\bar{t}_k) &\leq E_k^T\rho^k \quad \forall k \in \mathcal{K}. \end{aligned}$$

Since the dual linear programs have the same optimal objective value, $s(\bar{v})^T\bar{v} - w(\bar{t})^T\bar{t} = 0$. □

In the next section, these two lemmas are used to extend the theory of toll pricing to elastic demand traffic assignment.

9.3 TOLL SET FOR THE ELASTIC DEMAND CASE

The aim is to obtain system optimal flows by perturbing the costs in the user problem. Let $s_\beta(v) = s(v) + \beta$ be the perturbed cost map where β is a toll vector. Then the perturbed user problem is:

Find $(\bar{v}, \bar{t}) \in V$ such that

$$(s(\bar{v}) + \beta)^T(v - \bar{v}) - w(\bar{t})^T(t - \bar{t}) \geq 0, \quad \forall (v,t) \in V. \qquad (\text{UOPT}_\beta\text{-ED})$$

Let U_β^* be the set of **tolled equilibrium solutions**, i.e., those $(\bar{v}, \bar{t}) \in V$ that satisfy the above variational inequality, and let S^* be the optimal solution set for SOPT-ED. Any β satisfying $\emptyset \neq U_\beta^* \subseteq S^*$ will be called a **valid toll vector**. The set of all such vectors denoted by $\mathcal{T} := \{\beta | \emptyset \neq U_\beta^* \subseteq S^*\}$ is called the **toll set**. For a given vector $(\bar{v}, \bar{t}) \in V$, $W(\bar{v}, \bar{t}) = \{\beta | (\bar{v}, \bar{t}) \in U_\beta^*\}$ is the set of all tolls for which $(\bar{v}, \bar{t})$ is a solution of UOPT$_\beta$-ED. Using Lemma 2, $W(\bar{v}, \bar{t})$ is the polyhedron given by the β part of the following linear inequality system in β and ρ^k variables:

$$\begin{aligned} s(\bar{v}) + \beta &\geq A^T\rho^k \quad \forall k \in \mathcal{K} \\ w_k(\bar{t}_k) &\leq E_k^T\rho^k \quad \forall k \in \mathcal{K} \\ (s(\bar{v}) + \beta)^T\bar{v} &= w(\bar{t})^T\bar{t}. \end{aligned}$$

With these definitions, the following theorem gives expressions for $\mathcal{T}$ in terms of the polyhedra $W(\bar{v}, \bar{t})$.

Theorem 1 *The toll set can be characterized by*

$$\mathcal{T} = \cup_{(\bar{v},\bar{t}) \in S^*} \{\beta \in W(\bar{v}, \bar{t}) | U_\beta^* \subseteq S^*\}.$$

If s and w are strictly monotone, then $\mathcal{T} = \cup_{(\bar{v},\bar{t}) \in S^*} W(\bar{v}, \bar{t})$.
If S^* is the singleton (v^*, t^*), then $\mathcal{T} = W(v^*, t^*)$.

Proof of this theorem is omitted since the results follow from arguments similar to those in (Hearn and Ramana, 1998) where expressions for $\mathcal{T}$ are given in the fixed demand case.

For a given $(\bar{v}, \bar{t})$, a valid toll β generates total toll revenue $\beta^T \bar{v}$. Interestingly, this quantity is *constant* for all $\beta \in \mathcal{T}$ as shown in the corollary below. Larsson and Patriksson (1998) also make this observation. Examples in (Hearn and Ramana, 1998; Hearn and Ramana, 1997) show this is clearly not true in the fixed demand case, and, in fact, one of the original motivations for that work was to determine tolls that are less expensive than MSCP tolls (Bergendorff, 1995; Bergendorff, Hearn and Ramana, 1997).

Corollary 1 *The total toll revenue is equal to $w(\bar{t})^T \bar{t} - s(\bar{v})^T \bar{v}$ for any valid toll $\beta \in \mathcal{T}$.*

Proof: Follows from Lemma 2, since $(s(\bar{v}) + \beta)^T \bar{v} = w(\bar{t})^T \bar{t}$. □

Valid tolls are also restricted in that total costs are the same on utilized routes, as the equilibrium principle dictates for the UOPT$_\beta$-ED problem. To state precisely, let r_k be a route for commodity k and γ_{ar_k} be 1 if link a is on route r_k and zero otherwise.

Corollary 2 *At the UOPT$_\beta$-ED solution $(\bar{v}, \bar{t})$ the total generalized cost on any route, for any commodity k, satisfies the inequality $\sum_{a \in A} \gamma_{ar_k}(s_a(\bar{v}_a) + \beta_a) \geq w_k(\bar{t}_k)$. For any route with positive flow, the inequality holds as an equality. Therefore the costs on utilized routes are constant for any commodity.*

Proof: For $a = (i, j)$ and $k = (o, d)$, on a given route r_k, $s_a(\bar{v}_a) + \beta_a \geq \rho_j^k - \rho_i^k$ and $w_k(\bar{t}_k) \leq \rho_o^k - \rho_d^k$. By summing over a route, the inequality in the corollary statement is obtained. Further, the complementarity conditions force the inequality to hold as an equality when there is flow on the route. The conclusion of the corollary then follows. □

This corollary shows clearly that every commodity reaches an equilibrium exactly when the route costs, including tolls, equals the user benefit at the final demand level. MSCP tolls are easily computed by the formula $\beta_{MSCP} = \nabla s(v^*) v^*$ whenever (v^*, t^*) solves SOPT-ED, hence $\mathcal{T}$ is nonempty. However, there are disadvantages to MSCP tolls. A primary disadvantage, recognized in the theory of *second-best* tolls (McDonald, 1995; Verhoef, 1999), is that some links may not be available for tolling. In fact, there is a positive component of β_{MSCP} for any congested link with flow, implying the need for many toll booths. Also, MSCP tolls do not allow the possibility for *subsidies*, or negative tolls, on some links. The toll pricing framework of the following section is a prescription for defining $\mathcal{T}$ and then choosing a particular $\beta \in \mathcal{T}$ based on some secondary criteria.

9.4 THE TOLL PRICING FRAMEWORK

The toll pricing framework from (Hearn and Ramana, 1998) extends readily given the results above. Assume uniqueness of the solutions for SOPT-ED and for UOPT-ED. Uniqueness will follow, for example, if both s and w are strictly monotone and $s(v)^T v$ is strictly convex. The framework then consists of the following steps:

Step 1: Solve SOPT-ED to obtain the system optimal solution (v^*, t^*). The algorithm of Evans (1976) can be used for this step.

Step 2: Define the toll set which is the β part of the polyhedron $W(v^*, t^*)$:

$$\begin{aligned}(s(v^*) + \beta) &\geq A^T \rho^k \quad \forall k \in \mathcal{K} \\ w_k(t_k^*) &\leq E_k^T \rho^k \quad \forall k \in \mathcal{K} \\ (s(v^*) + \beta)^T v^* &= w(t^*)^T t^*.\end{aligned}$$

Step 3: Define and optimize an objective function over the toll set, possibly intersected with other constraints. (See Table 9.1 for examples.)

Table 9.1 Alternative Optimization Formulations

TOLL	OBJECTIVE FUNCTION	EXTRA CONSTRAINTS
MINREV	$\min_{(\beta,\rho)} \beta^T v^*$	
MINMAX	$\min_{(z,\beta,\rho)} z$	$z \geq \beta_a, \quad \forall a \in \mathcal{A}, \beta \geq 0$
MINTB	$\min_{(z,\beta,\rho)} \sum_{a \in \mathcal{A}} z_a$	$\beta_a \leq M z_a \quad \forall a \in \mathcal{A}, z_a \in \{0,1\}, \beta \geq 0$ M a large constant

As in the fixed demand case, various objectives in step 3 lead to linear programs or linear integer programs. For the elastic demand problem, two natural choices are the MINTB and MINMAX objectives in (Hearn and Ramana, 1998). The first of these aims to minimize the number of toll booths and the second minimizes the maximum toll on any one link. The objective of minimizing total revenue from nonnegative tolls (MINSYS) is of less interest due to corollary 1. MINREV is similar to MINSYS, but tolls are free to be negative as well as positive. It is worth noting that the step 3 objectives can lead to problems with nonunique solutions. Hence further objectives might be possible in specific problems. Table 9.1 summarizes the choices suggested.

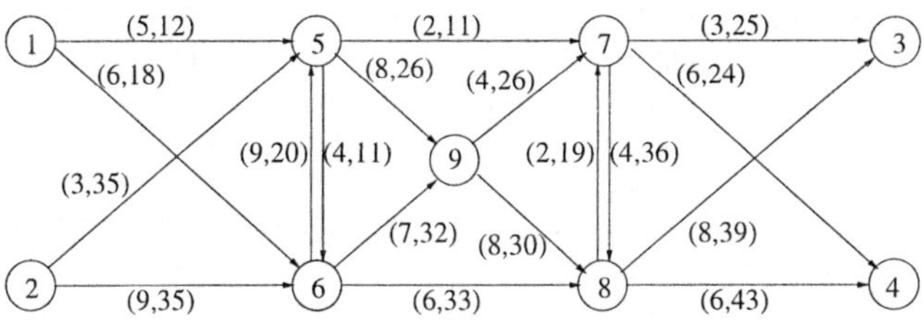

Figure 9.1 The Nine Node Network

To compare the formulations given above as well as provide comparison with MSCP tolls, we have modified the nine node example from (Hearn and Ribera, 1980; Bergendorff, Hearn and Ramana, 1997; Hearn and Ramana, 1998) which has cost data similar to large-scale traffic assignment problems. The nine node network has 18 links and all links have cost functions with the same structure: $s_a(v) = s_a(v_a) = T_a(1 + 0.15(v_a/C_a)^4)$ where T_a is a measure of travel time when there is zero flow and C_a is the practical capacity of link a. It is depicted in Figure 9.1. The tuple near link a is (T_a, C_a). There are four OD-pairs: (1,3), (1,4), (2,3) and (2,4). The demand functions between these OD-pairs are, $t_{(1,3)} = t_{(1,3)}(c_{(1,3)}) = 10 - 0.5c_{(1,3)}$,

$t_{(1,4)} = t_{(1,4)}(c_{(1,4)}) = 20 - 0.5c_{(1,4)}$, $t_{(2,3)} = t_{(2,3)}(c_{(2,3)}) = 30 - 0.5c_{(2,3)}$ and $t_{(2,4)} = t_{(2,4)}(c_{(2,4)}) = 40 - 0.5c_{(2,4)}$.

Table 9.2 Demand and Generalized Cost $(t^*, w(t^*))$ for the Nine Node ED Problem

OD-PAIR	SOPT-ED SOLUTION	UOPT-ED SOLUTION
(1,3)	(0.000, 20.000)	(0.151, 19.698)
(1,4)	(9.696, 20.607)	(10.698, 18.605)
(2,3)	(19.476, 21.047)	(20.672, 18.656)
(2,4)	(28.239, 23.523)	(29.232, 21.537)

Note first that SOPT-ED and UOPT-ED have different demands (Table 9.2). The total demand in the system problem is 57.411 and it is 60.753 in the user problem, so there is a 5.5% difference. This table also contains the generalized costs at the optimal demand levels.

Table 9.3 The Nine Node Problem – SOPT-ED and UOPT-ED Solutions

	SOPT-ED SOLUTION			UOPT-ED SOLUTION		
Link	v_a^*	$s_a(v_a^*)$	$v_a^* s_a(v_a^*)$	v_a^U	$s_a(v_a^U)$	$v_a^U s_a(v_a^U)$
1-5		5.000			5.000	20.000
1-6	9.696	6.076	58.914	10.849	6.119	66.380
2-5	31.715	3.303	104.769	34.458	3.423	117.940
2-6	15.999	9.059	144.938	15.446	9.051	139.805
5-6		9.000			9.000	
5-7	17.978	4.140	74.433	26.442	12.016	317.730
5-9	13.738	8.094	111.188	8.016	8.011	64.215
6-5		4.000			4.000	
6-8	25.696	6.331	162.677	26.295	6.363	167.308
6-9		7.000			7.000	
7-3	19.476	3.166	61.657	20.823	3.217	66.979
7-4	12.239	6.061	74.180	13.785	6.098	84.063
7-8		2.000			2.000	
8-3		8.000			8.000	
8-4	25.696	6.115	157.125	26.144	6.123	160.078
8-7		4.000		0.151	4.000	0.604
9-7	13.738	4.047	55.594	8.016	4.005	32.108
9-8		8.000			8.000	
User Benefit			2544.75			2613.50
System Cost			1005.474			1217.21
Net User Benefit			1539.284			1396.285

Table 9.3 provides optimal flows for the SOPT-ED and UOPT-ED problems. Although the UOPT-ED solution has higher user benefit than the SOPT-ED solution, it also has a higher user cost due to increased traffic on the network. Thus the net user benefit (NUB) in the system problem is greater. Examining the individual link flow values, notice that they are within 10% of each other on almost all links of the network, but not on links (5,7), (5,9) and (9,7). The *total* flow between nodes 5 and 7 for both user and system problems is within 10%, but the link flows differ substantially. Relative to the system solution, in the user problem (5,7) is over utilized while (5,9) and (9,7) are under utilized. Therefore, systems efficiency is increased by diverting traffic on (5,7) to the route 5-9-7. This can be done by making (5,7) less attractive, i.e., increasing the cost by tolling (5,7).

Table 9.4 The Nine Node Problem — Alternative Tolls

Link	**MSCP**	**MINREV**	**MINMAX**	**MINTB**
1-5			8.000	
1-6	0.303	2.085	2.085	0.067
2-5	1.214	-7.444		2.018
2-6	0.236	2.018	2.018	
5-6		6.218		
5-7	8.561		8.000	8.000
5-9	0.374	-3.953		
6-5				
6-8	1.323			
6-9				
7-3	0.663	17.882	2.438	0.420
7-4	0.243	17.462	2.018	
7-8		15.408		
8-3			2.320	
8-4	0.459			2.018
8-7			0.716	
9-7	0.187	-4.047		
9-8		9.408		
$\beta^T v^*$	268.519	268.519	268.519	268.519
$\beta^T v^*$**/(NUB) (%)**	17.44	17.44	17.44	17.44
Toll Booths	10	10	8	5

Table 9.4 contains alternative toll vectors for the nine node problem obtained with an optimization modeling package (GAMS/CPLEX) implementation of the Toll Pricing Framework. (Tolls are expressed in time units; Arnott and Small (1994) discuss the conversion to dollars based on studies in the U.S.) For comparison, the MSCP tolls are also listed. As expected, tolls on the link (5,7) are high for the tolling schemes with positive tolls, namely, MSCP, MINMAX and MINTB. However, MINREV, which allows negative tolls, rewards travel on the 5-9-7 route with subsidies

in order to achieve the SOPT-ED solution. By Corollary 1, all tolling schemes produce the same toll revenue. For this example the revenue is 268.519, which is 17.44% of net user benefit. MSCP and MINREV tolls require 10 toll booths. MINMAX has an objective value of 8.00 with eight toll booths. It happens that MINTB obtains the same result (i.e., the maximum toll is 8.00) with only five toll booths. Thus it could be argued that the MINTB solution is best in that it achieves the SOPT-ED solution and is cheapest to implement.

To illustrate Corollary 2, consider the routes 2-6-8-4 and 2-5-7-4. The delay functions on the first route give a total cost of 21.505, while on the second this total is only 13.504. However, the tolls on the first route are 2.018 and they are 10.018 on the second. Thus the total route cost is 23.523, in agreement with the total generalized cost in Table 9.2. There is no incentive for additional 2-4 trip demand. Any increase would result in lower user benefit and higher costs since the demand and cost functions are strictly monotone.

9.5 SUMMARY

The toll pricing methodology previously introduced extends readily to the elastic demand case when the system problem maximizes net user benefit. It offers clear advantages over marginal social cost pricing and heuristic tolling schemes. Further, it is anticipated that standard optimization software will be capable of solving modest sized urban networks. This expectation is based on prior experience with real data reported in (Hearn and Ramana, 1998; Hearn and Ramana, 1997). Alternative tolls were calculated for a Stockholm model with 417 nodes, 963 links and fixed demand of 272,873 rush hour trips using the GAMS/CPLEX combination, and the methodology of this paper is only marginally more complicated. Implementations and customized network algorithms for larger problems are important computational issues to be addressed in future research.

Acknowledgment *This research was partially supported by* NSF *grant* DMII-9634849.

References

Arnott, R. and K. Small (1994). “The Economics of Traffic Congestion,” *American Scientist*, Vol. 82, pp. 446-455.

Beckmann, M., C. McGuire, and C. Winston (1956). *Studies in the Economics of Transportation,* Yale University Press, New Haven. CT.

Bergendorff, P. (1995). *The Bounded Flow Approach to Congestion Pricing,* Univ. of Florida Center for Applied Optimization report and Master’s thesis, Royal Institute of Technology, Stockholm.

Bergendorff, P., D. W. Hearn and M. V. Ramana (1997). "Congestion Toll Pricing of Traffic Networks,” *Network Optimization,* P. M. Pardalos, D. W. Hearn and W. W. Hager (Eds.), Springer-Verlag series Lecture Notes in Economics and Mathematical Systems, pp. 51-71.

“Commuting,” *The Economist*, September 5th, 1998, pp. 3-18.

Evans, S. P. (1976). “Derivation and Analysis of Some Models for Combining Trip Distribution and Assignment,” *Transportation Research,* Vol. 10, pp. 37-57.

Florian, M. (1984). "An Introduction to Network Models Used in Transportation Planning," in *Transportation Planning Models* edited by M. Florian, Elsevier Science Publishers, pp. 137-152.

Florian, M. and D.W. Hearn (1995). "Network Equilibrium Models and Algorithms," Chapter 6 of *Handbooks in Operations Research and Management Science, 8: Network Routing,* M. O. Ball, T. L. Magnanti, C. L. Monma and G. L. Nemhauser (Eds.), North-Holland.

Gartner, N. H. (1980). "Optimal Traffic Assignment with Elastic Demands: A Review Part I: Analysis Framework; and Part II: Algorithmic Approaches," *Transportation Science*, Vol. 14, No. 2, pp. 174-207.

Hearn, D. W. (1980). "Bounding Flows in Traffic Assignment Models," Research Report 80-4, Dept. of Industrial and Systems Engineering, Univ. of Florida, Gainesville, FL.

Hearn, D. W. and J. Ribera (1980). "Bounded Flow Equilibrium Problems by Penalty Methods," *Proceedings of the IEEE International Conference on Circuits and Computers*, ICCC-80 Vol. 1, pp. 162-166.

Hearn, D. W. and M. V. Ramana (1997). "Toll Pricing Models for Traffic Networks," *Proceedings of 8th IFAC Symposium on Transportation,* Chania, Greece.

Hearn, D. W. and M. V. Ramana (1998). "Solving Congestion Toll Pricing Models," *Equilibrium and Advanced Transportation Modeling,* P. Marcotte and S. Nguyen (Eds.), Kluwer Academic Publishers, pp. 109-124.

Larsson, T. and M. Patriksson (1998). "Side Constrained Traffic Equilibrium Models—Traffic Management Through Link Tolls," *Equilibrium and Advanced Transportation Modeling,* P. Marcotte and S. Nguyen (Eds.), Kluwer Academic Publishers, pp. 125-151.

McDonald, J. F. (1995). "Urban Highway Congestion: An Analysis of Second-Best Tolls," *Transportation*, Vol. 22, pp. 353-369.

Patriksson, M. (1994). *The Traffic Assignment Problem — Models and Methods,* VSP, Utrecht.

Pigou, A. C. (1920). *The Economics of Welfare*, Macmillan, New-York.

Sheffi, Y. (1985). *Urban Transportation Networks. Equilibrium Analysis with Mathematical Programming Methods,* Prentice Hall, Englewood Cliffs, NJ.

Verhoef, E. T. (1999). "Second-Best Congestion Pricing in General Static Transportation Networks with Elastic Demands," *Research Report*, Free University Amsterdam.

Yang, H. and H. Huang (1998). "Principle of Marginal-Cost Pricing: How does it Work in a General Road Network?," *Transportation Research-A*, Vol. 32, No. 1, pp. 45-54.

Yang, H. and Lam, W. H. K. (1996). "Optimal Road Tolls under Conditions of Queueing and Congestion," *Transportation Research-A*, Vol. 30, No. 5, pp. 319-332.

Wardrop, J. G. (1952). "Some Theoretical Aspects of Road Traffic Research," *Proceedings of the Institution of Civil Engineers*, Volume I, Part II, pp. 325-378.

10 A DECISION SUPPORT METHODOLOGY FOR STRATEGIC TRAFFIC MANAGEMENT

Torbjörn Larsson
Jan T. Lundgren
Clas Rydergren
Michael Patriksson

Abstract: We propose a methodology for decision support in strategic traffic management. The methodology is based on an integrated model of traffic assignment and management decisions and its core is a traffic equilibrium model which assumes that the travellers choose their routes in accordance with Wardrop's principle. The management goals, regarding traffic flows and travel times in the network, are presumed to be described by constraints. It is also possible to specify a set of admissible actions in the network for achieving the goals; examples of such actions are changes in link capacities and the introduction of monetary tolls. The proposed approach constitutes a systematic methodology for finding appropriate changes in the traffic network in order to fulfill the management goals. We present a two-stage procedure for finding approximate solutions to the model.

10.1 INTRODUCTION

Strategic traffic management aims at the improvement of the functionality of the traffic network. This functionality can, for example, be expressed in terms of the traffic flows and the travel times on links in the network. Traffic management decisions are often preceded by the formulation (explicitly or implicitly) of targets, or *goals*, for the utilization of the traffic network. The management goals may be influenced by environmental, safety, practical, political, economical and other considerations. Examples of such goals are: maximal allowed traffic flows on links, maximal travel

M. Gendreau and P. Marcotte (eds.), Transportation and Network Analysis: Current Trends, 147–164.

times between locations using public transportation, maximal exhaust fume emissions in a certain area, and a sufficient proportion of the trips between an origin and a destination using the public transportation mode.

To reach the management goals, *actions* are taken that result in an altered behavior of the travelers. Any action, or combination of actions, which will alter the generalized travel times or costs (from now on referred to as travel times) for the travelers can, in principle, be used for this purpose. Examples of actions that can be used are: adjustments in the setting of traffic signal green times for public and private transportation, altered ticket prices for public transportation, adjustments in speed limits, and modifications in the street network. Another possible action is the introduction of road pricing.

Each action that is implemented causes an adjustment of the perceived travel times; the set of actions which are admissible to implement will hence result in a set of *admissible travel time adjustments.*

The traffic management process of today is often based on repeated scenario analyses, as follows. First, a management scheme, which is defined by a set of actions, is constructed, and the actions are included in the network model through adjustments of the travel time functions. An equilibrium package is then used to evaluate the scheme, through a comparison of the resulting flows and travel times with the management goals. If the goals are not satisfied, then ad-hoc changes in the actions are made, followed by a reevaluation of the management scheme. This trial-and-error procedure is repeated until the desired goals are reached, if possible.

We propose a *methodology* for the direct derivation of appropriate actions (or, rather, travel time adjustments). It is based on a model which integrates the management decisions with a traffic equilibrium model; this methodology is therefore a *systematic* and *automatic* methodology for strategic traffic management. The integrated model is a hierarchical (bilevel) model in which the management decision variables are at the upper level, and the travellers' decision variables are at the lower level.

The proposed methodology requires that the goals can be formulated as restrictions, or *goal constraints*, in an equilibrium assignment model. The methodology also presumes that the range of possible actions can be translated into constraints on the travel time adjustments.

There may, of course, be alternative feasible actions that can be used to enforce the fulfillment of the management goals. The objective of the upper level, which is a function of the travel time adjustments and the link flows, discriminates between such alternative actions.

Our methodology is the basis for a decision support tool. The tool is intended to be used by traffic planners for determining the actions that need to be taken in the network in order to reach the management goals.

We propose a numerical procedure for finding an approximate solution to the bilevel management model. This procedure has two stages. First, we need to establish whether or not the goal constraints are too restrictive, that is, if the goals are consistent with the travel demand. An inconsistency corresponds to too ambitious management goals, which whence must be revised. When the management goals and the travel demand have been found to be consistent, we are ready to solve a goal constrained traffic equilibrium problem.

Second, we establish whether or not the goals and the set of feasible travel time adjustments are consistent, that is, if the goals can be attained by the means available. In the case of inconsistency, the goals (or, the set of feasible travel time adjustments) have to be revised. When the goals and the feasible adjustments are verified to be consistent, a feasible travel time adjustment optimal with respect to the secondary objective is finally identified.

The contribution of this paper is a mathematical decision support methodology for a general strategic traffic management problem. The decision support methodology is based on earlier research on side-constrained traffic assignment (Larsson and Patriksson, 1998 and Larsson and Patriksson, 1999) and on toll pricing models (Hearn and Ramana, 1998). The focus of this paper is on the use of these earlier results in a realistic setting. Furthermore, details about solution strategies and inconsistency checks for each of the stages of the methodology are presented.

In the following section, we discuss the three parts of the traffic management model: the goal constraints, the set of feasible travel time adjustments, and the traffic equilibrium model. We present detailed examples of how both goals and restrictions on the travel time adjustments can be formulated and used in traffic management models. The traffic management problem is then formulated as a bilevel optimization problem. In Section 10.3 we present a solution procedure for finding an approximate solution to the traffic management model. The solution procedure has two stages. Each of the stages amounts to performing a consistency check (which is formulated as a convex optimization problem) and solving a convex optimization problem. A small numerical example is used to exemplify the two stages. We also present a decision support tool framework that is based on the proposed methodology. In Section 10.4, dual solution algorithms for the optimization problems in the two stages are presented.

In the first sections, our presentation is restricted to the case where the travelers choose their routes in accordance with the Wardrop principle for fixed demand traffic equilibrium. Extensions to the cases of elastic demands and combined assignment and mode split are discussed in Section 10.5.

10.2 FORMULATING THE TRAFFIC MANAGEMENT MODEL

The traffic management model consists of three parts: the mathematical formulation of management goals, the mathematical formulation of restrictions on the travel time adjustments, and a traffic equilibrium model. These three parts are discussed in this section.

10.2.1 Goal constraints

The manager's goals are often natural to formulate as restrictions on the traffic flows or on the travel times on sets of links. To incorporate the manager's goals in an equilibrium model we presume that the goals can be mathematically formulated as constraints. Let the traffic network be defined by the nodes set $\mathcal{N}$ and link set $\mathcal{A}$ and let the vector $f \in \mathbb{R}^{|\mathcal{A}|}$ with elements f_a represent the flow on link $a \in \mathcal{A}$. Each goal is described or quantified by a *goal constraint* $g_k(f) \leq 0$ where g_k is a convex function and $k \in \mathcal{K}$, with $\mathcal{K}$ being the set of goal constraints. The feasible traffic flows with respect to the management goals is the set

$$\mathcal{G} = \{f \mid g_k(f) \leq 0, \quad k \in \mathcal{K}\}.$$

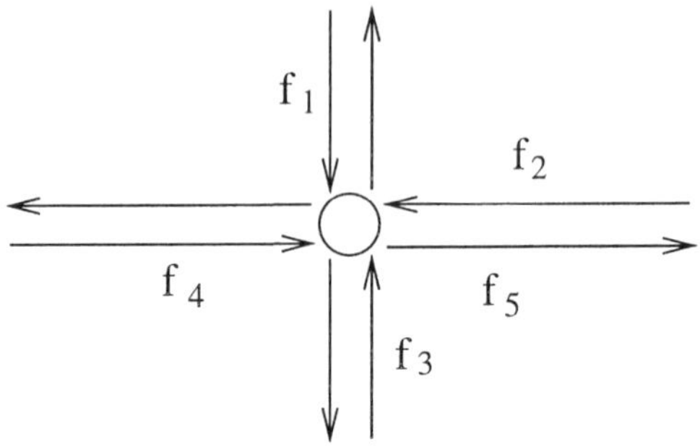

Figure 10.1 Simple intersection.

We next give examples of management goals modeled by constraints of this type.

Flow restriction examples. (a) Suppose that our goal is to restrict the total flow on the road defined by link 2, f_2, and link 5, f_5, shown in Figure 10.1. The restriction that the maximal total flow is b_1 units is imposed by formulating the constraint $g_1(f) = f_2 + f_5 - b_1 \leq 0$.

(b) Suppose that we would like to restrict the flow through the intersection shown in Figure 10.1. The inflow is calculated as the sum over the links 1 to 4. To restrict the flow to b_2 units, we use the constraint $g_2(f) = f_1 + f_2 + f_3 + f_4 - b_2 \leq 0$.

Travel time restriction example. A restriction that the total travel time on links 4 and 5 may not be larger than b_3 time units can be formulated as $g_3(f) = t_4(f_4) + t_5(f_5) - b_3 \leq 0$, where t_a is the (convex) travel time function for link a.

10.2.2 Restrictions on the travel time adjustments

Each of the possible actions in the network is associated with a travel time adjustment, which represent the change in the travel time perceived by the travelers on the link affected by the action. The restrictions that define the possible actions may therefore be translated into restrictions formulated in terms of the travel time adjustments. Let the vector $\rho \in \mathbb{R}^{|\mathcal{A}|}$, with elements ρ_a, represent the travel time adjustments caused by the action taken by the manager, and let the admissible travel time adjustments be described by the constraints $w_l(\rho) \leq 0$ for $l \in \mathcal{L}$, where each w_l is a convex function. The set of travel time adjustment restrictions is then given by the set

$$\mathcal{P} = \{\rho \mid w_l(\rho) \leq 0, \quad l \in \mathcal{L}\}.$$

Travel time adjustment restriction example. Consider the network in Figure 10.1. Assume that it is impossible to implement any action that alters the travel time on links 3 or 4, and that the travel time adjustment of an implementable action on link 1 should be at least a_1 and at most a_2 time units (which for example can be the maximal acceptable travel time adjustment that can be implemented by a traffic signal on the link). This scenario can be formulated by the constraints $a_1 \leq \rho_1 \leq a_2$ and $\rho_3 = \rho_4 = 0$.

10.2.3 User equilibrium

Assume that the travel demand for every origin-destination pair in the network is known and independent of the travel times. Let d_{pq} denote the travel demand from the origin p to the destination q, and let $\mathcal{C}$ be the set of origin-destination pairs (p, q). Let $\mathcal{R}_{pq}$ denote the (non-empty) set of routes in pair (p, q) and h_{pqr} denote the flow on route $r \in \mathcal{R}_{pq}$. The (compact) set of feasible route flows, $\mathcal{H}$, is then given by the solutions to the system

$$\sum_{r \in \mathcal{R}_{pq}} h_{pqr} = d_{pq}, \qquad \forall (p,q) \in \mathcal{C}, \tag{10.1a}$$

$$h_{pqr} \geq 0, \qquad \forall r \in \mathcal{R}_{pq}, \quad \forall (p,q) \in \mathcal{C}. \tag{10.1b}$$

Let $\delta_{pqra} = 1$ if route r from p to q includes link a, and 0 otherwise. The link flow on link a is then calculated as

$$f_a = \sum_{(p,q) \in \mathcal{C}} \sum_{r \in \mathcal{R}_{pq}} \delta_{pqra} h_{pqr}, \qquad \forall a \in \mathcal{A}. \tag{10.2}$$

Let $\mathcal{F}$ be the set of link flows that might result from the route flows in $\mathcal{H}$, that is, the link flows that satisfies constraints (10.1a), (10.1b) and (10.2). The traffic equilibrium is assumed to follow the Wardrop conditions (e.g., Patriksson, 1994). These conditions state that every traveler chooses to travel in the fastest possible way. The travel time on link a is given by $t_a(f_a)$, where t_a is an increasing, positive and continuous function. (The more general case where the travel time is given by $t_a(f)$ is commented in Section 10.5.) The travel time of a route $r \in \mathcal{R}_{pq}$ is given by $\sum_{a \in \mathcal{A}} \delta_{pqra} t_a(f_a)$, and the equilibrium conditions can be then formulated as

$$h_{pqr} > 0 \implies \sum_{a \in \mathcal{A}} \delta_{pqra} t_a(f_a) = \pi_{pq}, \qquad \forall (p,q) \in \mathcal{C}, \quad \forall r \in \mathcal{R}_{pq},$$

$$h_{pqr} = 0 \implies \sum_{a \in \mathcal{A}} \delta_{pqra} t_a(f_a) \geq \pi_{pq}, \qquad \forall (p,q) \in \mathcal{C}, \quad \forall r \in \mathcal{R}_{pq},$$

where π_{pq} is the minimum route travel time in pair (p, q). As is well known, the equilibrium link flow solution can be found by solving the convex optimization problem

$$\min_{f \in \mathcal{F}} \sum_{a \in \mathcal{A}} \int_0^{f_a} t_a(s)\, ds. \tag{TAP}$$

In the case when the travel times on the links can be adjusted, and then given by $t_a(f_a) + \rho_a$, the equilibrium conditions are given by

$$h_{pqr} > 0 \implies \sum_{a \in \mathcal{A}} \delta_{pqra}(t_a(f_a) + \rho_a) = \pi_{pq}, \qquad \forall (p,q) \in \mathcal{C}, \quad \forall r \in \mathcal{R}_{pq}, \tag{10.4a}$$

$$h_{pqr} = 0 \implies \sum_{a \in \mathcal{A}} \delta_{pqra}(t_a(f_a) + \rho_a) \geq \pi_{pq}, \qquad \forall (p,q) \in \mathcal{C}, \quad \forall r \in \mathcal{R}_{pq}. \tag{10.4b}$$

For a given vector of travel time adjustments, ρ, the equilibrium link flow solution that satisfies (10.4) can be found as a solution to an optimization problem of the form (TAP), but where the link travel times are given by $t_a(f_a) + \rho_a$. We denote this optimization problem by (TAP$_\rho$).

We also note that the system (10.4) can be reformulated as

$$\sum_{a \in \mathcal{A}} \delta_{pqra}(t_a(f_a) + \rho_a) - \pi_{pq} \geq 0, \qquad \forall r \in \mathcal{R}_{pq}, \quad \forall (p,q) \in \mathcal{C},$$

$$\sum_{r \in \mathcal{R}_{pq}} h_{pqr} \left(\sum_{a \in \mathcal{A}} \delta_{pqra}(t_a(f_a) + \rho_a) - \pi_{pq} \right) = 0, \qquad \forall (p,q) \in \mathcal{C}.$$

Using that $h_{pqr} \geq 0$, for all $(p,q) \in \mathcal{C}$ and $r \in \mathcal{R}_{pq}$, this system can be further rewritten as

$$\sum_{a \in \mathcal{A}} \delta_{pqra}(t_a(f_a) + \rho_a) - \pi_{pq} \geq 0, \qquad \forall r \in \mathcal{R}_{pq}, \quad \forall (p,q) \in \mathcal{C}, \tag{10.5a}$$

$$\sum_{a \in \mathcal{A}} (t_a(f_a) + \rho_a) f_a - \sum_{(p,q) \in \mathcal{C}} \pi_{pq} d_{pq} = 0. \tag{10.5b}$$

The latter system is used in the following sections to define the set of travel time adjustments that can be used to reach the management goals.

10.2.4 The traffic management model

The primary traffic management goal is to find feasible travel time adjustments (that is, $\rho \in \mathcal{P}$) that makes the resulting equilibrium flow feasible with respect to the goal constraints (that is, $f \in \mathcal{G}$).

To specify a secondary management goals an objective function $\varphi(f, \rho)$ is introduced. We give four examples of secondary goals. The first is to minimize the flow-weighted travel time adjustment, which is obtained by minimizing the function

$$\varphi(f, \rho) = \sum_{a \in \mathcal{A}} f_a \rho_a. \tag{10.6}$$

(In the case where each ρ_a is a monetary link toll, this function describes the total toll revenue.) The second example is to minimize the flow-weighted absolute travel time adjustment:

$$\varphi(f, \rho) = \sum_{a \in \mathcal{A}} f_a |\rho_a|. \tag{10.7}$$

This objective corresponds to minimizing the perceived travel time change. The third is to minimize a weighted absolute travel time adjustment,

$$\varphi(f, \rho) = \sum_{a \in \mathcal{A}} \bar{v}_a |\rho_a|, \tag{10.8}$$

where $\bar{v}$ is a vector of non-negative weights. The fourth is to minimize the system travel time, which corresponds to the choice

$$\varphi(f, \rho) = \sum_{a \in \mathcal{A}} (t_a(f_a) + \rho_a) f_a. \tag{10.9}$$

The traffic management problem can then be formulated as the bilevel model (e.g., Migdalas, 1995)

$$\min \varphi(f, \rho) \tag{10.10a}$$

subject to

$$f \in \mathcal{G}, \tag{10.10b}$$

$$\rho \in \mathcal{P}, \tag{10.10c}$$

$$f \text{ solves } (TAP_\rho). \tag{10.10d}$$

The interaction between the manager and the travelers is here described as a Stackelberg game, where the manager is the leader and where the travellers are followers, that is, they reevaluate their route choices after any change in the traffic network. This model is discussed in the more general setting of a mathematical problem with equilibrium constraints in Dirkse and Ferris (1998).

The bilevel problem (10.10) is in general non-convex and difficult to solve. Several heuristics have been proposed for the solution of similar bilevel problems, for example: genetic algorithms in Cree et al. (1998), simulated annealing in Huang and Bell (1998), and by approximating the bilevel problem by a nonlinear non-convex problem in Chen et al. (1998). Marcotte (1986) and Marcotte and Marquis (1992) describe a heuristic for finding a solution to the bilevel problem by solving a convex optimization problem, and present a worst-case analysis of this approach.

The special case of (10.10) that is obtained without congestion is discussed in Labbé et al. (1998). They show when this special case can be reduced to a linear bilevel problem, and how it may be solved, in the special case of a single origin destination pair where the route choice or link usage of the followers is known a priori.

Compared to most bilevel problems, ours has special characteristics. Our overall decision problem includes both primary and secondary goals of which the former are taken into account in the lower level, while the latter appear at the upper level. Because of this, the lower level is the most significant one, while the upper is of less importance. This characteristic motivates our solution approach, which is a sequential two-stage procedure where we first solve a lower level problem, and then an upper level problem.

In the next section we present this two-stage procedure for finding an approximate solution to (10.10). In the first stage a goal constrained traffic assignment problem is solved, following the ideas in Larsson and Patriksson (1998), the result being a constrained equilibrium flow f^*. Solving the goal constrained traffic assignment problem problem can be interpreted as solving (10.10) for the case where $\varphi \equiv 0$ and $\mathcal{P} \equiv \mathbb{R}^{|\mathcal{A}|}$. The second stage may similarly be interpreted as that of solving (10.10) with f fixed to f^*.

10.3 SOLUTION FRAMEWORK

We next present a solution framework for finding an approximate solution to the bilevel traffic management model. The first subsection describes a decision support system; in the following subsections, the details of the solution approach are presented.

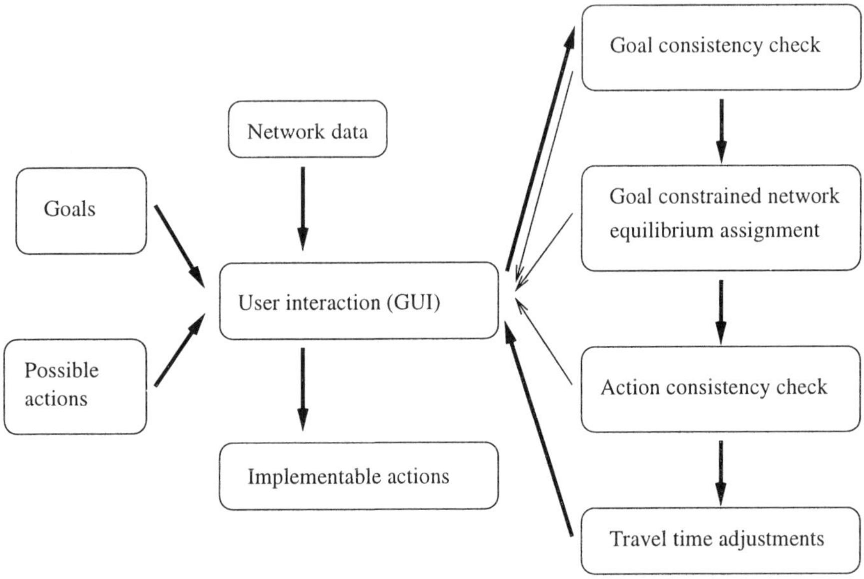

Figure 10.2 Relations between goals, actions, user interaction and the optimization problems. The bold arrows describe the main information flows.

10.3.1 Decision support framework

Our solution approach for the traffic management problem is intended to be used in combination with a user interface (e.g., a graphical user interface, GUI). From this user interface it is possible to specify the management goals and the feasible travel time adjustments. User interaction is used to support the methodology in several stages of the solution procedure, see Figure 10.2.

The goals and the network model are first run through the consistency check, and the interface reports on the consistency between the travel demands and the management goals. In the case of inconsistency, the manager gets information about which constraints that result in the inconsistency and how much the goals are violated. When consistency has been ascertained the goals are used in the goal constrained assignment, which produces a goal constrained equilibrium flow which satisfies the goals. The availability of a feasible travel time adjustment that reproduces these link flows is examined in a second consistency check procedure. In the case of inconsistency, the goals (or the set of actions) have to be revised. Information about how the goals may be revised is available from the consistency check.

Finally, a travel time adjustment problem is solved, which gives travel time adjustments that are feasible and make the goals fulfilled. From these travel time adjustments the manager may finally identify a set of actions that can be implemented in the network in order to reach the management goals.

In the following two subsections we make the assumption that the optimization problems under consideration are consistent. The next two subsections discuss the consistency checks, and the information obtained from the checks that may be used to revise the management goals.

10.3.2 *Goal constrained traffic equilibrium*

In the first stage of the approximate method for the model (10.10), we find a link flow solution f^* and a vector of travel time adjustments ρ, such that $f^* \in \mathcal{G}$ and f^* solves (TAP_ρ). This is done by solving the goal constrained (or side constrained) traffic equilibrium problem

$$\min_{f \in \mathcal{F}} \sum_{a \in \mathcal{A}} \int_0^{f_a} t_a(s)\, ds \tag{10.11a}$$

subject to

$$g_k(f) \leq 0, \qquad \forall k \in \mathcal{K}. \tag{10.11b}$$

The possible inconsistency of this goal constrained assignment problem is discussed in Section 10.3.4. In Larsson et al. (1997), a dual algorithm for the special case of capacitated traffic assignment (where the constraints $g_k(f) \leq 0$ for $k \in \mathcal{K}$ specifies upper bounds on the link flows) is presented; the solution algorithm presented in Section 10.4.1 is a extension of that scheme.

Denote an optimal solution to problem (10.11) by f^* and a set of optimal values for the dual variables associated with the constraints (10.11b) by β^*. One set of travel time adjustments that satisfies the system (10.5) for $f = f^*$ can then be derived from the Karush-Kuhn-Tucker conditions for the problem (10.11), is given by

$$\rho_a = \sum_{k \in \mathcal{K}} \beta_k^* \frac{\partial g_k(f^*)}{\partial f_a}, \quad a \in \mathcal{A}, \tag{10.12}$$

(see Larsson and Patriksson, 1999, Theorem 2.1). These travel time adjustments are not necessarily feasible, that is, $\rho \notin \mathcal{P}$ might hold. The dual solution β^* is, in general, not unique (Larsson and Patriksson, 1998, Example 7.5.3). Nor is the solution set to the system (10.5) necessarily a singleton.

10.3.3 *Finding the travel time adjustments*

If the adjustments computed by (10.12), are not feasible with respect to the set $\mathcal{P}$, or non-optimal with respect to our secondary goals, then we wish to find an alternative set of adjustments which reproduces the traffic flows f^*, which belongs to the set $\mathcal{P}$, and which optimizes the objective of (10.10). This is made in stage two of the solution approach.

The set of travel time adjustments that produces the link flows f^* is defined by the system (10.5), with $f = f^*$. We incorporate the travel time restrictions $\rho \in \mathcal{P}$ to the system (10.5), by adding the constraints $w_l(\rho) \leq 0$ for $l \in \mathcal{L}$. We finally introduce the restriction that no link in the network can get a negative generalized travel time when the travel time adjustment is added. This restriction is imposed by putting a lower bound $\rho_a^{lb} := -t_a(f_a^*)$ on the individual travel time adjustments, so that $\rho_a \geq \rho_a^{lb}$ must hold for all $a \in \mathcal{A}$. From the equilibrium conditions (10.4) we observe that these bounds imply that $\pi_{pq} \geq 0$ for all $(p, q) \in \mathcal{C}$. The problem of finding a set of travel time adjustments that reproduces the flows f^* and optimizes the function $\varphi(f^*, \rho)$ is then given by

$$\min_{\rho} \varphi(f^*, \rho) \tag{10.13a}$$

subject to

$$\sum_{a \in \mathcal{A}} \delta_{pqra}(t_a(f_a^*) + \rho_a) - \pi_{pq} \geq 0, \quad \forall r \in \mathcal{R}_{pq}, \quad \forall (p,q) \in \mathcal{C}, \tag{10.13b}$$

$$\sum_{a \in \mathcal{A}} (t_a(f_a^*) + \rho_a) f_a^* - \sum_{(p,q) \in \mathcal{C}} \pi_{pq} d_{pq} = 0, \tag{10.13c}$$

$$w_l(\rho) \leq 0, \qquad \forall l \in \mathcal{L}. \tag{10.13d}$$

The constraints (10.13b), (10.13c) and (10.13d) are not necessarily consistent. This issue is discussed in Section 10.3.5.

The optimization problem (10.13) is convex. An inherent difficulty of this problem is, however, that the number of constraints in (10.13b) is, in general, very large. Hearn and Ramana (1998) discuss the problem (10.13) in a context where the link flow vector f^* is system-optimal and the problem is stated for the case where the objective (10.13a) is to minimize the total system time. Several kinds of restrictions on the travel time adjustments (in their context, link tolls), including non-convex ones, are discussed. No solution algorithm that utilizes the problem structure is given. In Larsson and Patriksson (1998), a schematic description of a solution algorithm for the problem (10.13) is given. This scheme was used as a starting point for the solution procedure to be presented in Section 10.4.2.

Numerical example. This example demonstrates results from first solving the goal constrained traffic assignment problem and then the travel time adjustment problem. The experiment has been made using the 9-nodes, 18-links test network in Hearn and Ramana (1998).

The management goals are formulated as link flow capacities on the links 1–6, 5–7 and 5–9. The traffic flow on link 1–6 is restricted to 20 units, and link 5–7 and link 5–9 are restricted to 15 units each. The goal constrained traffic equilibrium problem was solved using the algorithm presented in Section 10.4.1. The goal constrained equilibrium link flows are given in the third column of Table 10.1.

The travel time adjustments computed using (10.12) are given under column A in Table 10.1. (The restrictions $\rho_a \geq -t_a(f_a^*)$ for $a \in \mathcal{A}$ are inactive in this solution.)

The travel time adjustments that minimize the total perceived travel time adjustment, that is, the objective (10.7), are given in column B. The travel time adjustments that are restricted to $\max\{-t_a(f_a^*), -3\} \leq \rho_a \leq 10$ for all $a \in \mathcal{A}$ and minimize the total perceived travel time are given in column C. When the total system time is minimized, that is, when objective (10.9) is used, the corresponding travel time adjustments are given in column D.

The objective values for (10.7) and (10.9) are given in the rows Min. adj. and Min. sys. respectively.

10.3.4 Goal consistency

The traffic management goals are formulated as goal constraints and added to the equilibrium assignment model. The goals may be overly ambitious, making them impossible to satisfy together with the travel demand. Consequently, we obtain an *infeasible* goal constrained equilibrium assignment problem (10.11). Goals that are so restrictive that they do not allow the travel demand to be fulfilled are clearly too optimistic or poorly formulated, and therefore need to be revised. To proceed with

Table 10.1 Numerical results for the Hearn and Ramana network.

Link	User eq. flow	Goal flow	ρ A	B	C	D
1–5	8.159	10.000	0.000	0.000	−3.000	−3.000
1–6	21.840	20.000	7.006	0.000	1.006	1.006
2–5	47.373	26.680	0.000	7.006	3.000	−0.001
2–6	22.627	43.320	0.000	0.000	0.000	−3.000
5–6	0.000	6.680	0.000	−7.007	−3.000	−3.000
5–7	27.842	15.001	18.868	10.853	10.000	9.867
5–9	27.691	14.998	8.015	0.000	2.015	2.015
6–5	0.000	0.000	0.000	0.000	0.000	0.000
6–8	44.467	50.984	0.000	0.000	−2.107	−3.000
6–9	0.000	19.016	0.000	−1.008	−3.000	−3.000
7–3	38.160	40.000	0.000	0.000	0.000	−3.000
7–4	17.373	9.016	0.000	1.008	3.761	0.000
7–8	0.000	0.000	0.000	0.000	0.000	0.000
8–3	1.840	0.000	0.000	0.000	0.000	0.000
8–4	42.627	50.984	0.000	0.000	0.000	−3.000
8–7	0.000	0.000	0.000	0.000	0.000	0.000
9–7	27.691	34.014	0.000	0.000	−2.868	−3.000
9–8	0.000	0.000	0.000	0.000	0.000	0.000
Min. adj.			543.4	424.8	626.3	963.4
Min. sys.			3171.1	2920.5	2629.9	2061.0

the solution approach we then need information about how the goals may be modified in order to make them more conform with the travel demand.

The consistency check amounts to solving a (convex) optimization problem for finding a point in $\mathcal{F}$ that minimizes the maximal level of violation of the side constraints, that is, to finding

$$\min_{f \in \mathcal{F}} \max_{k \in \mathcal{K}} \bar{v}_k g_k(f),$$

where $\bar{v}_k$, $k \in \mathcal{K}$ are positive. The weights may be chosen to reflect the importance of the individual goals. This problem can be solved with a dual solution algorithm which is similar to the one presented for the goal constrained traffic assignment problem (10.11). The optimal solution, y^*, is feasible point in (TAP$_\rho$), or the optimal objective value gives us a measure of how difficult the side constraints are to satisfy. The optimal value is non-positive for the case of consistency, and positive if no feasible flow in $\mathcal{F}$ that is conform with the goal constraints can be found. In case of inconsistency, the values of the goal constraints $g_k(y^*)$ for $k \in \mathcal{K}$ will of course provide information about the level of attainment of the management goals.

10.3.5 Action consistency

The question of inconsistency arises also in stage two of the proposed solution approach. When the restrictions on the travel time adjustments, is introduced in the problem (10.13), it may become infeasible, although the system (10.13b) and (10.13c) is always consistent. In order to deal with this ussue, we consider the system

$$\sum_{a\in\mathcal{A}} \delta_{pqra}(t_a(f_a^*) + \mu_a) - \pi_{pq} \geq 0, \quad \forall r \in \mathcal{R}_{pq}, \quad \forall (p,q) \in \mathcal{C}, \tag{10.14a}$$

$$\sum_{a\in\mathcal{A}} (t_a(f_a^*) + \mu_a)\, f_a^* - \sum_{(p,q)\in\mathcal{C}} \pi_{pq} d_{pq} = 0, \tag{10.14b}$$

$$w_l(\nu) \leq 0, \qquad \forall l \in \mathcal{L}. \tag{10.14c}$$

which is equivalent to the system (10.13b), (10.13c) and (10.13d) if $\mu = \nu$ is also imposed. One way of checking the consistency is then to find

$$\min_{\mu,\pi,\nu} \max_{a\in\mathcal{A}} \bar{v}_a |\mu_a - \nu_a| \tag{10.15}$$

subject to (10.14a), (10.14b) and (10.14c), where $\bar{v}_a$, $a \in \mathcal{A}$ is some positive weights. The weights are chosen to reflect the importance of the adjustment restriction on individual links. This problem can be solved by a dual algorithm which is similar to the one presented for the travel time adjustment problem (10.13). The optimal objective value is obviously positive if and only if the problem (10.13) is inconsistent. In such a situation the goals are unattainable by the actions that can be taken, and the goals or set of actions need to be revised. In order to give information about how the goals may be revised, we solve the problem (TAP$_\rho$) with $\rho = \nu$, which will result in a link flow solution y^*. Then, we compute the values of the goal constraints $g_k(y^*)$ for $k \in \mathcal{K}$, which will provide information about the level of violation of the management goals.

10.4 OUTLINE OF ALGORITHMS

In this section we present solution algorithms for the (consistent) convex optimization problems (10.11) and (10.13), which arise in the two stages of the solution approach. The algorithms are similar in that they solve the corresponding Lagrangian dual problem using a subgradient method.

10.4.1 Solution algorithm for the goal constrained traffic assignment problem

We outline a dual solution algorithm for the (consistent) goal constrained traffic assignment problem (10.11). We treat the goal constraints separately in the solution algorithm depending on whether these are bound, linear or nonlinear constraints. Let $\mathcal{K}_1 \subseteq \mathcal{K}$ include the linear goal constraints, and $\mathcal{K}_2 \subseteq \mathcal{K}$ the nonlinear ones. The solution algorithm is thus described for a version of the problem where the feasible set of link flows in the problem (10.11) is defined by (10.1a), (10.1b), (10.2) and

$$\mathcal{G} := \left\{ f \;\middle|\; f_a \leq f_a^{ub},\, a \in \mathcal{A};\; \sum_{a\in\mathcal{A}} e_{ak} f_a - b_k \leq 0,\, k \in \mathcal{K}_1;\; g_k(f) \leq 0,\, k \in \mathcal{K}_2 \right\} \tag{10.16}$$

where $f_a^{ub} \in \Re_+ \cup \{+\infty\}$ for each $a \in \mathcal{A}$, $e_{ak} \geq 0$ for each $a \in \mathcal{A}$ and $k \in \mathcal{K}_1$, and $g_k(f) = \sum_{a\in\mathcal{A}} g_{ak}(f_a)$, $k \in \mathcal{K}_2$ (that is, each g_k is separable), and $dg_{ak}(f_a)/df_a \geq 0$ for $f_a \geq 0$ and $k \in \mathcal{K}_2$.

In our numerical approach we dualize the equality constraints (10.2) with multipliers α_a and the constraints (10.11b) for $k \in \mathcal{K}_1 \cup \mathcal{K}_2$ with multipliers β_k. The Lagrangian dual problem then is

$$\max_{\alpha,\beta} \Psi(\alpha, \beta) = \Psi_1(\alpha, \beta) + \Psi_2(\alpha, \beta)$$

subject to

$$\alpha_a \geq t_a(0), \qquad \forall a \in \mathcal{A},$$
$$\beta_k \geq 0, \qquad \forall k \in \mathcal{K}_1 \cup \mathcal{K}_2,$$

where

$$\Psi_1(\alpha, \beta) = \min_{0 \leq f \leq f^{ub}} \sum_{a \in \mathcal{A}} \left(\int_0^{f_a} t_a(s)ds - \alpha_a f_a \right) + \sum_{k \in \mathcal{K}_2} \beta_k \sum_{a \in \mathcal{A}} g_{ak}(f_a),$$

and

$$\Psi_2(\alpha, \beta) = \min_{h \in \mathcal{H}} \sum_{a \in \mathcal{A}} \sum_{(p,q) \in \mathcal{C}} \sum_{r \in \mathcal{R}_{pq}} \delta_{pqra} \left(\alpha_a + \sum_{k \in \mathcal{K}_1} \beta_k e_{ak} \right) h_{pqr}.$$

Based on our assumptions it can be shown that the bounds $\alpha_a \geq t_a(0)$ can be added without changing the optimal value of Ψ. The former problem, the computation of Ψ_1, is a convex problem which is separable by links. The latter problem, the computation of Ψ_2, is solved by finding a fastest route in each origin-destination pair. Since, from our assumptions, $\alpha_a + \sum_{k \in \mathcal{K}_1} \beta_k e_{ak} \geq 0$ holds for all $a \in \mathcal{A}$, these routes can be found by Dijkstra's algorithm. Denote an optimal solution in terms of route flows to this problem by $h(\alpha, \beta)$, and in terms of link flows by $y_a(\alpha, \beta) = \sum_{(p,q) \in \mathcal{C}} \sum_{r \in \mathcal{R}_{pq}} \delta_{pqra} h_{pqr}(\alpha, \beta)$ for all $a \in \mathcal{A}$.

We search for optimal values on the dual variables (α, β) by means of a subgradient method (e.g., Larsson et al. 1999). Under additional regularity conditions on the constraints of the problem (10.11) the method converges to a dual optimal solution (α^*, β^*), see Larsson and Patriksson (1998).

Denote the dual solution in iteration i by (α^i, β^i). Since we are searching for a primal optimal solution, we generate an ergodic (averaged) sequence of the link flows $y^i = y(\alpha^i, \beta^i)$, starting from some iteration $i_0 + 1$. For each $a \in \mathcal{A}$, the ergodic link flow denoted $\hat{y}_a^i$ is, at iteration $i \geq i_0 + 1$, updated as

$$\hat{y}_a^i = y_a^i, \qquad i = i_0 + 1,$$
$$\hat{y}_a^i = \frac{i - i_0 - 1}{i - i_0} \hat{y}_a^{i-1} + \frac{1}{i - i_0} y_a^i, \qquad i = i_0 + 2, i_0 + 3, \ldots$$

which can be interpreted as an average of the aggregated fastest route flows in iteration $i_0 + 1$ to i. The sequences $\{\hat{y}_a^i\}$ converge to the set of optimal link flows in problem (10.11); see Larsson et al. (1999) for a demonstration.

10.4.2 Solution algorithm for the travel time adjustment problem

Here we present a dual solution algorithm for a version of the (consistent) travel time adjustment problem (10.13), where the restrictions on the travel time adjustments are assumed to be linear.

The solution algorithm assumes that finite upper bounds ρ_a^{ub} on the travel time adjustments ρ_a are present. To ensure that every dual subproblem solution is bounded we add the (redundant) constraints $\pi_{pq} \leq \pi_{pq}^{ub}$, where

$$\pi_{pq}^{ub} = \min_{r \in \mathcal{R}_{pq}} \left\{ \sum_{a \in \mathcal{A}} \delta_{pqra}(t_a(f_a^*) + \rho_a^{ub}) \right\}, \qquad \forall (p,q) \in \mathcal{C}.$$

These values are found by solving a fastest route problem from every origin to every destination, using link travel times $t_a(f_a^*) + \rho_a^{ub}$.

We dualize the constraints (10.13b) and (10.13d) with multipliers γ_{pqr} and σ_l, respectively. The Lagrangian dual problem is

$$\max_{\gamma,\sigma} \Phi(\gamma,\sigma) \tag{10.17a}$$

subject to

$$\gamma_{pqr} \geq 0, \qquad \forall (p,q) \in \mathcal{C}, \quad \forall r \in \mathcal{R}_{pq}, \tag{10.17b}$$
$$\sigma_l \geq 0, \qquad \forall l \in \mathcal{L}, \tag{10.17c}$$

where

$$\Phi(\gamma,\sigma) = \min_{\rho,\pi} \varphi(f^*,\rho) + \sum_{(p,q)\in\mathcal{C}} \sum_{r\in\mathcal{R}_{pq}} \gamma_{pqr} \left[\pi_{pq} - \sum_{a\in\mathcal{A}} \delta_{pqra}(t_a(f_a^*) + \rho_a) \right] + \sum_{l\in\mathcal{L}} \sigma_l w_l(\rho) \tag{10.17d}$$

subject to

$$\sum_{a\in\mathcal{A}} (t_a(f_a^*) + \rho_a) f_a^* - \sum_{(p,q)\in\mathcal{C}} d_{pq}\pi_{pq} = 0, \tag{10.17e}$$
$$0 \leq \pi_{pq} \leq \pi_{pq}^{ub}, \qquad \forall r \in \mathcal{R}_{pq}, \quad \forall (p,q) \in \mathcal{C}, \tag{10.17f}$$
$$-t_a(f_a^*) \leq \rho_a^{lb} \leq \rho_a \leq \rho_a^{ub}, \qquad \forall a \in \mathcal{A}. \tag{10.17g}$$

The subproblem (10.17d)–(10.17g) is a convex bounded variable knapsack problem solved by a specialized simplex-like technique. (For the secondary objective (10.7) and (10.8) a variable transformation is also necessary.) Denote the subproblem solution by $(\rho(\gamma,\sigma), \pi(\gamma,\sigma))$.

The dual problem of finding an optimal vector (γ,σ) is solved using a subgradient method. The sequence $\{\hat{\rho}^i\}$ is constructed similarly to the sequence $\{\hat{y}^i\}$ in the previous section.

Since the set of routes $\mathcal{R}_{pq}$, $(p,q) \in \mathcal{C}$ in general is very large, the solution algorithm for the dual problem must invoke a route (that is, constraint) generation procedure. The set of generated routes are included in the sets $\hat{\mathcal{R}}_{pq} \subset \mathcal{R}_{pq}, (p,q) \in \mathcal{C}$. New routes/constraints are generated from the current averaged solution $\hat{\rho}^i$, by finding the fastest route from every origin to every destination in the network using link travel times $t_a(f_a^*) + \hat{\rho}_a^i$, and these routes are then added to the sets $\hat{\mathcal{R}}_{pq}$ if not already present therein.

10.5 EXTENSIONS

The proposed methodology has this far been described for the case of fixed demand, but it can be extended to more complex equilibrium models without any principal difficulties. The two extensions considered here are cases of elastic demand and combined modal split and traffic assignment.

The case where the travel time functions are non-separable complicates stage one of our procedure considerably. In this case, it may not be possible to find the goal constrained traffic equilibrium problem by solving a convex optimization problem. Instead the goal constrained traffic equilibrium can be stated and solved as a variational inequality problem with side constraints. However, the model and algorithm for stage two, the travel time adjustment problem, do not require any modifications.

10.5.1 Elastic demands

Let the demands be given by functions D_{pq}, $(p,q) \in \mathcal{C}$, each of which is non-negative, strictly decreasing, upper bounded and continuous in the travel time π_{pq}. Let $d_{pq} = D_{pq}(\pi_{pq})$. The goal constrained traffic assignment problem for the case of elastic demand can then be stated as

$$\min_{f,d} \sum_{a \in \mathcal{A}} \int_0^{f_a} t_a(s)ds - \sum_{(p,q) \in \mathcal{C}} \int_0^{d_{pq}} D_{pq}^{-1}(s)ds$$

subject to

$$\begin{aligned} &(f,d) \in \mathcal{F}, \\ &g_k(f) \leq 0, \qquad k \in \mathcal{K}. \end{aligned}$$

If the goals are formulated by constraints which are consistent internally, then this goal constrained traffic assignment problem is always consistent. It can be solved using a dual solution algorithm similar to the one presented in Section 10.4.1 (cf. Larsson et al., 1997). Denote the (unique) solution by (f^*, d^*).

The set of travel time adjustments which reproduces the solution (f^*, d^*) is in this case of defined by the system

$$\sum_{a \in \mathcal{A}} \delta_{pqra}(t_a(f_a^*) + \rho_a) - \pi_{pq} \geq 0, \quad \forall r \in \mathcal{R}_{pq}, \ \forall (p,q) \in \mathcal{C}, \tag{10.18a}$$

$$\sum_{r \in \mathcal{R}_{pq}} h_{pqr}^* \left(\sum_{a \in \mathcal{A}} \delta_{pqra}(t_a(f_a^*) + \rho_a) - \pi_{pq} \right) = 0, \quad \forall (p,q) \in \mathcal{C}, \tag{10.18b}$$

$$\pi_{pq} - D_{pq}^{-1}(d_{pq}^*) \geq 0, \quad \forall (p,q) \in \mathcal{C}, \tag{10.18c}$$

$$d_{pq}^*(\pi_{pq} - D_{pq}^{-1}(d_{pq}^*)) = 0, \quad \forall (p,q) \in \mathcal{C}, \tag{10.18d}$$

where h^* is any vector of route flows that is consistent with the travel demands d^* and the link flow f^*. From (10.18d) we observe that the travel time π_{pq} is uniquely determined in a pair (p,q) with $d_{pq}^* > 0$. By the summation of the constraints (10.18b) and (10.18d), and summing over $(p,q) \in \mathcal{C}$, this system can be equivalently stated as

$$\sum_{a \in \mathcal{A}} \delta_{pqra}(t_a(f_a^*) + \rho_a) - \pi_{pq} \geq 0, \quad \forall r \in \mathcal{R}_{pq}, \ \forall (p,q) \in \mathcal{C}, \tag{10.19a}$$

$$\pi_{pq} - D_{pq}^{-1}(d_{pq}^*) \geq 0, \quad \forall (p,q) \in \mathcal{C}, \tag{10.19b}$$

$$\sum_{a \in \mathcal{A}} (t_a(f_a^*) + \rho_a) f_a^* - \sum_{(p,q) \in \mathcal{C}} d_{pq}^* D_{pq}^{-1}(d_{pq}^*) = 0. \tag{10.19c}$$

From (10.19c) we observe that the sum $\sum_{a \in \mathcal{A}} f_a^* \rho_a$ is constant. This observation leads us to believe that the degree of freedom in the system (10.19) is, in general, smaller than the degree of freedom in the corresponding system for the fixed demand case.

The travel time adjustment problem is defined as the problem of minimizing a secondary objective under the constraints (10.19) and (10.13d). A solution method for the travel time adjustment problem for the case of elastic demand can be based on the algorithm in Section 10.4.2.

10.5.2 Car/Public transport travel mode split

In this section we present an extension to a model with combined mode split and traffic assignment. The model considered has two modes: car and public transport. Congestion caused by the interaction between the modes is not considered. A more detailed description of this model can be found, for example, in Sheffi (1985, Chapter 6.4). Let d_{pq} denote the (fixed) total travel demand in pair (p,q) and let p_{pq} denote the (variable) number of travellers using the public transport mode. We consider the mode split functions

$$p_{pq} = d_{pq} \frac{1}{1 + e^{\Theta(\hat{\pi}_{pq} - \pi_{pq})}}, \qquad \forall (p,q) \in \mathcal{C},$$

where Θ is a positive parameter and $\hat{\pi}_{pq}$ is the travel time for the public transport mode in pair (p,q). The choice of a logit formula for the mode split implies that p_{pq} is always positive, that is, some proportion of the total number of travellers choose to use the public transport alternative, irrespective of its travel time. From this mode split function we can derive the excess demand function

$$W_{pq}(p_{pq}) = \frac{1}{\Theta} \ln \frac{p_{pq}}{d_{pq} - p_{pq}} + \hat{\pi}_{pq}, \qquad \forall (p,q) \in \mathcal{C}.$$

The mode split and equilibrium assignment problem can be formulated as

$$\min \sum_{a \in \mathcal{A}} \int_0^{f_a} t_a(s) ds + \sum_{(p,q) \in \mathcal{C}} \int_0^{p_{pq}} W_{pq}(s) ds$$

subject to

$$\sum_{r \in \mathcal{R}_{pq}} h_{pqr} + p_{pq} = d_{pq}, \qquad \forall (p,q) \in \mathcal{C},$$

$$h_{pqr} \geq 0, \qquad \forall (p,q) \in \mathcal{C}, \quad \forall r \in \mathcal{R}_{pq}.$$

It can be solved using a dual solution algorithm similar to the one presented in Section 10.4.1 (cf. Larsson et al., 1997). Denote the (unique) solution by (f^*, p^*).

We now introduce τ_{pq} as the adjustments made to the public transport travel time in origin destination pair (p, q), and denote adjustments on the links in the car network by ρ_a. The set of travel time adjustments which reproduces the solution (f^*, p^*) is here defined by the system

$$\sum_{a\in\mathcal{A}} \delta_{pqra}(t_a(f_a^*) + \rho_a) - \pi_{pq} \geq 0, \qquad \forall r \in \mathcal{R}_{pq}, \quad \forall (p,q) \in \mathcal{C},$$

$$\sum_{r\in\mathcal{R}_{pq}} h_{pqr}^* \left(\sum_{a\in\mathcal{A}} \delta_{pqra}(t_a(f_a^*) + \rho_a) - \pi_{pq} \right) = 0, \qquad \forall (p,q) \in \mathcal{C},$$

$$W_{pq}(p_{pq}^*) - (\hat{\pi}_{pq} + \tau_{pq}) \geq 0, \qquad \forall (p,q) \in \mathcal{C},$$

$$p_{pq}^*(W_{pq}(p_{pq}^*) - (\hat{\pi}_{pq} + \tau_{pq})) = 0, \qquad \forall (p,q) \in \mathcal{C}.$$

This system can be reformulated as

$$\sum_{a\in\mathcal{A}} \delta_{pqra}(t_a(f_a^*) + \rho_a) - \pi_{pq} \geq 0, \qquad \forall r \in \mathcal{R}_{pq}, \quad \forall (p,q) \in \mathcal{C}, \tag{10.21a}$$

$$W_{pq}(p_{pq}^*) - (\hat{\pi}_{pq} + \tau_{pq}) \geq 0, \qquad \forall (p,q) \in \mathcal{C}, \tag{10.21b}$$

$$\sum_{a\in\mathcal{A}} (t_a(f_a^*) + \rho_a) f_a^* - \sum_{(p,q)\in\mathcal{C}} \left[\sum_{r\in\mathcal{R}_{pq}} h_{pqr}^* \pi_{pq} + p_{pq}^* \left(W_{pq}(p_{pq}^*) - (\hat{\pi}_{pq} + \tau_{pq}) \right) \right] = 0. \tag{10.21c}$$

The travel time adjustment problem for the case of combined mode split and traffic assignment is defined as the problem of minimizing a secondary objective under the constraints (10.21) and (10.13d). Again, a solution method for this problem can be based on the algorithm in Section 10.4.2.

10.6 CONCLUSIONS

In this paper we have outlined a decision support methodology for traffic management which is based on a bilevel optimization approach to the management problem. This methodology generates changes in the traffic network that are adequate for reaching the management goals. The travel time adjustments in the network, or the corresponding implementable actions, are provided automatically. This is in contrast with a traditional scenario analysis strategy, in which a series of tentative travel time adjustments is generated by a heuristic trial-and-error technique, and evaluated through the use of a traditional traffic equilibrium package.

The solution technique for the bilevel problem consists of two stages, and solution algorithms for each of the stages are discussed in detail. Extensions for elastic demands and to combined assignment and mode split were also discussed.

Future research includes a detailed analysis of the computational performance of the solution algorithms as well as evaluations of alternative algorithms.

References

Chen, M., Bernstein, D.H., Chien, S.I.J., and Mouskos, K.C. (1998). "A simplified formulation of the toll design problem," Paper submitted for presentation and publication to the Transportation Research Board, National Research Council, July.

Cree, N.D., Maher, M.J., and Paechter, B. (1998). "The continuous equilibrium optimal network design problem: a genetic approach," *In: Transportation Networks: Recent Methodological Advances*, Ed: Bell, M.G.H., 163–174, Amsterdam: Pergamon.

Dirkse, S.P., and Ferris, M.P. (1998). "Traffic modeling and variational inequalities using GAMS." Ed: Ph. L. Toint, M. Labbé, K. Tanczos, and G. Laporte, Operations Research and Decision Aid Methodologies in Traffic and Transportation Management, NATO ASI Series F, 136-163. Berlin: Springer-Verlag.

Hearn D.W., and Ramana M.V. (1998). "Solving congestion toll pricing models," In: *Equilibrium and Advanced Transportation Modelling,* Ed: Marcotte P. and Nguyen S., New York: Kluwer Academic Publishers, 109–124.

Huang, H.J., and Bell, M.G.H. (1998). "Continuous equilibrium network design problem with elastic demand: derivative-free solution methods," *In: Transportation Networks: Recent Methodological Advances*, Ed: Bell, M.G.H., 175–193, Amsterdam: Pergamon.

Labbé, M., Marcotte, P., and Savard, G. (1998). "A Bilevel Model of Taxation and Its Application to Optimal Highway Pricing," *Management Science*, 44, 1608–1622.

Larsson, T., Liu, Z., and Patriksson, M. (1997). "A dual scheme for traffic assignment problems," *Optimization*, 42, 323–358.

Larsson, T., and Patriksson, M. (1998). "Side constrained traffic equilibrium models — traffic management through link tolls," In: *Equilibrium and Advanced Transportation Modelling*, Ed: Marcotte, P., and Nguyen, S., New York: Kluwer Academic Publishers, 125–151.

Larsson, T., and Patriksson, M. (1999). "Side constrained traffic equilibrium models — analysis, computation and applications," *Transportation Research*, 33B, 233–264.

Larsson, T., Patriksson, M., and Strömberg, A.-B. (1999). "Ergodic, primal convergence in dual subgradient schemes for convex programming," *Mathematical Programming*, 86, 283–312.

Marcotte, P. (1986). "Network Design Problem with Congestion Effects: A Case of Bilevel Programming," *Mathematical Programming*, 34, 142–162.

Marcotte, P., and Marquis, G. (1992). "Efficient implementation of heuristics for the continuous network design problem," *Annals of Operations Research*, 34, 163–176.

Migdalas, A. (1995). "Bilevel programming in traffic planning: models, methods and challenge," *Journal of Global Optimization*, 7, 381–405.

Patriksson, M. (1994). "The Traffic Assignment Problem — Models and Methods," Utrecht: VSP, BV, The Netherlands.

Sheffi, Y. (1985). "Urban Transport Networks: Equilibrium Analysis with Mathematical Programming Methods," Englewood Cliffs, NJ: Prentice-Hall.

11 COLUMN GENERATION METHOD FOR NETWORK DESIGN

Nelson Maculan

Marcos M. Passini

José André M. Brito

Abdel Lisser

Abstract: This work deals with the Two-architecture Network Problem, which consists in satisfying growing traffic demand by superimposing rings on an existing meshed network. The goal of this work is to solve the proposed mathematical model via column generation techniques. Firstly a complete model is proposed, then it is simplified to become more manageable. A branch-and-price approach is presented with all its problem-dependent issues. The numerical experience reported consists of two cases, one of 7 nodes and 11 edges, the other with 10 nodes and 36 edges, both drawn from real European telecommunications networks.

INTRODUCTION

In recent years, studies about network dimensioning and survivability focused mainly on ring based and mesh based architectures. Most of these works treated both models separately. A very good introduction to the design of a meshed telecommunications network can be found in (Minoux, 1987), while (Vachani et al., 1996) and (Alain Sutter, 1998) deal with rings, and (Lisser et al., 1995) with transmission networks survivability.

At the present time, several companies deploy SDH networks based on rings, but there are many meshed networks still in operation or even under construction, as different traffic scenarios usually lead to different design solutions. Specifically, in large metropolitan areas, it is easy to identify clusters of nodes (banks, communication companies, etc.) generating large data traffic among themselves. These clusters are

M. Gendreau and P. Marcotte (eds.), Transportation and Network Analysis: Current Trends, 165–179.

identified as cliques or quasi-cliques in the graph of calls. Meanwhile, peripheral areas of the same network may present little traffic. If the metropolitan area under study is already served by a meshed network, it is possible to deal with the growth of traffic demand by superimposing self-healing rings on the existing mesh. This idea, called the *two-architecture* network design solution, was introduced in (Lisser et al., 1996b) and (Lisser et al., 1996a). Once the nodes forming the cluster have been identified, we search for rings to interconnect them. This can be interesting, in terms of costs, as an alternative to improve the capacity of the network. This approach can also prove useful in circumventing the high costs of cross-connect equipments used for point-to-point networks, since rings only require add-drop multiplexers.

This describes the Two-architecture Network Problem. Our work consists of modelling the problem and bringing some elements of answer. We have already considered models for designing telecommunications networks with rings and meshed circuits, see (Passini, 1996), (Lisser et al., 1996b), (Lisser et al., 1996a).

Sections 11.1 and 11.2 present a mathematical model derived from (Lisser et al., 1996a), defining variables, constraints and the objective function. For the solution of this model, we consider in Section 11.3 its linear relaxation, to be solved by column generation techniques (Chvátal, 1983). Feasible solution acquirement is presented in Section 11.4. In Section 11.5 we comment on the *branch-and-price* method presented in (Savelsbergh, 1997) where a column generation technique is used. Numerical experiments on two real telecom networks are considered in Section 11.6. Finally, we present our conclusions.

11.1 MATHEMATICAL MODEL

In the mathematical model to be proposed, we consider a support graph on which we study the enlargement of the meshed network capacity and also the building of rings. After this, we define the data, the constraints and the objective function for a network design problem.

Some of the cited works, (Passini, 1996), (Lisser et al., 1996b), (Lisser et al., 1996a), have looked at rings as subsets of nodes, without considering their real path through the network. This may lead to infeasible rings, due to technical constraints like maximum length and maximum distance between adjacent nodes. So we introduce the concept of *support cycle*: a cycle in the network graph acting as a layer for a ring. Each ring must have a support cycle and every node belonging to a ring must belong to its support cycle.

Throughout the following definitions, let *cluster* be the subset of nodes effectively belonging to one or more rings in the final solution. A subset of nodes that are candidates to belong to the cluster is provided with the data, see V' below.

11.1.1 Data

- $G = (V, E)$, a non-oriented support graph;
- $F_{ij} > 0$, $[i,j] \in D$, $D = \{[i,j] | i < j,\ i \in V,\ j \in V\}$, demand between vertices i and j;
- $g_i > 0$, the cost of each multiplexer installed in $i \in V$;

- The capacities that can be installed on each edge: multiples of $2, 34, 155$ with respective costs, $c_{(2)} > 0$, $c_{(34)} > 0$, $c_{(155)} > 0$. We assume that: $\frac{c_{(155)}}{155} \leq \frac{c_{(34)}}{34} \leq \frac{c_{(2)}}{2}$;

- $CR > 0$, the capacity of each ring;

- $CV_i > 0$, the maximum capacity of vertex $i \in V$, for the meshed part;

- $d_{ij} > 0$, the length of each edge $[i, j] \in E$;

- $DIST > 0$, the maximum length of each ring;

- $RING \in \mathbb{Z}_+$, the maximum number of nodes in a ring;

- $V' \subseteq V$, the set of vertices that are candidates to be in the cluster;

- $CE_u \geq 0$, the current capacity of edge $u \in E$;

- a_k^{ij}, the demand satisfied by ring k, between i and j, where $[i, j] \in D$, $k = 1, 2, \ldots, K$, where K is the total number of feasible rings;

- $0 \leq f_{ij} \leq F_{ij}$, where f_{ij} is a part of the demand that might be routed by a ring;

- h_k, the cost of ring k.

11.1.2 *Auxiliary Data*

- $G_1 = (V, A)$, an oriented graph, where $A = \{(i, j), (j, i) | [i, j] \in E\}$;

- $G_2 = (V', A')$, where $A' = \{(i, j), (j, i) | [i, j] \in E \text{ and } i, j \in V'\}$, an oriented subgraph formed by the candidate nodes;

- $G_3 = (V'', A'')$, where $V'' = V' \cup \{\odot\}$ and $A'' = \{(\odot, i), (i, \odot) | i \in V'\} \cup A'$, a graph over which we formulate the rings.

11.1.3 *Variables*

- $r_i \in \{0, 1\}$, $r_i = 1$ if vertex i is chosen to have a multiplexer, $r_i = 0$ otherwise;

- $v_u^{(2)}, v_u^{(34)}, v_u^{(155)} \in \mathbb{Z}_+, u \in E$, the capacities of type 2, 34 and 155 to be installed on edge u;

- $z_{\alpha\beta}^{ij} \geq 0$, flow through the arc $(\alpha, \beta) \in A'$ associated to the demand between i and j;

- $x_k \in \mathbb{Z}_+$, the number of rings k.

11.1.4 Constraints

$$\sum_\beta z^{ij}_{i\beta} - \sum_\beta z^{ij}_{\beta i} + \sum_{k=1}^{K} a^{ij}_k x_k \geq F_{ij},\ [i,j] \in D \tag{11.1}$$

$$\sum_\beta z^{ij}_{\alpha\beta} - \sum_\beta z^{ij}_{\beta\alpha} = 0,\ \alpha \in V - \{i,j\},\ [i,j] \in D \tag{11.2}$$

$$\sum_\beta \sum_{[i,j]\in D} z^{ij}_{\alpha\beta} \leq CV_\alpha,\ \alpha \in V \tag{11.3}$$

$$\sum_\beta \sum_{[i,j]\in D} z^{ij}_{\beta\alpha} \leq CV_\alpha,\ \alpha \in V \tag{11.4}$$

$$\sum_{[i,j]\in D} (z^{ij}_{\alpha\beta} + z^{ij}_{\beta\alpha}) \leq CE_{\alpha\beta} + 2v^{(2)}_{\alpha\beta} + 34v^{(34)}_{\alpha\beta} + 155v^{(155)}_{\alpha\beta},\quad [\alpha,\beta] \in E \tag{11.5}$$

$$x_k \in \mathbb{Z}_+,\ k = 1,2,\ldots,K \tag{11.6}$$

$$v^{(2)}_{\alpha\beta},\ v^{(34)}_{\alpha\beta},\ v^{(155)}_{\alpha\beta} \in \mathbb{Z}_+,\ [\alpha,\beta] \in E \tag{11.7}$$

$$z^{ij}_{\alpha\beta} \geq 0,\ (\alpha,\beta) \in A',\ [i,j] \in D \tag{11.8}$$

The constraints (11.1) and (11.2) guarantee that the demands F_{ij} are satisfied. The inequalities (11.3) and (11.4) limit the capacity of each vertex α. The expression (11.5) represents the capacity constraint for edge $[\alpha,\beta]$. The constraints (11.6) – (11.8) define the sets to which the decision variables must belong.

11.1.5 Objective Function

$$\text{minimize} \sum_{k=1}^{K} h_k x_k + \sum_{[\alpha,\beta]\in E} (c_{(2)} v^{(2)}_{\alpha\beta} + c_{(34)} v^{(34)}_{\alpha\beta} + c_{(155)} v^{(155)}_{\alpha\beta}) \tag{11.9}$$

This objective function represents the sum of ring building and edge expansion costs.

Minimizing (11.9) under the constraints (11.1) – (11.8) is an extremely hard combinatorial optimization problem for the problems we are to deal with. However, it is possible to define the simplified model that follows.

11.2 SIMPLIFIED MODEL

To create a simpler model, we remove the vertex capacity constraints (11.3) and (11.4). We also assume that $2v^{(2)}_{\alpha\beta} + 34v^{(34)}_{\alpha\beta} + 155v^{(155)}_{\alpha\beta} = v_{\alpha\beta}$ and $c_{(2)} v^{(2)}_{\alpha\beta} + c_{(34)} v^{(34)}_{\alpha\beta} + c_{(155)} v^{(155)}_{\alpha\beta} = cv_{\alpha\beta}$, where the continuous variable $v_{\alpha\beta}$ represents the capacity expansion of edge $[\alpha,\beta]$ and c the unit cost of this expansion. The problem we shall solve may be written in the following form:

$$(P):\ \text{minimize} \sum_{k=1}^{K} h_k x_k + \sum_{[\alpha,\beta]\in E} cv_{\alpha\beta} \tag{11.10}$$

under the constraints:

$$\sum_{\beta} z^{ij}_{i\beta} - \sum_{\beta} z^{ij}_{\beta i} + \sum_{k=1}^{K} a^{ij}_k x_k \geq F_{ij}, \ [i,j] \in D \tag{11.11}$$

$$\sum_{\beta} z^{ij}_{\alpha\beta} - \sum_{\beta} z^{ij}_{\beta\alpha} = 0, \ \alpha \in V - \{i,j\}, \ [i,j] \in D \tag{11.12}$$

$$\sum_{[i,j]\in D} (z^{ij}_{\alpha\beta} + z^{ij}_{\beta\alpha}) \leq CE_{\alpha\beta} + v_{\alpha\beta}, \ [\alpha,\beta] \in E \tag{11.13}$$

$$x_k \in \mathbb{Z}_+, \quad k = 1, 2, \ldots, K \tag{11.14}$$

$$v_{\alpha\beta} \geq 0, \quad [\alpha,\beta] \in E \tag{11.15}$$

$$z^{ij}_{\alpha\beta} \geq 0, \quad (\alpha,\beta) \in A', \ [i,j] \in D \tag{11.16}$$

A first step towards solving (P) consists in solving its linear relaxation, henceforth called $(\bar{P})$. We also introduce the slack variables $w_{ij} \geq 0$ and $s_{\alpha\beta} \geq 0$ respectively in (11.11) and (11.13). We thus obtain:

$$(\bar{P}) : \text{ minimize} \sum_{k=1}^{K} h_k x_k + \sum_{[\alpha,\beta]\in E} c v_{\alpha\beta} \tag{11.17}$$

under the constraints:

$$\sum_{\beta} z^{ij}_{i\beta} - \sum_{\beta} z^{ij}_{\beta i} + \sum_{k=1}^{K} a^{ij}_k x_k - w_{ij} = F_{ij}, \ [i,j] \in D \tag{11.18}$$

$$\sum_{\beta} z^{ij}_{\alpha\beta} - \sum_{\beta} z^{ij}_{\beta\alpha} = 0, \ \alpha \in V - \{i,j\}, [i,j] \in D \tag{11.19}$$

$$\sum_{[i,j]\in D} (z^{ij}_{\alpha\beta} + z^{ij}_{\beta\alpha}) - v_{\alpha\beta} + s_{\alpha\beta} = CE_{\alpha\beta}, \ [\alpha,\beta] \in E \tag{11.20}$$

$$x_k \in R_+, \quad k = 1, 2, \ldots, K \tag{11.21}$$

$$v_{\alpha\beta} \geq 0, \quad s_{\alpha\beta} \geq 0, \ [\alpha,\beta] \in E \tag{11.22}$$

$$w_{ij} \geq 0, \quad [i,j] \in D \tag{11.23}$$

$$z^{ij}_{\alpha\beta} \geq 0, \quad (\alpha,\beta) \in A', \ [i,j] \in D \tag{11.24}$$

Let us denote by $val(\cdot)$ the optimal value of the objective function of $(\cdot)$. The value $val(\bar{P})$ provides a lower bound for $val(P)$. We will solve $(\bar{P})$ using a column generation method, see (Dantzig and Wolfe, 1960), (Gilmore and Gomory, 1961), (Gilmore and Gomory, 1963).

11.3 COLUMN GENERATION

The linear program $(\bar{P})$ may be solved by the simplex method, using column generation in linear programming, see also (Chvátal, 1983) and (Lasdon, 1970). Let u be the row vector of simplex multipliers associated with a primal-feasible solution. We have that $u^T \in \mathbb{R}^\delta$, where δ is equal to the number of rows from constraints (11.18) to (11.20).

The column vector a_k, whose elements are equal to a_k^{ij} for the constraints (11.18) and zero for all the others, may be known *a priori* or generated as required.

To know if a column a_k, not presently in the basis matrix, is a candidate to enter it, we must compute the reduced cost $\bar{h}_k = h_k - ua_k$ associated with column a_k. If $\bar{h}_k < 0$, a_k enters the new basis and we must find the column that leaves the basis (primal simplex method). Let

$$\bar{h}_l = h_l - ua_l = \min_{k=1,\dots,K} \{h_k - ua_k\}.$$

If $\bar{h}_l \geq 0$, no column a_k will be a candidate to enter the basis.

11.3.1 An Automatic Method

Since a column a_k is associated with a ring k, we will attempt to design a feasible ring k in order to generate column a_k.

We add a dummy node $\odot$ to G_2 from where we create an arc ending at each candidate node (this defines G_3). Then we inject a "virtual flow" (thus called to avoid confusion with the "real" demand flow between nodes) in $\odot$ and extract it in the same place. By forbidding bifurcations or self-crossings we force the virtual flow to follow a unique path in the subgraph of candidates. Of course, the first node after and the last before $\odot$ must be adjacent for the cycle to be closed. In this way, we create a cycle in the subgraph of candidates.

This will be done over the graph G_3, so we include in the sets A' and A'' all arcs $(i,j) \notin A'$, $i \neq j$, $i \in V'$, $j \in V'$. The distance between i and j for these last arcs will be $DIST + 1$.

We introduce the following variables:

- $t_i \in \{0,1\}$, $t_i = 1$ if the vertex i is chosen to belong to a support cycle, $t_i = 0$ otherwise.

- $y^l_{\alpha\beta} \geq 0$, virtual flow through the arc $(\alpha,\beta) \in A''$ associated with the virtual flow from $\odot \in V''$ to $l \in V'$ through a ring.

- $w_{\alpha\beta} \in \{0,1\}$, $w_{\alpha\beta} = 1$ if the arc $(\alpha\beta) \in A''$ belongs to a ring, $w_{\alpha\beta} = 0$ otherwise.

- $p_{\alpha\beta} = w_{\alpha\odot} . w_{\odot\beta}, \alpha \neq \beta, \alpha \neq 0, \beta \neq 0.$

- $q_{ij} \in \{0,1\}$, $q_{ij} = 1$ if there is a flow from i to j through a ring and $q_{ij} = 0$ otherwise, $i,\ j \in V'$ and $i < j$.

The ring design model is:

$$\sum_{\beta} y^l_{\odot\beta} - \sum_{\beta} y^l_{\beta\odot} = t_l, \; l \in V', \tag{11.25}$$

$$\sum_{\beta} y^l_{\alpha\beta} - \sum_{\beta} y^l_{\beta\alpha} = 0, \; \alpha \in V' - \{l\}, \; l \in V', \tag{11.26}$$

$$\sum_{\beta} w_{\odot\beta} = \sum_{\beta} w_{\beta\odot} = 1, \tag{11.27}$$

$$\sum_{\beta} w_{\alpha\beta} = \sum_{\beta} w_{\beta\alpha} = t_\alpha, \; \alpha \in V', \tag{11.28}$$

$$0 \le y^l_{\alpha\beta} \le w_{\alpha\beta}, l \in V', (\alpha, \beta) \in A'', \tag{11.29}$$

$$p_{\alpha\beta} \ge 0, \; p_{\alpha\beta} \le w_{\alpha\odot}, \; p_{\alpha\beta} \le w_{\odot\beta}, \; (\alpha, \beta) \in A', \tag{11.30}$$

$$w_{\alpha\odot} + w_{\odot\beta} - 1 \le p_{\alpha\beta}, \; (\alpha, \beta) \in A', \tag{11.31}$$

$$w_{\alpha\beta} \in \{0, 1\}, \; (\alpha, \beta) \in A'', \tag{11.32}$$

$$\sum_{(\alpha,\beta)\in A'} d_{\alpha\beta} w_{\alpha\beta} + \sum_{(\alpha,\beta)\in A'} d_{\alpha\beta} p_{\alpha\beta} \le DIST, \tag{11.33}$$

$$q_{ij} \in \{0, 1\}, \;\; q_{ij} \le t_i, \;\; q_{ij} \le t_j, \; i < j, \; i, j \in V', \tag{11.34}$$

$$1 \ge r_i \ge q_{ij}, i < j, \; i, j \in V', \tag{11.35}$$

$$1 \ge r_j \ge q_{ij}, i < j, \; i, j \in V', \tag{11.36}$$

$$\sum_{i<j, \; i,j \in V'} f_{ij} q_{ij} \le CR, \tag{11.37}$$

$$\sum_{i \in V'} t_i \le RING. \tag{11.38}$$

After this, we can write: $h_k = \sum_{i \in V'} g_i r_i$ and $a^{ij}_k = f_{ij} q_{ij}$.

Let u^{ij} be the element of vector u associated with the constraint (11.18) related to the demand between i and j, $[i, j] \in D$ and $i, j \in V'$. We can define the objective function of the problem that will generate a_k as follows:

$$h_k - u a_k = \sum_{i \in V'} g_i r_i - \sum_{i<j, \; i,j \in V'} u^{ij} a^{ij}_k = \sum_{i \in V'} g_i r_i - \sum_{i<j, \; i,j \in V'} u^{ij} f_{ij} q_{ij}.$$

So we minimize

$$\sum_{i \in V'} g_i r_i - \sum_{i<j, \; i,j \in V'} u^{ij} f_{ij} q_{ij} \tag{11.39}$$

under constraints (11.25)–(11.38).

Constraints (11.25) to (11.33) and (11.38) define the feasible cycles. (11.25) and (11.26) balance the virtual flow as explained above; if $t_i = 1$ then a virtual flow is injected in $\odot$ to build the cycle; (11.27) states that $\odot$ is visited just once, (11.28) binds the presence of an arc in the cycle with the t_i's, (11.29) forces an arc to be present if flow passes through it; (11.30) and (11.31) linearize the condition $p_{\alpha\beta} = w_{\alpha\odot}.w_{\odot\beta}$. It should be noted that p just acts as a w for the arc that "closes" the ring between the first and the last nodes, both connecting directly to $\odot$ as explained above. (11.33) limits the total length of the cycle. To summarize, constraints (11.25) to (11.33) define a feasible cycle and (11.34) to (11.36) place a ring over the cycle in a way to minimize the objective function (11.39), which will give the reduced cost of the column associated to the generated ring.

If we do not have any $u^{ij} > 0$, there will not be any column associated to rings with a negative reduced cost.

For $|V'| = 30$, the number of 0–1 variables goes beyond 2000. We may rewrite the model with less 0–1 variables, but it will remain very hard to solve for practical instances. For this reason, other techniques must be looked for.

11.3.2 *Feasible Cycles Generation*

The cycles over graph G must be shorter than or equal to $DIST$; we can generate some of them by search methods.

The method we used finds all elementary circuits over an oriented graph. For our graph $G_1 = (V, A)$, it has complexity $O(|V| + [|C| + 1]|A|)$, where C is the set of elementary circuits in the graph G_1, see (Szwarcfiter and Lauer, 1975).

Fortunately, the problems we consider involve sparse graphs. The length of the support cycles is bounded by the distance $DIST$. Thanks to this condition, we were able to generate all cycles for two examples where $|V| = 7$ and $|V| = 10$, respectively. For the first case, the number of feasible cycles is equal to 15, and for the second, 165, see (Brito, 1999).

11.3.3 *A Semi-automatic Method*

As we already noted, the reduced cost associated with a column will be of the form $h = \sum_i g_i r_i - \sum_i \sum_j u^{ij} a^{ij}$. We want to generate a column having a negative reduced cost. Given $g_i > 0$, $r_i \geq 0$, $a^{ij} \geq 0$, for h to have a negative value, we must have $u^{ij} > 0$. Let us define $\bar{D} = \{[i,j] | \ u^{ij} > 0\}$ and $\bar{V} = \{i, j | \ [i,j] \in \bar{D}\}$. If $\bar{D} = \phi$, no column associated with a ring will be part of the new basis for $(\bar{P})$.

We may *a priori* generate support cycles whose total length is inferior to $DIST$. Among these cycles we take one containing the vertices of $\bar{V}$. Let us now suppose $|\bar{V}| = p$. Let $s(1) < s(2) < \cdots < s(p-1) < s(p)$ and $\bar{V} = \{s(1),\ s(2),\ \ldots\ , s(p-1),\ s(p)\}$. We suppose that the vector u is associated with to a feasible primal basis of $(\bar{P})$ such that the reduced costs of the columns corresponding to variables $z^{ij}_{\alpha\beta}$ and $s_{\alpha\beta}$ are nonnegative.

The problem that generates a column associated with this ring may be written as follows.

$$(PA): \text{ minimize } \sum_{i=1}^{p} g_{s(i)} r_{s(i)} - \sum_{i=1}^{p-1} \sum_{j=i+1}^{p} u^{s(i)s(j)} a^{s(i)s(j)}$$

under the constraints:

$$M_{s(i)} = \sum_{j=i+1}^{p} a^{s(i)s(j)}, \ i = 1, 2, \ldots, p-1$$

$$0 \leq M_{s(i)} \leq (\sum_{j=i+1}^{p} f_{s(i)s(j)}) r_{s(i)}, \ i = 1, 2, \ldots, p-1$$

$$\sum_{i=1}^{p-1} M_{s(i)} \leq CR$$

$$0 \leq a^{s(i)s(j)} \leq f_{s(i)s(j)} r_{s(j)}, \; i = 1, 2, \ldots, p-1, \; j = i+1, i+2, \ldots, p$$

$$r_{s(i)} \in \{0, 1\}, \; i = 1, 2, \ldots, p.$$

If we do not have any cycle to support $\bar{V}$, we can search for a set $D_0 \subseteq \bar{D}$, such that there is a feasible cycle for $V_0 = \{i, j | \; [i, j] \in D_0\}$.

If $val(PA) \geq 0$ and $\bar{V} = V_0$ we have found an optimal solution for $(\bar{P})$.

When $val(PA) \geq 0$ and $\bar{V} \neq V_0$, we can do it another way, to be detailed later. If $val(PA) < 0$, the column generated will enter the new basis and another simplex iteration will be performed.

11.3.4 An Alternative Method

We can also assume the existence of a feasible cycle, and with $\bar{D}$ and $\bar{V}$ defined as above, we solve problem (PA) and get as solution $\bar{r}_i \in \{0, 1\}, \; i = 1, 2, \ldots, n$. If $val(PA) \geq 0$, we have found an optimal solution for $(\bar{P})$.

If $val(PA) < 0$, we will verify the existence of a feasible cycle to support $V_1 = \{i \mid \bar{r}_i = 1\}$ in the two following ways:

- All feasible cycles have been listed in a former step. If there is no cycle to support V_1, we eliminate this solution by introducing the following constraint in (PA), see (Maculan et al., 1998a):

$$\sum_{i=1}^{n} (1 - 2\bar{r}_i) r_i \geq 1 - \sum_{i=1}^{n} \bar{r}_i. \tag{11.40}$$

  The inequality (11.40) removes the point $\bar{r} \in \{0, 1\}^p$ from the set $\{0, 1\}^p$. We can also verify that (11.40) is a facet for the convex hull of the set $\{0, 1\}^p - \{\bar{r}\}$, see (Maculan et al., 1998a).

- The feasible cycles are not yet listed. Let us note $\bar{V}_1 = V' - V_1$ and consider $\alpha \in V_1$. We must solve the following problem:

$$(VC): \text{ minimize } \sum_{(i,j) \in A'} d_{ij} w_{ij}$$

  under the constraints

$$\sum_j y_{\alpha j} - \sum_j y_{j\alpha} = |V_1| - 1$$

$$\sum_j y_{ij} - \sum_j y_{ji} = -1, \; i \in V_1 - \{\alpha\}$$

$$\sum_j y_{ij} - \sum_j y_{ji} = 0, \; i \in \bar{V}_1$$

$$\sum_j w_{ij} = \sum_j w_{ji} = 1, \ i \in V_1$$

$$\sum_j w_{ij} = \sum_j w_{ji}, \ i \in \bar{V}_1$$

$$0 \leq y_{ij} \leq (|V_1| - 1)w_{ij}, \ (i,j) \in A'$$

$$\sum_{(i,j)\in A'} w_{ij} \leq RING$$

$$w_{ij} \in \{0,1\}.$$

This problem finds a cycle in G containing the vertices of V_1 and having minimal length. If $val(VC) \leq DIST$, the $\bar{w}_{ij}$, $(i,j) \in A'$, solutions of (VC), will define a feasible cycle for V_1. If $val(VC) > DIST$, there is no cycle for $\bar{V}$. The solution $\bar{r}_i$, $i = 1, 2, \ldots, p$, will be eliminated by introducing in (VC) the constraint (11.40) previously defined.

11.3.5 A Heuristic for Column Generation

An algorithm producing for (PA) a feasible solution $\hat{r}_i$, $i \in \bar{V}$ and $\hat{a}^{ij}$, $[i,j] \in \bar{D}$ will be able to give a column that is a candidate to enter the new basis of $(\bar{P})$; for this there has to be a support cycle for $\hat{r}_i$, $i \in \bar{V}$ and $\sum_{i\in\bar{V}} g_i\hat{r}_i - \sum_{[i,j]\in\bar{D}} u^{ij}\hat{a}^{ij} < 0$. See the algorithm in Figure 11.1.

$u^{i(1)j(1)} f_{i(1)j(1)} \geq u^{i(2)j(2)} f_{i(2)j(2)} \geq \cdots \geq u^{i(p)j(p)} f_{i(p)j(p)} > 0$;
$\hat{h} := 0$; $\bar{h} := g_{i(1)} + g_{j(1)} - u^{i(1)j(1)} f_{i(1)j(1)}$;
$D := \{[i(1), j(1)]\}$; $V := \{i(1), j(1)\}$; `CAP` $:= f_{i(1)j(1)}$;
`for` $k = 2$ `to` p `do`
  `if (CAP` $+ f_{i(k)j(k)} \leq CR$`)` `and` $(i(k) \in V)$ `and` $(j(k) \in V)$ `then`
    $\hat{h} := \bar{h} - u^{i(k)j(k)} f_{i(k)j(k)}$;
  `if (CAP` $+ f_{i(k)j(k)} \leq CR$`)` `and` $(i(k) \in V)$ `and` $(j(k) \notin V)$ `then`
    $\hat{h} := \bar{h} + g_{j(k)} - u^{i(k)j(k)} f_{i(k)j(k)}$;
  `if (CAP` $+ f_{i(k)j(k)} \leq CR$`)` `and` $(i(k) \notin V)$ `and` $(j(k) \in V)$ `then`
    $\hat{h} := \bar{h} + g_{i(k)} - u^{i(k)j(k)} f_{i(k)j(k)}$;
  `if (CAP` $+ f_{i(k)j(k)} \leq CR$`)` `and` $(i(k) \notin V)$ `and` $(j(k) \notin V)$ `then`
    $\hat{h} := \bar{h} + g_{i(k)} + g_{j(k)} - u^{i(k)j(k)} f_{i(k)j(k)}$;
  `if` $(\bar{h} < 0)$ `and` $(\hat{h} > \bar{h})$ `then`
    `STOP (we take the solution associated to` $\bar{h}$`);`
  $\bar{h} := \hat{h}$;
  `CAP:= CAP` $+ f_{i(k)j(k)}$;
  $V := V \cup \{i(k), j(k)\}$; $D := D \cup \{[i(k), j(k)]\}$;
`end`

Figure 11.1 Heuristic column generation algorithm

If $\bar{h} \geq 0$, then we could not find a column with negative reduced cost. Otherwise, V and D define a potential ring, and we have to search for a cycle to support it.

11.4 GETTING A FEASIBLE SOLUTION FOR (P)

It is always important to get good feasible solutions for a combinatorial optimization problem. For this, we may build an *ad hoc* heuristic, based on our specific knowledge about the problem, or try to develop an approximation scheme through which we can determine, *a priori,* the distance, in the worst case, between the solution reached by this scheme and the optimum, see (Campello and Maculan, 1994), (Hochbaum, 1997), or even use metaheuristics (Glover et al., 1995).

We have chosen to develop an *ad hoc* method. We start by considering the largest feasible cycles, and we assign to them the flows associated with demands to build the rings. There may be no ring for some pair of nodes of graph G; in this case the demand between these nodes will be routed through the meshed network, see (Brito, 1999).

11.5 BRANCH-AND-PRICE METHOD

In order to solve the mathematical model with integer variables, we can consider *branch-and-price* methods, see (Barnhart et al., 1998) and (Maculan et al., 1998a).

Once $(\bar{P})$ has been solved, we check the values taken by the variables x_k. If all these values are integers, we have an optimal solution for (P). Otherwise, there should be a variable x_l such that $\bar{x}_l \notin \mathbb{Z}$ in the optimum solution of (P). The *branch-and-price* method we propose to solve (P) may be described as follows.

To start the algorithm, we consider two problems descending from $(\bar{P})$:

- $(\bar{P}_1) := (\bar{P}) \cap \{x_l \leq \lfloor \bar{x}_l \rfloor\}$;
- $(\bar{P}_2) := (\bar{P}) \cap \{x_l \geq \lfloor \bar{x}_l \rfloor + 1\}$,

where $\lfloor \alpha \rfloor$ represents the largest integer less than or equal to α.

In order to solve $(\bar{P}_1)$, we may use the dual simplex method, see (Dantzig, 1963), (Chvátal, 1983), because the optimal basis solution of $(\bar{P})$ is dual-feasible for $(\bar{P}_1)$ and $(\bar{P}_2)$.

We start by solving $(\bar{P}_1)$ by the dual simplex method, considering already generated columns to solve $(\bar{P})$; this problem will be named $(\bar{P'}_1)$. This may be easily done, see (Chvátal, 1983) page 157, or (Maculan et al., 1999).

At this point, the columns were generated ignoring other feasible columns. Considering these columns may lead to an infeasible $(\bar{P'}_1)$, even if $(\bar{P}_1)$ is feasible. In this case, we use a two-phase primal simplex method see (Maculan et al., 1998a). Once we have solved $(\bar{P'}_1)$, we solve $(\bar{P}_1)$ by a primal simplex method using column generation. For this we must delete the column associated with the variable x_l in order to prevent its return in the basis. This elimination is carried on introducing a constraint of type (11.40) in (PA), where the $\bar{r}_i$, $i = 1, 2, \ldots, p$, define the column associated to x_l.

The resolution of $(\bar{P}_2)$ is similar to that of $(\bar{P}_1)$, but we do not need to eliminate the column associated to x_l. For the descendants of $(\bar{P}_1)$ and $(\bar{P}_2)$, the preceding comments are still valid.

A problem of type $(\bar{P}_i)$ is fathomed when there are no more descendants: its optimal solution is feasible for (P), or $val(\bar{P}_i)$ is greater than or equal to the best feasible solution of (P), or $(\bar{P}_i)$ is infeasible.

11.6 NUMERICAL EXPERIMENTS

We present numerical results for two instances of the problem, both networks from a European metropolitan area. The program was written in the C programming language, calling the XPRESS-MP library. All experiments have been carried on a Pentium III 450 MHz machine. Table 11.1 shows some statistics on the networks and the resulting problems.

Table 11.1 Networks statistics

Nodes	Edges	Demands	Cycles[a]	Matrix[b]	Density
7	11	10	15	65×231	3.10%
10	22	36	165	291×1606	0.68%

[a] Feasible support cycles
[b] Master problem

The costs we adopted throughout the experiments were 1 for an active node in a ring and 2 for a unit edge expansion in the mesh. This is indeed an approximation for the real costs of a multiplexer, used to connect a node to a ring, and a pair of terminals to be installed in SDXCs.

Solving (P) for the 10-nodes network, we obtained an objective function value of 152, corresponding to 76 edge expansions. We started the column generator, whose output were 88 new columns with negative reduced cost, and the solution value dropped to 24.403. Adding these columns to (P), we launched a branch-and-bound procedure which produced an integer solution, of cost 29. From these 88 generated columns, 7 were related to non-zero variables, corresponding to 7 rings with capacity 12 in the final solution. All mesh expansions were reduced to zero.

Having obtained the results of the branch-and-bound with the generated columns, we launched the branch-and-price method described in Section 11.5. In both cases, we were able to find an optimum solution.

We considered two different values for the ring capacity. These results, together with the ones for the 7-nodes network, are summarized in Tables 11.2 and 11.3.

CONCLUSIONS

The main advantage of our method is the generation of new ring constraints. An extension of this study considering ring combinations is considered in (Maculan et al., 1998b).

Acknowledgements This work was partially supported by CNPq, Pronex, France Telecom and FAPERJ.

Table 11.2 Results on the 7-nodes-network

Ring capacity	**12**	**16**
Cost of a pure mesh network	50	50
Linear solution with rings	10.067	8.740
CPU time (sec.)	9	8
Integer solution with rings	12	10
Generated columns	25	24
Active columns	2	2
Total ring cost	10	10
Total mesh cost	2	0
CPU time (sec.)	1	1
Optimum (Branch-and-price)	11	10
Generated columns	43	50
Active column	3	2
Total ring cost	11	10
Total mesh cost	0	0
Open nodes	3	4
Closed nodes	3	5
CPU time (sec.)	10	7

Table 11.3 Results on the 10-nodes-network

Ring capacity	**12**	**16**
Cost of a pure mesh network	152	152
Linear solution with rings	24.403	21.958
CPU time (sec.)	48	57
Integer solution with rings	29	27
Generated columns	88	78
Active Columns	7	6
Total Ring Cost	29	27
Total Mesh Cost	0	0
CPU time (sec.)	6	3
Optimum (Branch-and-price)	27	24
Generated columns	4450	11104
Active column	6	6
Total ring cost	27	24
Total mesh cost	0	0
Open nodes	281	436
Closed nodes	540	685
CPU time (hours)	2	2.5

References

Sutter, A. and F. Vanderbeck, L. W.: 1998, 'Optimal Placement of Add/Drop Multiplexers: Heuristic and Exact Algorithms'. *Operations Research* **46**(5), 719–728.

Barnhart, C., E. L. Johnson, G. L. Nemhauser, M. W. P. Savelsbergh, and P. H. Vance: 1998, 'Branch-and-price: Column Generation for Solving Huge Integer Programs'. *Operations Research* **46**, 316–329.

Brito, J. A. M.: 1999, 'Um Modelo de Otimização para Dimensionamento de uma Rede de Telecomunicações'. Tese de mestrado, COPPE, Universidade Federal do Rio de Janeiro, Rio de Janeiro, Brasil.

Campello, R. E. and N. Maculan: 1994, *Algoritmos e Heurísticas: Desenvolvimento e Avaliação de Performance.* Niterói, Brasil: EDUFF.

Chvátal, V.: 1983, *Linear programming.* New York / San Francisco: W. H. Freeman and Company.

Dantzig, G. and P. Wolfe: 1960, 'Decomposition principle for linear programming'. *Operations Research* **8**, 101–111.

Dantzig, G. B.: 1963, *Linear programming and extensions.* New Jersey, USA: Princeton University Press.

Gilmore, P. and R. Gomory: 1961, 'A Linear Programming Approach to the Cutting Stock Problem'. *Operations Research* **9**, 849–859.

Gilmore, P. and R. Gomory: 1963, 'A linear programming approach to the cutting stock problem - part II'. *Operations Research* **11**, 863–888.

Glover, F., J. P. Kelly, and M. Laguna: 1995, 'Genetic Algorithms and Tabu Search: Hybrids for Optimization'. *Computers & Operations Research* **22**, 111–134.

Hochbaum, D. S. (ed.): 1997, *Approximation algorithms for NP-hard problems.* PWS Publishing Company.

Lasdon, L. S.: 1970, *Optimization Theory for Large Systems.* New York, USA: Macmillan.

Lisser, A., N. Maculan, and M. Passini: 1996a, 'A Mathematical Model for Solving Design Problems of Telecommunication Networks'. In: *INFORMS.* Atlanta, USA.

Lisser, A., N. Maculan, and M. Passini: 1996b, 'Modelling and Solving Design Problems of Telecommunication Network Using Two Different Architectures'. In: *Conférence Internationale en Recherche Opérationnelle.* Marrakech, Maroc.

Lisser, A., R. Sarkassian, and J.-P. Vial: 1995, 'Survivability in Transmission Telecommunication Networks'. Technical report, NT/PAA/ATR/ORI, CNET, France Télécom. To appear in *Investigación Operativa.*

Maculan, N., M. Fampa, and P. Michelon: 1999, 'Programação Linear e Inteira'. Notes - COPPE/Universidade Federal do Rio de Janeiro.

Maculan, N., M. Passini, and I. Loiseau: 1998a, 'Column-Generation in Integer Linear Programming'. Technical report, COPPE, Universidade Federal do Rio de Janeiro, Rio de Janeiro, Brasil.

Maculan, N., M. Passini, and A. Silvério: 1998b, 'Caracterização de um Emaranhado de Anéis em uma Rede de Telecomunicações'. Technical report, COPPE, Universidade Federal do Rio de Janeiro, Rio de Janeiro, Brasil.

Minoux, M.: 1987, 'Network Synthesis and Dynamic Network Optimization'. *Annals of Discrete Mathematics* **31**, 283–324.

Passini, M. M.: 1996, 'Um Modelo de Otimização Combinatória para o Dimensionamento de uma Rede Urbana de Telecomunicações'. Master's thesis, COPPE, Universidade Federal do Rio de Janeiro.

Savelsbergh, M.: 1997, 'A Branch-and-price Algorithm for the Generalized Assignment Problem'. *Operations Research* **46**, 831–841.

Szwarcfiter, J. and P. Lauer: 1975, 'A Search Strategy for the Elementary Cycles of a Directed Graph'. *BIT* **16**, 192–204.

Vachani, R., A. Shulman, and P. Kubat: 1996, 'Multicommodity Flows in Ring Networks'. *INFORMS Journal on Computing* **8**, 235–242.

12 COMPUTING FIXED POINTS BY AVERAGING

Thomas L. Magnanti
Georgia Perakis

Abstract: Averaging methods for solving fixed point problems combine the underlying fixed point map T with some "well-behaved" map g. The map g might, for example, be contractive or might be a nonexpansive map whose fixed points include those of the original map T. One class of averaging methods (inside averaging) averages any current iterate with its image under the map g, and then applies the map T. Another averaging method (outside averaging) first applies the maps T and g and then takes averages. When g is the identity map, outside averaging averages a given point with its image under the map T. In this paper we summarize a number of known results concerning these averaging methods, including (i) a general averaging framework that approximates the original fixed point problem with a trajectory of averaged (parametrized) fixed point subproblems, and (ii) a procedure for following these trajectories approximately to ease the computations.

12.1 INTRODUCTION

Fixed point problems arise in numerous practical and methodological contexts, including application areas as diverse as transportation, communication networks, mechanical systems, electric power systems, regional planning, and economic equilibria. We study the solution of fixed point problems defined on a closed, convex (constraint or ground) subset K of R^n. Given a map $T : K \subseteq R^n \to K$, the fixed point problem $FP(T, K)$ seeks

$$x^* \in K \subseteq R^n \text{ satisfying } T(x^*) = x^*. \tag{12.1}$$

The fixed point problem is closely related with the variational inequality problem, which for a given a map $f : K \subseteq R^n \to R^n$, seeks

$$x^* \in K \subseteq R^n \text{ satisfying } f(x^*)^t(x - x^*) \geq 0, \quad \forall x \in K. \qquad VI(f,K) \tag{12.2}$$

M. Gendreau and P. Marcotte (eds.), Transportation and Network Analysis: Current Trends, 181–198.

For any positive constant ρ, the projection operator $T = Pr_K(I - \rho f)$ is one possible way to model a variational inequality as a fixed point problem. Therefore, any algorithm for solving fixed point problems becomes an algorithm for solving variational inequality problems as well.

The central idea of averaging has been (i) to combine the underlying fixed point map with an auxiliary map (with special properties) so that the resulting hybrid map is better behaved than the original map, and (ii) to be able to recover the solution of the original problem from the solution to the problem with the hybrid map (perhaps by solving a sequence of hybrid maps). The methodology for achieving this objective is averaging, in various ways, the given map and the auxiliary map. The literature on solving fixed point problems by averaging contains a large array of algorithmic contributions. Examples include Bauschke (1994), Browder (1965, 1967), Browder and Petryshn (1967), Dunn (1973), Halpern (1967), Kaniel (1971), Epelman, Magnanti and Perakis (1998), and Magnanti and Perakis (1997, 1998a, 1998b).

As is well known, when the fixed point map T is a contraction, the classical function iteration $x_{k+1} = T(x_k)$ converges to a fixed point solution. Even when applied to well-behaved maps such as rotations, this algorithm need not converge. If, however, we choose x_{k+1} as the average of the current point and its image as in $x_{k+1} = \frac{1}{2}x_k + \frac{1}{2}T(x_k)$, then for rotations at least, the algorithm does converge. This observation prompts several natural questions:

- When does averaging converge?
- Is it possible to develop efficient averaging methods or use averaging to improve the convergence behavior of known algorithms?
- Are there different ways to average, and if so, are they related?
- Is there a unifying underlying theory of averaging that will synthesize known applications of this methodology?

This review offers partial answers to these questions. It is intended to provide a broad and intuitive summary of research results, many developed in our own prior research. The cited references contain more details including proof of any results.

12.2 A GENERAL AVERAGING FRAMEWORK

Example 1 To motivate the key idea of averaging, consider the rotation $T(x^1, x^2) = (x^2, -x^1)$ shown in Figure 12.1. When applied to this map, the classical function iteration cycles and so does not find the fixed point solution. How might we remedy this behavior? As a preliminary idea, suppose we attempt to "regularize" the map by defining a new map $F(x) = \frac{1}{2}x + \frac{1}{2}T(x)$. Note that $\|F(x)\| = \frac{\sqrt{2}}{2}\|x\|$. The new map is a contraction since $\|F(x)\| \leq a\|x\|$ for some constant $0 < a < 1$, and so, as shown in Figure 12.1, the classical (Banach) function iteration $x_{k+1} = F(x_k)$ converges to its unique fixed point. The map F has the same fixed points as the map T and so by finding a fixed point of this map we find a fixed point of T.

Geometrically, when applying the map F, function iteration chooses the next iterate x_{k+1} to be the midpoint of the segment connecting the current iterate x_k and its image $T(x_k)$. More generally, for any constant $0 < \lambda < 1$, we could define the

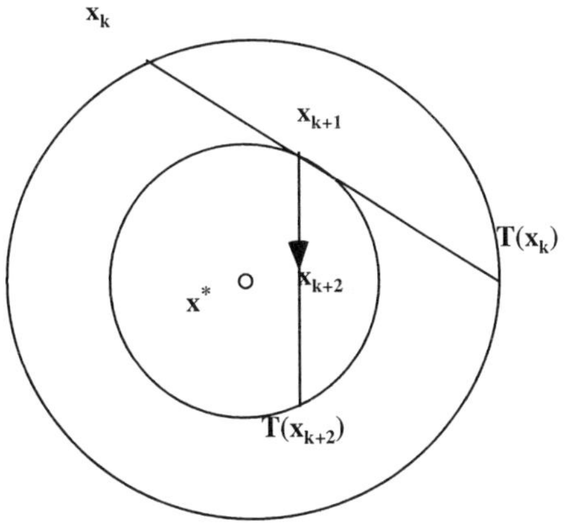

Figure 12.1 Averaging for a Rotation

contraction $F_\lambda(x) = (1-\lambda)x + \lambda T(x)$. When applied to this map, function iteration again will converge to a fixed point of the map T. Now function iteration moves a given fraction λ of the way along the segment connecting the point x and its image $T(x)$.

As a more general and flexible version of this approach, we might consider a sequence of maps $F_{\lambda_k}(x)$ and apply one step of function iteration to each map so that $x_{k+1} = (1-\lambda_k)x_k + \lambda_k T(x_k)$ (see Dunn (1973), and Magnanti and Perakis (1997)). In this case, at each step k, the algorithm moves a different fraction λ_k of the way along the segment connecting the point x_k and its image $T(x_k)$. For our rotation example, this scheme becomes $x_{k+1} = (1-\lambda_k)x_k + \lambda_k(x_k^2, -x_k^1)$. By choosing the parameters λ_k so that any iterate lies far "enough" from the previous iterate as well as far enough from the image of the previous iterate under the map T, this algorithm will also converge to a fixed point of the map T (see Figure 12.1). One possible choice is $\lambda_k = \frac{1}{k}$, and so $x_{k+1} = \frac{k-1}{k}x_k + \frac{1}{k}T(x_k)$, which implies that $x_{k+1} = \frac{1}{k}(x_0 + T(x_1) + T(x_2) + ... + T(x_k))$. Another possibility is to choose λ_k at each step by optimizing a potential function, for example, by minimizing the norm of the difference between the point $(1-\lambda_k)x_k + \lambda_k T(x_k)$ and its image under the map T (see Section 12.3.1).

Will a similar averaging scheme find a fixed point to more general fixed point maps T that do not have as nice a structure as a rotation? In Section 12.3.1 we characterize such schemes. To introduce other types of averaging, we next examine an example with multiple fixed point solutions.

Example 2 The fixed point problem defined by the map $T(x) = (2 - x^1, x^2)$ and the ground set $K = [0,2] \times [-1,1]$ has a set $FP(T) = \{(1, x^2) : x^2 \in [-1,1]\}$ of multiple solutions. Starting from the point $x_1 = (\frac{3}{2}, \frac{1}{2})$, the classical function iteration, $x_{k+1} = T(x_k)$, cycles by inducing the iterates $x_2 = T(x_1) = (\frac{1}{2}, \frac{1}{2})$ and $x_3 = T(x_2) = (\frac{3}{2}, \frac{1}{2}) = x_1$. Figure 12.2 illustrates this behavior.

To remedy this behavior, observe that by damping the map T by any constant (parameter) $0 < \lambda < 1$, we create a contractive map $\lambda T(x_\lambda)$. Moreover, the new

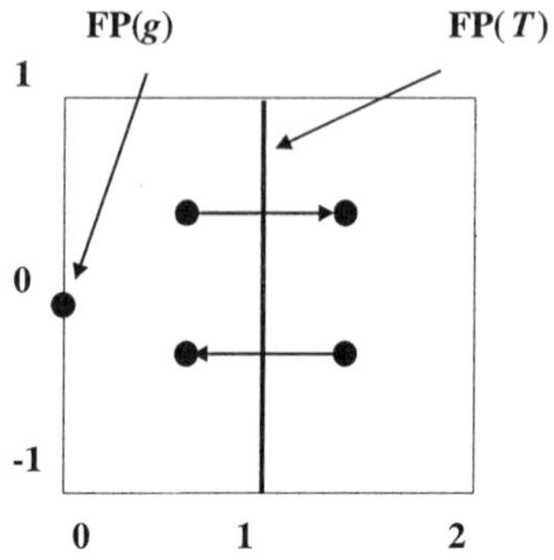

Figure 12.2 Cycling

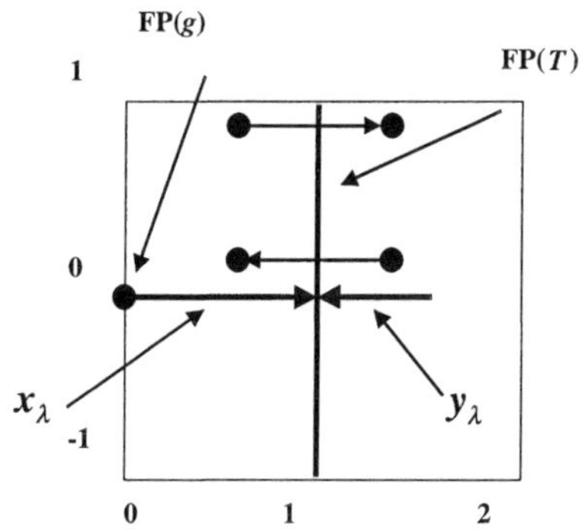

Figure 12.3 Averaging Trajectories

parametrized problem $x_\lambda = \lambda T(x_\lambda)$ has the unique fixed point solution $x_\lambda = (\frac{2\lambda}{1+\lambda}, 0)$ which approaches a fixed point of the map T as $\lambda \to 1$. This method is a special case of a more general averaging method that for any fixed constant $x_0 = (x_0^1, x_0^2) \in K$ and any constant $0 < \lambda < 1$ creates the fixed point problem $x_\lambda = (1-\lambda)x_0 + \lambda T(x_\lambda)$ which, in this example, defines a trajectory $x_\lambda = (\frac{2\lambda+(1-\lambda)x_0^1}{1+\lambda}, x_0^2)$ that again approaches a fixed point of the map T as $\lambda \to 1$. An even more general scheme uses any contractive map $g(x)$ and the averaging trajectory $x_\lambda = (1-\lambda)g(x_\lambda) + \lambda T(x_\lambda)$. Suppose, for example, that $g(x) = \frac{x}{2}$ which has a fixed point $x^* = (0,0)$. Then $x_\lambda = (\frac{4\lambda}{1+3\lambda}, 0)$. If we choose the contractive map as $g(x) = 0$ or as $g(x) = x_0$, then the general scheme becomes the damping method (due to Halpern (1967)) or the method for averaging with a constant (which is due to Wittmann (1992)). The most general scheme is due to Epelman, Magnanti and Perakis (1998). Figure 12.3 illustrates these trajectories.

The previous algorithmic methods take averages by weighting the given map with an auxiliary map. Alternatively, we might first average a given point with an auxiliary map and then apply the map T, leading to the averaging trajectory $y_\lambda = T((1-\lambda)g(y_\lambda) + \lambda y_\lambda)$ which for the choice $g(x) = \frac{x}{2}$ generates the iterates $y_\lambda = (\frac{4}{3+\lambda}, 0)$ (see Epelman, Magnanti and Perakis (1998)).

As λ approaches one, although the trajectories defined by the three x_λ schemes and the y_λ scheme we have considered are quite different, the end result is much the same. All of these methods remedy the cycling behavior of the classical function iteration. In fact, if $x_0^2 = 0$ for the second x_λ scheme, each trajectory approaches the same fixed point solution x^* of the original map T, namely, the minimum norm fixed point solution $x^* = Pr_{FP(T)}g(x^*) = (1,0)$. And even when $x_0^2 \neq 0$, the second x_λ scheme $x_\lambda = (1-\lambda)x_0 + \lambda T(x_\lambda)$ still approaches the fixed point solution $x^* = Pr_{FP(T)}g(x^*) = (1, x_0^2)$.

As an alternative, we could have adopted the approach of the first example by averaging with the identity map $g(x) = x$. However, in this case, for any nonzero parameter λ, the subproblem $x_\lambda = (1-\lambda)x_\lambda + \lambda T(x_\lambda)$ becomes the same as the original fixed point problem.

Computing trajectories of parametrized fixed points might be as hard as computing the original fixed point solution. For this reason, suppose we instead consider approximations of these trajectories by applying only one step of the classical function iteration toward the solution of each parametrized fixed point subproblem. In this example, we could consider the approximation $x_{k+1} = (1-\lambda_k)g(x_k) + \lambda_k T(x_k)$ which in the case of $g(x) = x/2$ becomes $x_{k+1} = (1-\lambda_k)(\frac{x_k^1}{2}, \frac{x_k^2}{2}) + \lambda_k(2 - x_k^1, x_k^2)$ or, in the case of the second trajectory, becomes $y_{k+1} = T((1-\lambda_k)g(y_k) + \lambda_k y_k) = (2-(1-\lambda_k)\frac{y_k^1}{2} - \lambda_k y_k^1, (1-\lambda_k)\frac{y_k^2}{2} + \lambda_k y_k^2)$. For appropriate choices of the step lengths (weights) λ_k (see Section 12.3.3), these trajectories converge to a fixed point of the map T. The previous examples suggest several questions:

Is the behavior in these examples a coincidence? Is there a general unifying averaging framework that generates trajectories of fixed points approaching a fixed point of the original problem and that includes the trajectories in the previous examples as special cases? What is the nature of the limit points of these trajectories? What choices of the step sizes λ_k ensure convergence? Are there efficient ways of choosing step sizes that will lead to more efficient schemes? How large a class of problems can we solve? What are the weakest possible conditions we might impose, weaker than nonexpansiveness?

12.2.1 A General Method

To develop a more general framework that contains each of the previous averaging schemes as special cases, suppose that we consider a scheme defined by a sequence of parameterized fixed point subproblems using a generalized averaging map F_λ. That is, we wish to solve the following fixed point subproblem for any choice of $0 < \lambda < 1$:

$$x_\lambda = F_\lambda(x_\lambda). \tag{12.3}$$

With the choice $F_\lambda(x_\lambda) = (1-\lambda)x_\lambda + \lambda T(x_\lambda)$, this generalized averaging method becomes averaging with the identity. More generally, when $F_\lambda(x_\lambda) = (1-\lambda)g(x_\lambda) + \lambda T(x_\lambda)$, it becomes outside averaging. As another special case, when $F_\lambda(x_\lambda) = T((1-\lambda)g(x_\lambda) + \lambda x_\lambda)$, the method becomes inside averaging.

We wish to show that as the parameter λ approaches one, the fixed points generated by general averaging (12.3) converge to a fixed point of the original map T. To do so, we need to impose some conditions on the generalized averaging problem (12.3) that will ensure that it approaches the original fixed point problem in some regular (continuous) sense. We are naturally led to impose the following conditions on the parameters $\lambda \in \Lambda \subseteq R$ and the function $F_\lambda(x) : \Lambda \times K \to K$:

C1. For all $\lambda \in \Lambda = [0,1]$, the parameterized problem $x_\lambda = F_\lambda(x_\lambda)$ has a solution.

C2. The function $F_\lambda(x)$ is a continuous function of λ and x.

C3. $FP(F_1) = FP(T)$.

The following preliminary result follows easily.

Proposition 1 *(Epelman, Magnanti and Perakis, 1998): When K is a closed and convex feasible region and the fixed point map T is continuous, then as the parameter λ approaches one, every limit point of the sequence $\{x_\lambda\}$, if one exists, is a fixed point solution of the map T.*

To guarantee that the sequence of iterates induced by the averaging framework converges and to also characterize the resulting limit point, we need to impose additional assumptions. Rather than attempting to analyze the most general form of the generalized averaging map F_λ, we will assume that it is a composite map of the form $F(G_\lambda)$. The generalized maps for inside and outside averaging are of this form: for inside averaging, $F = T$ and $G_\lambda = \lambda I + (1-\lambda)g$, while for outside averaging, $F = I$ and $G_\lambda = \lambda T + (1-\lambda)g$. We will also impose some conditions on the problems for limiting values of the parameter λ, namely, (i) at the limiting value of $\lambda = 1$, the fixed points of the outside map F include those of the original map T, that is, $FP(T) \subseteq FP(F)$, and (ii) as λ approaches the limit $\lambda = 1$, when evaluated at the fixed point x_λ of the composite map F_λ, the point $G_\lambda(x_\lambda)$ "looks like" x_λ (that is, $x_\lambda \approxeq G_\lambda(x_\lambda)$) so that $x_\lambda = F_\lambda(x_\lambda) = F(G_\lambda(x_\lambda)) \approxeq F(x_\lambda)$. Indeed, we will assume that $G_\lambda(x_\lambda)$ approaches x_λ faster than quadratically as λ approaches one. We also impose some modest conditions on the maps G_0 which corresponds to the map g in inside and outside averaging (conditions C6a and C6b formalize these conditions).

In Example 2, we noted that the trajectories generated by the various averaging schemes $x_\lambda = F_\lambda(x_\lambda)$ we considered each converged to a limit point $\bar{x}$ satisfying the condition $x^* = Pr_{FP(T)}g(x^*)$. Since for these examples $G_0(x) = g(x)$, they all satisfy the condition $x^* = Pr_{FP(T)}G_0(x^*)$ or, equivalently, they solve the variational inequality $(\bar{x} - G_0(\bar{x}))^t(x - \bar{x}) \geq 0$, for all $x \in FP(T)$. To establish this condition in general, we will impose a further technical condition on the problem, namely, an upper bounding (convexity like) condition on the map $G_\lambda(x)^t(x - x^*)$ stating that it always lies below the line segment with end points $G_1(x)^t(x-x^*)$ and $G_0(x)^t(x-x^*)$.

C4. The map F_λ is of the form $F_\lambda = F(G_\lambda)$, with $F : K \to K$, $G_\lambda : \Lambda \times K \to K$ continuous functions of x, and $\lambda \in \Lambda = [0,1]$. In addition,

- F is a nonexpansive map and $FP(F) \supseteq FP(T)$
- G_1 is a nonexpansive map and $FP(G_1) \supseteq FP(T)$
- $\lim_{\lambda \to 1} \frac{\|G_\lambda(x_\lambda) - x_\lambda\|^2}{1-\lambda} = 0$

C5. $G_\lambda(x)^t(x - x^*) \leq \lambda G_1(x)^t(x - x^*) + (1-\lambda)G_0(x)^t(x - x^*)$,
for all $\lambda \in [0,1]$, all $x \in K$ and for all $x^* \in FP(T)$.

C6a. $G_0(x)$ is a nonexpansive map whose fixed points contain those of the map T, or

C6b. $G_0(x)$ is a contractive map with contraction constant $a \in (0,1)$.

These additional conditions allow us to further characterize the limit points of the sequence induced by parametric fixed point problem (12.3).

Theorem 1 *(Epelman, Magnanti and Perakis, 1998): Let T be a nonexpansive fixed point map and K be a convex, closed ground set.*

1. *Suppose conditions C1–C6a are valid. Then as λ approaches one, every limit point $\bar{x}$ of the sequence $\{x_\lambda\}$ is a fixed point solution of $FP(T, K)$ satisfying the condition*

$$\bar{x} \in FP(T): \quad (\bar{x} - G_0(\bar{x}))^t(x - \bar{x}) \geq 0, \quad \text{for all} \quad x \in FP(T).$$

2. *Suppose conditions C1–C5 and C6b are valid. Then as λ approaches one, the sequence $\{x_\lambda\}$ converges to a fixed point solution x^* of $FP(T, K)$.*

Remark Theorem 1 considers two approaches. The first approach, described in part 1, considers averaging with a map G_0 that is nonexpansive but whose fixed point solutions contain those of the original map T. The second approach, described in part 2, relaxes this latter requirement by considering a "stronger" averaging map G_0 that is contractive but whose fixed point solutions do not necessarily relate to those of the orginal map T.

Example 3 Consider outside averaging $F_\lambda(x) = (1 - \lambda)g(x) + \lambda T(x)$ with a specific contractive map $g(x) = ax$ for some constant $0 < a < 1$. Let $F = I$ so that $F_\lambda(x) = G_\lambda(x)$. Since the map $g = G_0$ is contractive, part 2 of Theorem 1 implies that the limit of the sequence x_λ that the general averaging framework induces satisfies the condition $(1 - a)\,(\bar{x})^t(x - \bar{x}) \geq 0, \;\text{for all}\; x \in FP(T)$ which are the first order (necessary and sufficient) optimally condition for the minimum norm problem defined on the set of fixed points solutions of the original problem $FP(T, K)$.

It is instructive to consider several examples of the maps $F_\lambda(\cdot)$.

Averaging with the identity map: $F_\lambda(x) = (1 - \lambda)x + \lambda T(x)$. In this case $F = I$ and $G_\lambda = (1 - \lambda)I + \lambda T$. Dunn (1973), Magnanti and Perakis (1997, 1998a, 1998b) have studied this type of averaging.

Outside averaging with a contractive map: $F_\lambda(x) = (1 - \lambda)g(x) + \lambda T(x)$. In this case, $F = I$ and $G_\lambda = (1 - \lambda)g + \lambda T$.
Halpern (1967) and Browder (1965, 1967) introduced, and Bauschke (1994) and Wittmann (1992) further studied, a special case of this map (with the constant map $g(x) = c$, for some constant c).

Inside averaging with a contractive map: $F_\lambda(x) = T((1 - \lambda)g(x) + \lambda x)$. In this case, $F = T$ and $G_\lambda = (1 - \lambda)g + \lambda I$.
The next two examples use the notion of a proximal point map. Given a map H (in our case map $H = I - T$), a proximal point map is the resolvent of map H which is the map $J_{cH} = (I + cH)^{-1}$ (see Rockafellar (1976)).

Outside averaging with the proximal point map: $J_{c(I-T)}$ $F_\lambda(x) = \lambda(I + c(I - T))^{-1}(x)$. In this case, $F = I$ and $G_\lambda = \lambda J_{c(I-T)} = \lambda(I + c(I - T))^{-1}$.

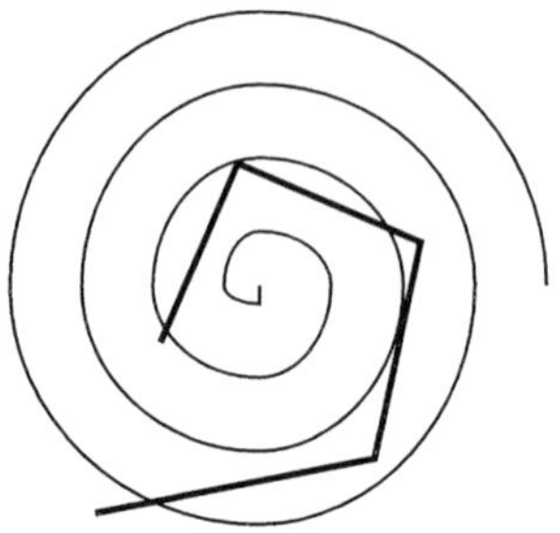

Figure 12.4 Approximate Parametric Trajectory

Inside averaging with the proximal point map: $J_{c(I-T)}\ F_\lambda(x) = (I + c(I - T))^{-1}(\lambda x)$. In this case, $F = J_{c(I-T)}$ and $G_\lambda = \lambda I$. Note that the last two methods are analogous to outside and inside averaging of the map $J_{c(I-T)}$ with $g \equiv 0$ respectively. Like in the inside and outside averaging methods, we could apply averaging with a more general contractive map g.

Convex combinations of any F_λ maps that satisfy the conditions C1-C6a or C1-C5 and C6b.

Averaging with a family of contractive maps. In the previous examples, we can replace the map g with a family of maps $g_\lambda : \Lambda \times K \to K$ that are contractive for every fixed λ and Lipschitz continuous for every fixed x.
Epelman, Magnanti and Perakis (1998) examine these examples in further detail.

12.3 AN APPROXIMATE AVERAGING FRAMEWORK

The general averaging framework discussed so far has one significant drawback: each step requires the solution of a parametrized fixed point subproblem. Unfortunately, this task might be as hard as solving the original fixed point problem. Consequently, rather than tracing out the trajectory of parametrized fixed point subproblem solutions exactly, we might attempt to following the trajectory only "closely" as in the following approximate averaging framework,

$$x_{k+1} = F_{\lambda_k}(x_k). \tag{12.4}$$

At each step, this method takes a single step of the function iteration procedure toward solving the parameterized fixed point subproblem and then alters the value of λ.

We will consider the application of this approximation procedure in several different contexts. In particular, we will first consider several variations of the first approach (that is, averaging with a nonexpansive map whose fixed points contain those of map T) described in part 1 of Theorem 1.

12.3.1 Averaging with the Identity Map

Perhaps the most straightforward type of approximate averaging $x_{k+1} = F_{\lambda_k}(x_k)$ is averaging with the identity map, which is a specific form of outside averaging with the map $g(x) = x$, so that

$$x_{k+1} = (1 - \lambda_k)\, x_k + \lambda_k T(x_k) = x_k + \lambda_k (T(x_k) - x_k).$$

This approach to averaging (line search procedures) has received much attention in the literature. Our next result characterizes the limit of the sequence that this type of averaging induces, as a fixed point solution, using a predetermined choice of step sizes. These step sizes ensure that the new iterate is far "enough" from the previous iterate and from its image under the fixed point map (that is, the step sizes do not approach the values zero or one too quickly).

Theorem 2 *(Dunn, 1973, Magnanti and Perakis, 1997): Consider a fixed point problem defined on a closed, convex subset K of R^n. Assume it has a solution. If the underlying map T is a nonexpansive map on K relative to the $\|.\|_G$ norm, the sequence $x_{k+1} = x_k + \lambda_k(T(x_k) - x_k)$ converges to a fixed point of the map T whenever $\sum_{k=1}^{\infty} \lambda_k(1 - \lambda_k) = +\infty$. This fixed point is also the limit of the projection of the points x_k on the set of fixed points of map T.*

As is easy to see in the rotation in Figure 12.1, if the step sizes approach zero or one too quickly, then averaging with the identity need not converge. The choice $\lambda_k = \frac{1}{k}$ is an example of step sizes satisfying the condition $\sum_{k=1}^{\infty} \lambda_k(1 - \lambda_k) = +\infty$. Magnanti and Perakis (1997) illustrate how Theorem 2 applies to the solution of variational inequality problems by using averaging with the identity map when applying projection, relaxation and linearization type of methods to solve the variational inequality problems. They invoke versions of a condition called strong-f-monotonicity (also refered to as co-coercivity), rather than the condition of strong monotonicity typically made in the literature. They also demonstrate similar results for solving nonlinear systems of equations.

Averaging by Optimizing Potentials. Averaging methods using predetermined choices of step sizes like $\lambda_k = \frac{1}{k}$ have one significant disadvantage: they often converge slowly to a solution. Therefore, we might pose the following question: *Can we compute step sizes λ_k so that the resulting algorithms will be more efficient?* To achieve this goal, we might consider a class of algorithms that compute step sizes λ_k and so $x_{k+1} = x_k(\lambda) \equiv x_k + \lambda_k(T(x_k) - x_k)$ at each step by optimizing a potential function g.

Adaptive Averaging Framework

$$\lambda_k = argmin_{\{\lambda \in S\}} g(x_k(\lambda)).$$

A natural choice of the potential $g(x_k(\lambda))$ might be $\|x_k(\lambda) - T(x_k(\lambda))\|^2$ which would find the point that lies closest to its image under the map T. In our discussion to follow, we provide several other examples for the potential $g(x_k(\lambda))$. We assume the step size search set S is a subset of R. Examples include $S = [0, 1]$, $S = R^+$ and $S = [-1, 0]$.

To better understand the underlying convergence behavior of these averaging schemes, we consider a generalization of the notion of nonexpansiveness of a fixed point map T. If x^* is a fixed point of the map T and $\bar{G}$ is a positive definite, symmetric matrix, for any sequence of step sizes $\{\lambda_k\}$, we define the quantity

$$\Lambda_k(x^*) \equiv \frac{\|x_k - x^*\|_{\bar{G}}^2 - \|T(x_k) - T(x^*)\|_{\bar{G}}^2}{\|x_k - T(x_k)\|_{\bar{G}}^2} + (1 - \lambda_k).$$

Observe that if map T is nonexpansive, then by definition the first term of $\Lambda_k(x^*)$ is nonnegative. If, additionally, $0 \leq \lambda_k \leq 1$, then the entire quantity $\Lambda_k(x^*)$ is nonnegative. Therefore, when $\lambda_k \in [0,1]$, the condition $\Lambda_k(x^*) \geq 0$ provides a generalization of the notion of nonexpansiveness of the fixed point map T, which makes a tradeoff between the choice of step sizes λ_k and nonexpansiveness of the map T. Two key conditions underlie the convergence of the sequence x_k that the adaptive averaging scheme induces:

C7. For any fixed point x^* of the map T, $\lambda_k \Lambda_k(x^*)$ is nonnegative.

C8. For any fixed point x^* of the map T, if $\lambda_k \Lambda_k(x^*)$ converges to zero, then every limit point of the sequence $\{x_k\}$ is a fixed point solution.

To satisfy conditions C7 and C8, we might either (i) restrict the line searches we perform to determine step sizes λ_k in a set $S \subseteq [0,1)$ or $(-1,0]$ so that our results apply for maps that satisfy a condition weaker than nonexpansiveness, or (ii) extend the line searches to a set $S \subseteq [0,1]$ or $[-1,0]$, but as a result we might need to impose stronger assumptions on the map T. For example if $\lambda_k = 1$ and the map T is a contraction, then the condition $\Lambda_k(x^*) = 0$ implies that $x_k = T(x_k)$ is a fixed point solution. Otherwise, the condition implies the contradiction $\|T(x_k) - T(x^*)\|_G^2 = \|x_k - x^*\|_G^2$.

A large class of maps T satisfy conditions C7 and C8, including those that for all $x \in K$ satisfy the condition

$$\|T(x) - T(x^*)\|_G^2 \leq A\|x - x^*\|_G^2 + B\|x - T(x)\|_G^2$$

for constants $B \in [0,1)$ and $A \in (0,1]$. When $A \in (0,1)$ and $B = 0$, the map T is contractive, and when $A = 1$ and $B = 0$, the map T is nonexpansive. When $A = 1$ and $B \in (0,1)$, the map T is nonexpansive-contractive. The latter class of maps includes examples of maps that expand such as $T(x) = -3x$ (see Magnanti and Perakis (1998a), for more details). Moreover, conditions C7 and C8 also apply to a class of maps T that always expand, that is, for all $x \in K$,

$$\|T(x) - T(x^*)\|_G^2 \geq \|x - x^*\|_G^2 + B\|x - T(x)\|_G^2,$$

for some constant $B > 1$. For this class of maps T, conditions C7 and C8 require step sizes $\lambda_k \in S \subseteq [-1,0]$.

The following definition describes a connection between the conditions C7 and C8 and the choice of potentials g.

Definition 1 *A potential g is compatible if the step sizes λ_k it generates, together with map T, satisfy conditions C7 and C8.*

The following theorem establishes a convergence result for the adaptive averaging framework.

Theorem 3 *(Magnanti and Perakis, 1998a): Consider iterates of the type $x_{k+1} = x_k + \lambda_k(T(x_k) - x_k)$, with step sizes λ_k chosen through the optimization of potentials g. If the map T has a fixed point solution x^*, then for potentials g that are compatible, the sequence of iterates $\{x_k\}$ converges to a fixed point solution.*

This theorem includes Theorem 2 as a special case (see Magnanti and Perakis (1998a)).

Examples of potentials g that are *compatible* and step size search sets S include:

- $g_1(x_k(\lambda)) = ||x_k(\lambda) - T(x_k(\lambda))||_G^2 + \beta||x_k(\lambda) - x_k(0)||_G^2$, $\beta \geq 0$ and a step size search set $S = [0,1]$.
- $g_2(x_k(\lambda)) = ||x_k(\lambda) - T(x_k(\lambda))||_G^2$ and a step size search set $S = R^+$.
- $g_3(x_k(\lambda)) = ||x_k(\lambda) - T(x_k(\lambda))||_G^2 - \beta||x_k(\lambda) - x_k(0)||_G^2$ and a step size search set $S = [0, c_1] \subseteq [0, 1)$.
- $g_4(x_k(\lambda)) = ||x_k(\lambda) - T(x_k(\lambda))||_G^2 - \beta||x_k(\lambda) - x_k(0)||_G.||x_k(\lambda) - x_k(1)||_G$, and a step size search set $S = R^+$.
- $g_5(x_k(\lambda)) = [(x_k(\lambda) - T(x_k(\lambda)))^t(x_k - T(x_k))]^2$, and a step size search set $S = R^+$.
- $g_6(x_k(\lambda)) = [(x_k(\lambda) - T(x_k(\lambda))^t G(x_k - T(x_k))]^2 - \beta[(x_k(\lambda) - x_k(0))^t G(x_k - T(x_k))].[(x_k(\lambda) - x_k(1))^t G(x_k - T(x_k))]$, and a step size search set $S = R^+$.
- $g_7(x_k(\lambda)) = -f(x_k(\lambda))^t(T(x_k(\lambda)) - x_k(\lambda)) - \frac{1}{2}||T(x_k(\lambda)) - x_k(\lambda)||_G^2$ and a step size search set $S = [0,1]$.

Magnanti and Perakis (1998a) provide additional details. Other choices of potentials can be found for example in Fukushima (1992), Wu, Florian and Marcotte (1993), and Zhu and Marcotte (1993).

For certain types of maps and some step sizes, the adaptive averaging framework provides a better rate of convergence than classical function iteration. To achieve an improved convergence rate, we need to choose step size λ_k "close" to the "optimal" step size $\lambda_k^* = \frac{(x_k - T(x_k))^t \bar{G}(x_k - x^*)}{||x_k - T(x_k)||_{\bar{G}}^2}$. The step size λ_k^* is "optimal" in the sense that it provides the largest possible decrease of $||x_k(\lambda) - x^*||_{\bar{G}}^2$. However, we cannot compute λ_k^* since it requires the knowledge of a fixed point solution x^*.

Proposition 2 *(Magnanti and Perakis, 1998a): For fixed point problems with maps T satisfying the condition*

$$|\lambda_k^* - 1| = |\frac{(x - T(x))^t \bar{G}(x - x^*)}{||x - T(x)||_{\bar{G}}^2} - 1| \geq c > 0, \quad \text{for all} \quad x \in K \tag{12.5}$$

adaptive averaging schemes with choices of step size λ_k lying with the range

$$\lambda_k^* - |\lambda_k^* - 1| + d \leq \lambda_k \leq \lambda_k^* + |\lambda_k^* - 1| - d, \tag{12.6}$$

with $0 < d < c$, have a better rate of convergence than function iteration.

Examples of maps T that give rise to step lengths satisfying the condition $|\lambda_k^* - 1| \geq c$ include "tightly" nonexpansive maps, that is, for all $x \in K$, $||T(x) - T(x^*)||_{\bar{G}}^2 \approx ||x - x^*||_{\bar{G}}^2$. Magnanti and Perakis (1998a) show that several of the examples of g presented previously generate step sizes within the range in the Proposition 2. They also show that the adaptive averaging framework also applies using inexact line searches.

A Generalization. Several of the examples of *compatible* potentials are of the functional form $g(x_k(a)) = P(x_k(a), T(x_k(a)))^2 - \beta h(a) P(x_k, T(x_k))^2$. To provide a convergence result in this more general setting, we need to impose conditions on the choices of the functions P and h.

We will choose $P : K \times K \to R$ as a function satisfying the following conditions:

- $P(x^*, T(x^*)) = 0$ if and only if x^* is a fixed point solution.
- $P(x, x) = 0$.
- $P(x, T(x)) \geq 0$, $P(x, x^*) \geq 0$ for each fixed point x^* of T.
- P is *convex* relative to the first component, that is, $P(., y)$ is convex for all fixed y, and for all points x_1, z_1 and $a \in R$, some constant $D_1 \geq 0$ satisfies the condition

$$P(ax_1 + (1-a)z_1, y) \leq aP(x_1, y) + (1-a)P(z_1, y) - D_1 a(1-a)P(z_1, x_1).$$

In the remainder of this paper we will refer to the function P as a *generalized norm.* Examples of such functions include $P(x, y) = ||x - y||_G^2$ and $P(x, y) = f(x)^t(x - y)$ or $P(x, y) = (f(x) - f(y))^t(x - y)$ for variational inequality problems $VI(f, K)$. The analogous definition of quantity $\Lambda_k(x^*)$ in this case becomes

$$\Lambda_k(x^*) \equiv \frac{P(x_k, x^*) - P(T(x_k), T(x^*))}{P(x_k, T(x_k))} + (1 - \lambda_k).$$

The results in this section generalize to include maps T that satisfy "nonexpansiveness-type" of conditions with respect to generalized norms P (see Magnanti and Perakis (1998b), for further details). These results also provide weaker convergence results for the Frank-Wolfe and the Steepest Descent methods for solving nonlinear programming problems and, more generally, a descent framework for solving variational inequality problems.

To complete our discussion of potentials $g(x_k(a)) = P(x_k(a), T(x_k(a)))^2 - \beta h(a)$ $P(x_k, T(x_k))^2$, we will impose further assumptions of the continuous function $h : R \to R^+$ and constant $\beta > 0$. Overall the assumption become:

- P is a generalized norm.
- $h(0) = 0$.
- For some point $\bar{\lambda} \in S \cap (0, 1)$, $h(\bar{\lambda}) > 0$.
- h is bounded from above on $S \cap [0, 1]$, that is, $0 < C = sup_{\lambda \in S \cap [0,1]} h(\lambda) < +\infty$.
- $1 < C\beta$.

In settings satisfying these assumptions, Theorem 2 remains valid. The previous results also apply to maps T satisfying "nonexpansiveness" conditions with respect to generalized norms P rather than $||.||_G$

Example 4 The fixed point problem with the map $T(x^1, x^2) = b(x^2, -x^1)$ is expansive when $b > 1$ and "tightly" nonexpansive when $b = 1$. For any choice of $b > 0$, these maps T are always nonexpansive-contractive. Observe that the classical function iteration will not solve the fixed point problem for this map unless the map T is contractive and so $0 < b < 1$. Although the classical function iteration successfully computes the solution $(0, 0)$ when $0 < b < 1$, the adaptive averaging schemes compute the solution faster. For example, when $b = 0.95$, from a starting point $(2, 2.5)$, the adaptive averaging framework with potential g_2 computes the solution (to 10 significant digits) in 67 iterations while the classical function iteration requires 479 iterations. When $b = 0.9$, the adaptive averaging framework with potential g_6 requires 63 iterations while function iteration requires 234. When $b = 0.7$, the adaptive averaging framework with potential g_6 requires 47 iterations while function iteration requires 70. We reach similar conclusions from a variety of starting points and a variety of different values of the parameter b.

Example 5 Bottom et al. (1999) have used fixed point formulations to model an anticipatory route guidance problem in the area of Intelligent Transportation Systems. They use the adaptive averaging framework to solve these problems. The authors used the Boston Central Artery network during the morning traffic period as a test example. Their empirical results showed that the adaptive averaging framework clearly outperforms averaging with step size $\lambda_k = 1/k$ at each iteration. Using the potential g_4 in the adaptive averaging framework proved to be slightly more effective than using averaging with constant step sizes. The classical functional iteration did not converge for these test problems.

12.3.2 Averaging with Nonexpansive Maps whose Fixed Points include those of the Original Map

In this section we examine a generalization of averaging with the identity map by considering an approximate averaging framework with a general averaging map F_λ that is nonexpansive and whose fixed points include those of the original map T. The map $F_\lambda = I + \lambda(T - I)$, for averaging with the identity, is a special case.

As before, we assume that $F_\lambda(x)$ is a continuous function of x and λ, mapping $\Lambda \times K \times K$ into K. We impose several conditions on F_λ that extend conditions C1-C6 in Section 12.2:

C9. $FP(F_1) = FP(T)$.

C10. $F_\lambda = F(G_\lambda)$

- F is a nonexpansive map and $FP(T) \subseteq FP(F)$.
- G_0 is a nonexpansive map and $FP(T) \subseteq FP(G_0)$.
- G_1 is a nonexpansive map and $FP(T) \subseteq FP(G_1)$.

C11. For all $y \in K$ and $\lambda \in [0, 1]$,
$G_\lambda(x^*)^t(y - x^*) \leq \lambda G_1(x^*)^t(y - x^*) + (1 - \lambda)G_0(x^*)^t(y - x^*)$, $\forall x^* \in FP(T)$.

C12. Either F is a contractive map and G_λ is a nonexpansive map: $||G_\lambda(x) - G_\lambda(x^*)||^2 \leq ||x - x^*||^2 \; \forall x \in K$ and $\forall x^* \in FP(T)$
or F is a nonexpansive map, $FP(G_1) = FP(T)$ and some $\gamma(\lambda) \in R^+$ satisfies the condition,
$||G_\lambda(x) - G_\lambda(x^*)||^2 \leq ||x - x^*||^2 - \gamma(\lambda)||x - G_1(x)||^2, \; \forall x \in K.$

C13. For some $L > 0$, $||G_\lambda(x) - G_{\bar{\lambda}}(x)|| \leq L|\lambda - \bar{\lambda}|, \; \forall x \in K, \; \forall \lambda, \bar{\lambda} \in \Lambda.$
The step sizes λ_k satisfy the conditions

C14. The $\lim_k \lambda_k = 1$ and $\sum_k \gamma(\lambda_k) = +\infty.$

As in Section 12.2, we will consider the general iteration

$$x_{k+1} = F_{\lambda_k}(x_k). \tag{12.7}$$

Condition **C12** holds, for example, when map G_1 is firmly nonexpansive, that is, for all $x \in K$, $||G_1(x) - G_1(x^*)||^2 \leq ||x - x^*||^2 - ||x - G_1(x)||^2$ (see Epelman, Magnanti and Perakis (1998)). The following theorem summarizes a convergence result.

Theorem 4 *(Epelman, Magnanti and Perakis, 1998): Conditions C9–C14 imply that the entire sequence $\{x_k\}$ induced by the averaging scheme (12.7) converges to a fixed point solution.*

This result applies to the algorithmic extensions of the examples we considered in Section 12.2.

Outside Averaging: $x_{k+1} = (1 - \lambda_k)g(x_k) + \lambda_k T(x_k).$

We require the map T to be *firmly nonexpansive*, the map g to be *nonexpansive*, $FP(g) \supseteq FP(T)$, and step sizes $\lambda_k \in (0,1)$ with $\sum_k \lambda_k(1 - \lambda_k) = +\infty.$
Observe that when $g(x) = x$, this scheme reduces to averaging with the identity.

Inside Averaging: $x_{k+1} = T((1 - \lambda_k)g(x_k) + \lambda_k x_k).$

We require the map T to be *contractive*, the map g to be *nonexpansive*, $FP(g) \supseteq FP(T)$, and step sizes $\lambda_k \in (0,1)$ with $\sum_k \lambda_k(1 - \lambda_k) = +\infty.$

Outside Averaging with the Proximal Point Mapping: $J_{c(I-T)}$ $x_{k+1} = (1 - \lambda_k)g(x_k) + \lambda_k(I + c(I - T))^{-1}(x_k).$

We require the map T to be *nonexpansive*, the map g to be *nonexpansive*, $FP(g) \supseteq FP(T)$, and step sizes $\lambda_k \in (0,1)$ with $\sum_k (1 - \lambda_k)^2 = +\infty.$

Inside Averaging with the Proximal Point Mapping: $J_{c(I-T)}$ $x_{k+1} = (I + c(I - T))^{-1}((1 - \lambda_k)g(x_k) + \lambda_k x_k).$
We require the map T to be *nonexpansive*, the map g to be *nonexpansive*, $FP(g) \supseteq FP(T)$, and step sizes $\lambda_k \in (0,1)$ with $\sum_k \lambda_k(1 - \lambda_k) = +\infty.$

Convex Combinations of each of the previous examples.
Epelman, Magnanti and Perakis (1998) examine these examples in further detail.

12.3.3 Averaging with Contractive Maps

So far, we have examined approximate trajectories for general averaging maps F_λ that are nonexpansive and whose fixed points include those of the original map T. In this subsection we examine the second approach in Theorem 1, that is, we consider a general averaging map F_λ that is contractive. The averaging map $F_\lambda(x)$ is again defined on the set $\Lambda \times K \to K$. Using a new parameter $\beta(\lambda)$, we will impose conditions C15-C17 on this function that are relaxed versions of conditions C1-C3.

C15. $FP(F_1) = FP(T)$.

C16. $||F_\lambda(x) - F_{\bar{\lambda}}(x)|| \leq |\lambda - \bar{\lambda}|L(x)$, $\forall x \in K$ and for some continuous function $L : K \to R_+$.

C17. $F_\lambda(x)$ is a continuous function of (x, λ), and is contractive for all $\lambda \in [0, 1)$, that is, for all x, $y \in K$, $||F_\lambda(x) - F_\lambda(y)|| \leq \beta(\lambda)||x - y||$, with $\beta(\lambda) \geq 0$, $0 < \beta(1) \leq 1$ and $\beta'(1) > 0$ and $\beta(\lambda) \leq \beta(1) + \beta'(1)(\lambda - 1)$.

Note that condition C17 is valid if $\beta''(\lambda) \leq 0$, that is, $\beta(\lambda)$ is a concave function. Consider step sizes λ_k satisfying:

C18. $\lambda_k \to 1$, $\sum_k \lambda_k = +\infty$, $\sum_k |\lambda_k - \lambda_{k+1}| < +\infty$.

Wittmann (1992) has used the same choice of step sizes for the special case of outside averaging with a constant map.

C19. $L(x)$ is bounded over the feasible set K (for example, the set K is bounded).

Observe that when $F_1(x) = T$, the condition $0 < \beta(1) \leq 1$ captures the nonexpansiveness of the map T. Moreover, conditions $\beta'(1) > 0$ and $\beta(\lambda) \leq \beta(1) + \beta'(1)(\lambda - 1)$ imply that $\beta(\lambda) < 1$ for all $\lambda \in [0, 1)$. Consequently, F_λ is a contractive map for every fixed $\lambda \in [0, 1)$ and F_λ is a nonexpansive map for every fixed $\lambda \in [0, 1]$. For inside and outside averaging, these assumptions imply the contractiveness of the map g, which implies that for every fixed $\lambda \in [0, 1]$, $F_\lambda(x)$ is a continuous function of x (see condition C17 above).

Several iterative schemes that are the algorithmic extensions of the examples that we considered in Section 12.2 satisfy properties C15-C17.

Outside Averaging: $x_{k+1} = (1 - \lambda_k)g(x_k) + \lambda_k T(x_k)$.

Inside Averaging: $x_{k+1} = T((1 - \lambda_k)g(x_k) + \lambda_k x_k)$.

Outside Averaging with the Proximal Point Mapping: $J_{c(I-T)}$ $x_{k+1} = \lambda_k(I + c(I - T))^{-1}(x_k)$.

Inside Averaging with the Proximal Point Mapping: $J_{c(I-T)}$ $x_{k+1} = (I + c(I - T))^{-1}(\lambda_k)x_k$.

Convex Combinations of each of the previous examples.

Epelman, Magnanti and Perakis (1998) analyze these schemes further. The following preliminary result is an easy consequence of these conditions.

Theorem 5 *(Epelman, Magnanti and Perakis, 1998): Conditions C15-C19 imply that every limit point, if any, of the sequence $\{x_k\}$ induced from the approximate averaging framework is a fixed point solution.*

This theorem ensures that the limit points of the sequence induced from the approximate averaging framework are, indeed, fixed points of T. Can we characterize the limiting fixed points for this approximate averaging framework, as we did in Section 12.2? To do so, we consider a stronger version of conditions C15-C19, obtained by replacing conditions C16 and C17 with the following conditions:

C16′ For all $x \in K$, $||G_\lambda(x) - G_{\bar{\lambda}}(x)|| \leq |\lambda - \bar{\lambda}|L(x)$, for some continuous function $L(x)$ of x.

C17′ $F_\lambda = F(G_\lambda)$.

- The map F is nonexpansive and $FP(F) \supseteq FP(T)$.
- The map G_1 is nonexpansive, $FP(G_1) \supseteq FP(T)$ and G_λ is a contraction for all $\lambda \in [0, 1)$, that is,

$$||G_\lambda(x) - G_\lambda(y)|| \leq \beta(\lambda)||x - y||, \ \forall x, y \in K$$

  with $\beta(\lambda) \geq 0$, $0 < \beta(1) \leq 1$, $\beta'(1) > 0$ and $\beta(\lambda) \leq \beta(1) + \beta'(1)(\lambda - 1)$.
- For all $y \in K$ and all $\lambda \in [0, 1]$,

$$G_\lambda(x^*)^t(y-x^*) \leq (1-\lambda)G_0(x^*)^t(y-x^*)+\lambda G_1(x^*)^t(y-x^*), \ \forall x^* \in FP(T).$$

Conditions C16′ and C17′ imply C16 and C17.

Theorem 6 *(Epelman, Magnanti and Perakis 1998): Conditions C15, C16′, C17′, C18, and C19 imply that the entire sequence $\{x_k\}$ induced by the iterative scheme (12.4) converges to a fixed point solution x^* satisfying the property $x^* = Pr_{FP(T)}G_0(x^*)$.*

12.3.4 Solving Mathematical Programming Problems by Averaging

The fixed point problem is closely related to several problem classes in mathematical programming: linear and nonlinear optimization as well as variational inequality problems. We will illustrate this connection by showing how averaging results apply to solving linear programs and other optimization problems.

As we noted in Section 12.1, we can solve variational inequalities defined over a closed, convex (constraint) set K in R^n, that is,

$$VI(f, K): \ \text{find} \ \ x^* \in K \subseteq R^n : \ f(x^*)^t(x - x^*) \geq 0, \ \forall x \in K. \tag{12.8}$$

by solving a fixed point problem over the set K using the problem map $T = Pr_K(I - \rho f)$ for some positive constant ρ. The next proposition shows that the averaging results we have considered apply to variational inequality problems. Let $\lambda_{min}(G)$ denote the minimum eigenvalue of a matrix G.

Proposition 3 *(Magnanti and Perakis, 1998a): Let G be a positive definite and symmetric matrix, $f : K \subseteq R^n \rightarrow R^n$ be a given function and ρ be a given constant satisfying the condition $\rho(\frac{\rho}{\lambda_{min}(G)} - 2l) \leq 0$. Suppose that either (i) f is strongly-f-monotone with constant $l > 0$, that is, for all $x \in K$, $(f(x) - f(x^*))^t(x - x^*) \geq l||f(x) - f(x^*)||^2$, or (ii) for all $x \in K$, $(f(x) - f(x^*))^t(x - x^*) \leq l||f(x) - f(x^*)||^2$ for some constant $l \neq 0$. Then the map $T = Pr_K^G(I - \rho G^{-1} f)$ is nonexpansive with respect to the $||.||_G$ norm.*

This proposition allows us to apply the averaging framework to solve variational inequalities defined by strongly-f-monotone maps as well a class of variational inequality problems that are nonmonotone.

Since the map $f(x) = c$ is strongly-f-monotone for any $l > 0$, Proposition 3 provides us with one means for applying averaging to solve linear programs

$$min_{P=\{x \in R^n:\ Ax \geq b, x \geq 0\}} c^t x$$

defined by an $m \times n$ matrix A and vectors c and b. We simply choose $K = P$ and $f = c$. However, the projection computation $T = Pr_K(x - \rho c)$ requires the solution of a quadratic optimization problem, which is computationally expensive relative to solving the linear program itself, and so this approach does not appear to be particularly attractive. Is there an alternate formulation of the problem that might be more attractive computationally?

Suppose we rewrite the linear optimization problem in the following well known form of a nonlinear "linear complementarity" system

$$c - A^t p = s, \quad x^t s = 0, \quad Ax - b \geq 0, \quad p^t(Ax - b) = 0, \quad x \geq 0, \quad s \geq 0, \quad p \geq 0,$$

which, in turn, is equivalent to the variational inequality problem $VI(f, R_+^{n+m})$, with $f(x,p) = \begin{bmatrix} 0 & -A^t \\ A & 0 \end{bmatrix} \begin{bmatrix} x \\ p \end{bmatrix} + \begin{bmatrix} c \\ -b \end{bmatrix}$. For this reformulation, the map f is always monotone. For any positive constant ρ, we can cast any linear optimization problem as the fixed point problem $FP(T, R_+^{n+m})$ with $T(x) = Pr_{R_+^{n+m}}(x - \rho f(Pr_{R_+^{n+m}}(x - \rho f(x))))$. Since $(Pr_{R_+^{n+m}}(z))_i = \max(0, z_i)$ for every component i, computing the map T is for this formulation is "easy." In fact, it is possible to show that the fixed point map T is nonexpansive. This observation allows us to use averaging results for solving linear optimization problems.

In closing, we note that we can use averaging in a variety of other ways to solve optimization problems. For example, by formulating the first order optimality conditions of nonlinear optimization problems $\min\{F(x) : x \in K\}$ as variational inequality problems $VI(\nabla F, K)$, under appropriate condition (e.g., convexity), we can again use Proposition 1 to formulate the problem as an fixed point problem. Another example of averaging is the noted method of Julia Robinson (see Robinson (1951)) for solving two person zero sum games (which can be cast as linear programs), though the connection between this method and the material presented in this paper remains an open question.

References

Bauschke H.H. (1994). The approximation of fixed points of compositions of nonexpansive mappings in Hilbert space. *Preprint.* Center for Experimental and Constructive Mathematics, Simon Fraser University, Burnaby, British Columbia, Canada V5A 1S6.

Bottom J., Ben-Akiva M., Bierlaire M., Chabini I., Koutsopoulos H. and Yang Q. (1999). Investigation of route guidance generation issues by simulation with DynaMIT. *Proceedings of 14th International Symposium on Transportation and Traffic Theory*, Jerusalem.

Browder F.E. (1965). Fixed-point theorems for noncompact mappings in Hilbert space. *Proc. Nat. Acad. Sci. U.S.A.*, 53: 1272-1276.

Browder F.E. (1967). Convergence of approximants to fixed points of nonexpansive nonlinear mappings in Banach spaces. *Archive for Rational Mechanics and Analysis*, 24:82–90.

Browder F. E. and Petryshn W.V. (1967). Construction of fixed points of nonlinear mappings in a Hilbert space. *Journal of Analysis and Applications*, 20:197-228.

Dunn J.C. (1973). On the recursive averaging processes and Hilbert space extensions of the contraction mapping principle. *Journal of the Franklin Institute*, 295:117-133.

Epelman M., Magnanti T.L. and Perakis G. (1998). Finding fixed points by averaging with well-behaved maps. *preprint.*

Fukushima M. (1992). Equivalent differentiable optimization problems and descent methods for asymmetric variational inequality problems. *Mathematical Programming.* 53: 99-110.

Halpern B. (1967). Fixed points of nonexpanding maps. *Bulletin of the American Mathematical Society*, 73:957–961

Kaniel S. (1971). Construction of a fixed point for contractions in Banach space. *Israel Journal of Mathematics*, 9:535–540.

Magnanti T.L. and Perakis G. (1997). Averaging schemes for variational inequalities and systems of equations. *Mathematics of Operations Research*, 22(3):568-587.

Magnanti T.L. and Perakis G. (1998a). Solving variational inequalities and fixed point problems by averaging and optimizing potentials. *preprint.*

Magnanti T.L. and Perakis G. (1998b). From Frank-Wolfe to steepest descent; a descent framework for solving asymmetric variational inequalities. *preprint.*

Rockafellar R.T. (1976). Monotone operators and the proximal point algorithm. *SIAM Journal on Control and Optimization*, 14(5):877–898.

Robinson J. (1951). An iterative method of solving a game. *Annals of Mathematics*, 54,2:296-301.

Wittmann R. (1992). Approximation of fixed points of nonexpansive mappings. *Archiv der Mathematik*, 58:486–491.

Wu J.H., Florian M., and Marcotte P. (1993)). A general descent framework for the monotone variational inequality problem. *Mathematical Programming*, 61:281–300.

Zhu D.L., and Marcotte P. (1993). Modified descent methods for solving the monotone variational inequality problem. *Operations Research Letters*, 14:111–120.

13 POLLUTION PERMITS FOR SPATIAL PRICE NETWORKS

Anna Nagurney

Abstract: In this paper, we develop a spatial price network equilibrium model with marketable pollution permits using the theory of variational inequalities. The model has the distinctive feature that it allows for the imposition of environmental quality standards for both the producers at the supply markets as well as on the transportation links. The permits are distinct and recognize that producers may emit pollutants that are different from those that are emitted during transportation. We provide qualitative properties of the equilibrium supply, shipment, demand, marginal cost of emission abatement, license, and license price pattern as well as a numerical scheme for the computation of the equilibrium pattern. We conclude with several numerical examples for illustrative purposes.

13.1 INTRODUCTION

Environmental pollution is a topic of increasing concern and has captured the interests of many constituencies on regional, national, and global levels. The development of appropriate environmental policy instruments for pollution or emission control is, hence, of both theoretical interest as well as practical importance.

In this paper, we consider the economic-incentive approach of marketable pollution permits as a mechanism for pollution amelioration. The idea of pollution permits can be traced to Crocker (1966) and Dales (1968) and was theoretically formulated as a mathematical problem in the case of firms which are fixed in location by Montgomery (1972). Montgomery (1972) explicitly recognized the spatial nature of pollution dispersion. Nagurney and Dhanda (1996) extended the basic framework of Montgomery's to include oligopolistic behavior of the firms and studied the problem both qualitatively and computationally as a variational inequality problem. They subsequently incorporated transaction costs (cf. Nagurney and Dhanda (1997a)) as well as technological investments into the models (cf. Nagurney and Dhanda (1997b)).

M. Gendreau and P. Marcotte (eds.), Transportation and Network Analysis: Current Trends, 199–220.

Recently, Nagurney, Ramanujam, and Dhanda (1998) introduced marketable pollution permits in the case of transportation networks. These permits were link-based. Subsequently, Nagurney (1999) proposed both origin/destination pair based as well as path-based permit systems. Permit systems for transportation networks, in contrast to those for firms, involve travelers which are mobile. The work established that as in the case of marketable pollution permits for firms, the environmental quality standard can be achieved provided that the initial license allocation is set accordingly.

In this paper, we turn to pollution permits in the context of spatial price networks. We present a model for marketable pollution permits in the structure of the spatial price network equilibrium problem (cf. Samuelson (1952), Takayama and Judge (1971), Florian and Los (1982), Dafermos and Nagurney (1987)). The model includes emission permits for both the producing or supply regions as well as for the transportation links or routes. Note that emissions discharged by producers may differ from those discharged by the shippers of the commodities and, consequently, a system of marketable pollution permits that handles both production and transportation emissions offers a more complete framework for environmental policy and decision-making.

In particular, the model that we develop allows the policy decision-maker to impose environmental quality standards on both the supply regions as well as on the shipments of the commodity to the demand regions. This model, hence, explicitly recognizes the duality of pollution generation by both producers as well as transporters and makes the distinction between the pollutant emissions that they may generate. It is clear that at regional/national/international levels both producers as well as transporters may add to pollution emissions and, therefore, a complete policy framework must address both of these generators. The system of marketable pollution permits presented in this paper provides flexibility on the part of the agents in the economy in that they can trade permits while guaranteeing that the imposed environmental standards will be met.

The paper is organized as follows. In Section 13.2, we develop the pollution permit spatial network equilibrium model, derive the equilibrium conditions, and formulate them as a variational inequality problem. We also present some qualitative properties of the model that may be useful for policy-makers.

In Section 13.3, we present additional qualitative results, specifically those that are needed for establishing the convergence of the algorithmic scheme in Section 13.4. In Section 13.5, we provide numerical results for illustrative purposes.

13.2 THE SPATIAL PRICE NETWORK MODEL WITH POLLUTION PERMITS

In this section, we develop a spatial network policy model in which pollution, due to emissions, generated by either the producers and/or the transporters is restrained by the incorporation of emission pollution licenses or permits. The market-based fee, or equilibrium license price, for these emission permits stimulates the reduction in production and/or transportation and, hence, pollution. Note that the system of marketable pollution permits allows those with advanced emission technology to sell off their permits to those that may need them. For simplicity of notation and presentation, we assume that there is only a single commodity produced, and that the producers emit a single pollutant whereas the shippers generate another pollutant while the commodity is being transported.

We now describe the spatial network model. The model is based on the spatial price equilibrium models of Samuelson (1952), Takayama and Judge (1971), Florian and Los (1982), and Dafermos and Nagurney (1987) (see also Nagurney (1987, 1999)). We consider m supply markets involved in the production of a homogeneous commodity and n demand markets. We assume that these markets are spatially separated. We denote a typical supply market by i and a typical demand market by j. Let s_i denote the supply at supply market i and let d_j be the demand at demand market j. We group the supplies into a column vector $s \in R^m_+$ and the demands into a column vector $d \in R^n_+$. Let Q_{ij} denote the nonnegative commodity shipment between supply and demand market pair (i,j), and group the commodity shipments into a column vector $Q \in R^{mn}_+$.

The commodity shipments and the supplies and demands must satisfy the following conservation of flow equations:

$$s_i = \sum_{j=1}^{n} Q_{ij}, \quad i = 1, \ldots, m \tag{13.1}$$

and

$$d_j = \sum_{i=1}^{m} Q_{ij}, \quad j = 1, \ldots, n, \tag{13.2}$$

that is, the supply at each supply market must be equal to the sum of the commodity shipments from that market to all the demand markets, and the demand at each demand market must be equal to the sum of the commodity shipments from all supply markets to each demand market.

We now describe the price and cost structure. Let π_i denote the supply price at supply market i and we let ρ_j the demand price at demand market j. We group the supply prices into a column vector $\pi \in R^m$ and the demand prices into a column vector $\rho \in R^n$. The unit transaction/transportation cost associated with shipping the commodity between supply market i and demand market j is denoted by c_{ij}. We group the transportation costs into a column vector $c \in R^{mn}$.

For a graphical depiction of the spatial network, see Figure 13.1.

We assume that the supply price at a supply market may, in general, depend upon the supplies of the commodity at every supply market, that is,

$$\pi = \pi(s) \tag{13.3}$$

where π is a smooth function.

Similarly, the demand price at a demand market may, in general, depend upon the demands for the commodity at every demand market, that is,

$$\rho = \rho(d) \tag{13.4}$$

where ρ is a smooth function.

The unit transaction/transportation cost, in turn, associated with shipping the commodity between a pair of supply and demand markets is assumed to be, in general, a function of the entire commodity shipment pattern, that is,

$$c = c(Q). \tag{13.5}$$

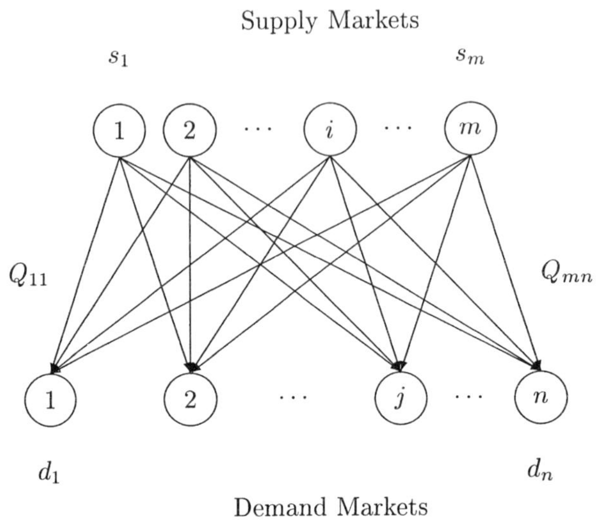

Figure 13.1 The spatial price network

In the absence of a marketable pollution permit system and under the assumption of perfect competition, the well-known spatial price equilibrium conditions are recalled here for easy reference:

Definition 1 (Spatial Price Equilibrium)
Following Samuelson (1952) and Takayama and Judge (1971), the supply, commodity shipment, and demand pattern (s^*, Q^*, d^*) *constitutes a spatial price equilibrium, if it is feasible with respect to* (13.1) *and* (13.2), *and for all pairs of supply and demand markets* (i, j), *it satisfies the conditions:*

$$\pi_i(s^*) + c_{ij}(Q^*) \begin{cases} = \rho_j(d^*), & \textit{if} \quad Q^*_{ij} > 0 \\ \geq \rho_j(d^*), & \textit{if} \quad Q^*_{ij} = 0. \end{cases} \tag{13.6}$$

Hence, if the commodity shipment between a pair of supply and demand markets is positive at equilibrium, then the demand price at the demand market must be equal to the supply price at the originating supply market plus the unit transaction/transportation cost. If the commodity shipment is zero in equilibrium, then the supply price plus the transaction/transportation cost can exceed the demand price.

It is well-known (cf. Florian and Los (1982) and Nagurney (1999) for a proof and additional references) that the spatial price equilibrium can be formulated as a variational inequality problem. Precisely, we have

Theorem 1 (Variational Inequality Formulation - Spatial Price Equilibrium)
A commodity supply, shipment, and demand pattern $(s^*, Q^*, d^*) \in K$ *is a spatial price equilibrium if and only if it satisfies the following variational inequality problem:*

$$\langle \pi(s^*), s - s^* \rangle + \langle c(Q^*), Q - Q^* \rangle - \langle \rho(d^*), d - d^* \rangle \geq 0, \quad \forall (s, Q, d) \in K, \tag{13.7}$$

where $K \equiv \{(s, Q, d) : (13.1)\text{–}(13.2) \textit{ hold}\}$ *and* $\langle \cdot, \cdot \rangle$ *denotes the inner product in Euclidean space.*

Recall that (cf. Samuelson (1952) and Takayama and Judge (1971)) in the case of symmetric and positive semidefinite Jacobian matrices of the supply price, transportation cost, and minus demand price functions $\left[\frac{\partial \pi}{\partial s}\right]$, $\left[\frac{\partial c}{\partial Q}\right]$, and $-\left[\frac{\partial \rho}{\partial d}\right]$, the spatial price equilibrium can be obtained as the solution of a convex minimization problem.

We now introduce the notation for the permit system. We first discuss the notation associated with the permits for the supply markets and then for the transportation links (or routes). We assume that both in the case of the producers and in the case of the transporters that the estimation of pollution can be modeled as the product of an emission factor times, respectively, the quantity of the commodity produced or the quantity of the commodity transported (see, for example, Glover and Brzezinski (1989), DeCorla-Souza, et al. (1995), Anderson, et al. (1996), and Allen (1996)). Our definition of the permits or licenses follows that of Montgomery (1972) who also, as we do, explicitly considered the spatial diffusion of pollution.

Let h_i denote the emission factor associated with supply market i and let l_i denote the number of license permits for supply market i that allows the suppliers to emit at a certain rate. Let l_i^0 denote the initial allocation of the licenses for supply market i. Similarly, let h_{ij} denote the emission factor associated with transporting the commodity from supply market i to demand market j and let l_{ij} denote the number of licenses or permits associated with transporting the commodity from i to j. Finally, let l_{ij}^0 denote the initial license allocation for the shipment route from i to j. We group the supply market emission licenses into the vector $l \in R_+^m$ and the transportation emission licenses into the vector $\lambda \in R_+^{mn}$.

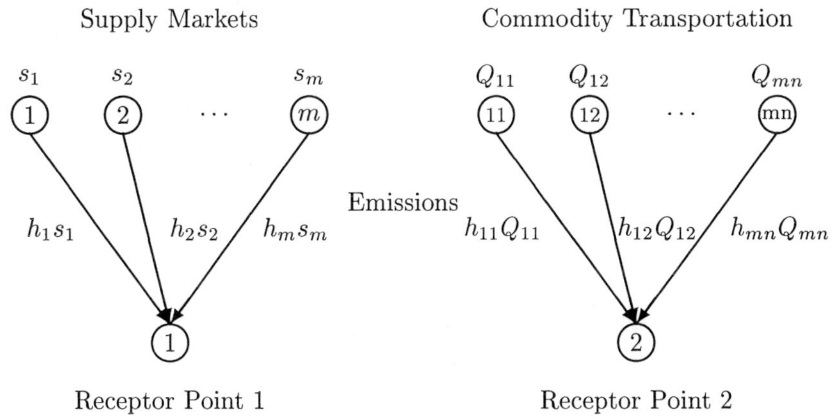

Figure 13.2 The emissions network for the spatial price problem

In Figure 13.2, we depict the emissions network for the production as well as the transportation problem. Note that in this model we assume that the producers, in effect, emit a distinct pollutant from the transporters. Moreover, we assume that there is a single receptor point for the production emissions and a single receptor point for the transportation emissions. The following inequalities guarantee, respectively, that the total amount of emissions from the supply markets, and from the shipment routes, will not exceed the amounts permitted by the license holdings:

$$h_i s_i \leq l_i, \quad \forall i \tag{13.8}$$

and

$$h_{ij} Q_{ij} \leq l_{ij}, \quad \forall i, j. \tag{13.9}$$

Moreover, the total number of licenses on the production side as well as on the transportation side cannot exceed the initial allocation of the respective licenses, that is, the following inequalities must also be satisfied:

$$\sum_{i=1}^{m} l_i \leq \sum_{i=1}^{m} l_i^0 \tag{13.10}$$

and

$$\sum_{i=1}^{m} \sum_{j=1}^{n} l_{ij} \leq \sum_{i=1}^{m} \sum_{j=1}^{n} l_{ij}^0. \tag{13.11}$$

We now discuss the price and cost structure associated with the marketable pollution permit system. Their relationship will become more apparent when we develop the equilibrium conditions. We let p_s denote the emission license price for the supply markets and we let p_Q denote the emission license price for the transportation of the commodity. Also, we let τ_i denote the marginal cost of emission abatement associated with supply region i and we let τ_{ij} denote the marginal cost of emission abatement associated with shipping the commodity from i to j. We group the supply market marginal costs of abatement into the vector $t \in R_+^m$ and the transportation marginal costs of abatement into the vector $\tau \in R_+^{mn}$.

In particular, we associate with the constraint (13.8), for each supply market i, the marginal cost of emission abatement τ_i, which, in equilibrium, must satisfy the following system: For each i:

$$h_i s_i^* \begin{cases} = & l_i^*, & \text{if } \tau_i^* > 0 \\ \leq & l_i^*, & \text{if } \tau_i^* = 0. \end{cases} \tag{13.12}$$

In other words, if the marginal cost of emission abatement, τ_i^*, is positive in equilibrium for a supply market i, then the emissions by that supply market are precisely equal to the pollution license holdings for that supply market; if the number of licenses exceeds the emissions of a supply market, then, in equilibrium, the marginal cost of abatement is zero.

The following condition must also be met at equilibrium: For each supply market i:

$$\tau_i^* \begin{cases} = & p_s^*, & \text{if } l_i^* > 0 \\ \leq & p_s^*, & \text{if } l_i^* = 0. \end{cases} \tag{13.13}$$

Hence, in equilibrium, a positive holding of licenses for a supply market implies that the marginal cost of abatement must be equal to the price of the license. However, if the price of the license exceeds the marginal cost of abatement, then the number of licenses for that supply market will be zero.

Similarly, we associate with constraint (13.9), for each transportation link or route (i, j), the marginal cost of emission abatement τ_{ij}, which, in equilibrium, must satisfy the following system: For each link (i, j):

$$h_{ij}Q^*_{ij} \begin{cases} = & l^*_{ij}, & \text{if } \tau^*_{ij} > 0 \\ \leq & l^*_{ij}, & \text{if } \tau^*_{ij} = 0. \end{cases} \tag{13.14}$$

Thus, if the marginal cost of emission abatement, τ^*_{ij}, is positive in equilibrium for the transportation link (i, j), then the emissions on that transportation link are precisely equal to the pollution license holdings for that link; if the number of licenses exceeds the emissions on that transportation link, then, in equilibrium, the marginal cost of abatement on that transportation link is zero.

The following condition must also be met at equilibrium: For each transportation link or route (i, j):

$$\tau^*_{ij} \begin{cases} = & p^*_Q, & \text{if } l^*_{ij} > 0 \\ \leq & p^*_Q, & \text{if } l^*_{ij} = 0. \end{cases} \tag{13.15}$$

Hence, in equilibrium, a positive holding of licenses for a transportation route implies that the marginal cost of emission abatement must be equal to the price of the license. However, if the price of the license exceeds the marginal cost of abatement, then the number of licenses for that link will be zero.

Finally, the equilibrium price, p^*_s, of a production emission license must satisfy the following equilibrium condition:

$$\sum_{i=1}^{m}(l^0_i - l^*_i) \begin{cases} = & 0, & \text{if } p^*_s > 0 \\ \geq & 0, & \text{if } p^*_s = 0. \end{cases} \tag{13.16}$$

Indeed, expression (13.16) corresponds to the well-known economic equilibrium conditions that state that, in equilibrium, if a price of a good (which in this case is the license) is positive, then the market for that good must clear, that is, the supply of the production licenses, which is equal to $\sum_{i=1}^{m} l^0_i$, must be equal to the demand for the production licenses in equilibrium, which is given by $\sum_{i=1}^{m} l^*_i$. On the other hand, if the price of a license is zero, then we may have an excess supply of the licenses.

Similarly, we must have for the transportation emission licenses that the transportation emission license price p^*_Q, in equilibrium, must satisfy the following condition:

$$\sum_{i=1}^{m}\sum_{j=1}^{n}(l^0_{ij} - l^*_{ij}) \begin{cases} = & 0, & \text{if } p^*_Q > 0 \\ \geq & 0, & \text{if } p^*_Q = 0. \end{cases} \tag{13.17}$$

We let $\mathcal{K}$ denote the feasible set such that $\mathcal{K} \equiv K \times R^{2m+2mn+2}_+$, since the licenses, the marginal costs of emission abatement, and the license prices are all nonnegative.

The cost of a commodity produced at supply market i and consumed at demand market j is not only equal to the supply price plus the transportation or transaction cost but it is now subject to the true cost of emissions and, hence, the *generalized* cost of a commodity produced at supply market i and shipped to demand market j, denoted by $\hat{c}_{ij}$, is given by:

$$\hat{c}_{ij}(s, Q\ t, \tau) = \pi_i(s) + h_i\tau_i + c_{ij}(Q) + h_{ij}\tau_{ij}.$$

Consequently, the spatial price equilibrium conditions now take the form: For each pair of supply and demand markets (i,j):

$$\hat{c}_{ij} = \pi_i(s^*) + h_i\tau_i^* + c_{ij}(Q^*) + h_{ij}\tau_{ij}^* \begin{cases} = & \rho_j(d^*), & \text{if } Q_{ij}^* > 0 \\ \geq & \rho_j(d^*), & \text{if } Q_{ij}^* = 0. \end{cases} \tag{13.18}$$

Indeed, equilibrium conditions (13.18) state that the price of a commodity at a demand market, if consumed, is equal to the supply price at the originating supply market plus the unit transportation cost plus the emissions costs incurred by the supply market and on the transportation link. Note that here the consumers "pay" for the costs of emissions associated with production and transportation in the form of the demand price at the demand market, if they purchase from that particular supply market.

In subsequent propositions we establish, first, that if a commodity is produced at a supply market, then the supply market emission license price in equilibrium is precisely equal to the marginal emission abatement cost at the supply market and, second, if a commodity is shipped between a pair of supply and demand markets, then the transportation emission license price in equilibrium is equal to the marginal emission cost of abatement associated with transporting the commodity on that link.

We are now ready to define the equilibrium state:

Definition 2 (Spatial Price Permit Market Equilibrium)
A vector $(s^*, Q^*, d^*, t^*, \tau^*, l^*, \lambda^*, p_s^*, p_Q^*) \in \mathcal{K}$ *is an equilibrium of the spatial price network equilibrium emission permits market model if and only if it satisfies the systems of equalities and inequalities* (13.12)–(13.18).

Remark 1 The spatial price permit market equilibrium can be reformulated as a convex minimization problem under the same assumptions given immediately following Theorem 1 and subject to constraints (13.8), (13.9), (13.10), and (13.11), the conservation of flow equations (13.1), (13.2), as well as the nonnegativity assumptions on the commodity shipments (and supplies and demands) and the licenses. Hence, τ_i then is the Lagrange multiplier associated with constraint (13.8) for firm i; τ_{ij}, in turn, is the Lagrange multiplier associated with constraint (13.9) for market pair (i,j); p_s is the Lagrange multiplier associated with constraint (13.10), and p_Q is the Lagrange multiplier associated with constraint (13.11). In this special case the equilibrium conditions coincide with the Kuhn-Tucker conditions.

In the environmental economics literature (cf. Montgomery (1972) and Tietenberg (1985), for example), where cost efficiency of various policy instruments is emphasized the Lagrange multipliers such as τ_i and τ_{ij} (cf. (13.12) and (13.14)) are given the interpretation of marginal costs of emission abatement. Of course, one could also give them the interpretation of the marginal value to the supplier or the shipper, respectively, of another unit of the allowed pollution rate. Conditions (13.13) and (13.15), in turn, then reveal that the cost of emission abatement for the suppliers (and for the shippers) is precisely equal to the price of the license to pollute, if the supplier (shipper) holds a positive amount of the license in equilibrium. Note that conditions (13.12) through (13.15) and the above interpretations are still relevant whether or not we have an optimization reformulation of the spatial price permit market equilibrium model. Conditions (13.16) and (13.17), in turn, are simply the economic equilibrium conditions governing the supply of and demand for the licenses.

We now derive the variational inequality formulation of the equilibrium conditions for the model. We then consider a special case.

Theorem 2 (Variational Inequality Formulation - Spatial Price Permit Market Model)
A vector of commodity supplies, shipments, demand, and marginal costs of emission abatement, license, and license prices $(s^*, Q^*, d^*, t^*, \tau^*, l^*, \lambda^*, p_s^*, p_Q^*) \in \mathcal{K}$ *is an equilibrium of the spatial price network equilibrium problem with marketable pollution permits if and only if it is a solution to the variational inequality problem:*

$$\sum_{i=1}^{m}(\pi_i(s^*)+h_i\tau_i^*)\times(s_i-s_i^*)+\sum_{i=1}^{m}\sum_{j=1}^{n}(c_{ij}(Q^*)+h_{ij}\tau_{ij}^*)\times(Q_{ij}-Q_{ij}^*)$$
$$-\sum_{j=1}^{n}\rho_j(d^*)\times(d_j-d_j^*)$$
$$+\sum_{i=1}^{m}(l_i^*-h_is_i^*)\times(\tau_i-\tau_i^*)+\sum_{i=1}^{m}\sum_{j=1}^{n}(l_{ij}^*-h_{ij}Q_{ij}^*)\times(\tau_{ij}-\tau_{ij}^*)$$
$$+\sum_{i=1}^{m}(p_s^*-\tau_i^*)\times(l_i-l_i^*)+\sum_{i=1}^{m}\sum_{j=1}^{n}(p_Q^*-\tau_{ij}^*)\times(l_{ij}-l_{ij}^*)$$
$$+\sum_{i=1}^{m}(l_i^0-l_i^*)\times(p_s-p_s^*)+\sum_{i=1}^{m}\sum_{j=1}^{n}(l_{ij}^0-l_{ij}^*)\times(p_Q-p_Q^*)\geq 0,$$
$$\forall(s,Q,d,t,\tau,l,\lambda,p_s,p_Q)\in\mathcal{K}. \tag{13.19}$$

Proof: We first establish that a solution to the equilibrium conditions (13.12)–(13.18) satisfies variational inequality (13.19).

Equilibrium conditions (13.18) imply that, for a fixed pair of markets (i, j):

$$(\pi_i(s^*)+h_i\tau_i^*+c_{ij}(Q^*)+h_{ij}\tau_{ij}^*-\rho_j(d^*))\times(Q_{ij}-Q_{ij}^*)\geq 0,\quad \forall Q_{ij}\geq 0. \tag{13.20}$$

Indeed, if $Q_{ij}^*=0$, then $(Q_{ij}-Q_{ij}^*)\geq 0$ and according to (13.18): $(\pi_i(s^*)+h_i\tau_i^*+c_{ij}(Q^*)+h_{ij}\tau_{ij}^*-\rho_j(d^*))\geq 0$ and, hence, (13.20) holds true. If $Q_{ij}^*>0$, then note that the term $(Q_{ij}-Q_{ij}^*)$ may be positive, negative, or zero, depending on Q_{ij}, but since, according to (13.18), in this case, $(\pi_i(s^*)+h_i\tau_i^*+c_{ij}(Q^*)+h_{ij}\tau_{ij}^*-\rho_j(d^*))=0$, inequality (13.20) again holds true.

Summing now (13.20) over all market pairs (i, j) and using (13.1) and (13.2) yields:

$$\sum_{i=1}^{m}(\pi_i(s^*)+h_i\tau_i^*)\times(s_i-s_i^*)+\sum_{i=1}^{m}\sum_{j=1}^{n}(c_{ij}(Q^*)+h_{ij}\tau_{ij}^*)\times(Q_{ij}-Q_{ij}^*)$$
$$-\sum_{j=1}^{n}\rho_j(d^*)\times(d_j-d_j^*)\geq 0,\quad \forall(s,Q,d)\in K. \tag{13.21}$$

From (13.12), in turn, we have that, for a fixed supply market i,

$$(l_i^*-h_is_i^*)\times(\tau_i-\tau_i^*)\geq 0,\quad \forall\tau_i\geq 0, \tag{13.22}$$

and, similarly, from (13.14) we have that, for a fixed transportation link (i,j):

$$(l_{ij}^* - h_{ij}Q_{ij}^*) \times (\tau_{ij} - \tau_{ij}^*) \geq 0, \quad \forall \tau_{ij} \geq 0. \tag{13.23}$$

Summing now (13.22) over all supply markets and (13.23) over all transportation links (i,j), and adding the resultants yields:

$$\sum_{i=1}^{m}(l_i^* - h_i s_i^*) \times (\tau_i - \tau_i^*) + \sum_{i=1}^{m}\sum_{j=1}^{n}(l_{ij}^* - h_{ij}Q_{ij}^*) \times (\tau_{ij} - \tau_{ij}^*) \geq 0. \tag{13.24}$$

Equilibrium condition (13.13), in turn, implies that, for a fixed supply market i

$$(p_s^* - \tau_i^*) \times (l_i - l_i^*) \geq 0, \quad \forall l_i \geq 0, \tag{13.25}$$

whereas equilibrium condition (13.15) implies that for a transportation link (i,j):

$$(p_Q^* - \tau_{ij}^*) \times (l_{ij} - l_{ij}^*) \geq 0, \quad \forall l_{ij} \geq 0. \tag{13.26}$$

Summing now inequality (13.25) over all supply markets and inequality (13.26) over all transportation links, and adding the resultants, yields:

$$\sum_{i=1}^{m}(p_s^* - \tau_i^*) \times (l_i - l_i^*) + \sum_{i=1}^{m}\sum_{j=1}^{n}(p_Q^* - \tau_{ij}^*) \times (l_{ij} - l_{ij}^*) \geq 0, \quad \forall (l,\lambda) \in R_+^{m+mn}. \tag{13.27}$$

Turning now to the equilibrium conditions that govern the prices, we note that equilibrium condition (13.16) implies that

$$\sum_{i=1}^{m}(l_i^0 - l_i) \times (p_s - p_s^*) \geq 0, \quad \forall p_s \geq 0, \tag{13.28}$$

whereas equilibrium condition (13.17) implies that

$$\sum_{i=1}^{m}\sum_{j=1}^{n}(l_{ij}^0 - l_{ij}^*) \times (p_Q - p_Q^*) \geq 0, \quad \forall p_Q \geq 0. \tag{13.29}$$

Summing now inequalities (13.28) over all supply markets and (13.29) over all supply and demand markets, we obtain

$$\sum_{i=1}^{m}(l_i^0 - l_i^*) \times (p_s - p_s^*) + \sum_{i=1}^{m}\sum_{j=1}^{n}(l_{ij}^0 - l_{ij}^*) \times (p_Q - p_Q^*) \geq 0, \quad \forall (p_s, p_Q) \in R_+^2. \tag{13.30}$$

Finally, summing inequalities (13.21), (13.24), (13.27), and (13.30) yields the variational inequality (13.19).

We now prove that a solution to (13.19) satisfies equilibrium conditions (13.12)–(13.18).

Let $(s^*, Q^*, d^*, t^*, \tau^*, l^*, \lambda^*, p_s^*, p_Q^*) \in \mathcal{K}$ be a solution to (13.19). Let $s = s^*, Q = Q^*, d = d^*, \tau = \tau^*, l = l^*, \lambda = \lambda^*$, and $p_s = p_s^*, p_Q = p_Q^*$, and make these substitutions into (13.19), yields:

$$\sum_{i=1}^{m}(l_i^* - h_i s_i^*) \times (\tau_i - \tau_i^*) \geq 0, \quad \forall t \in R_+^m,$$

which implies equilibrium condition (13.12).

Making the same substitutions into (13.19) as previously but now setting $t = t^*$ and making no substitution for τ, yields:

$$\sum_{i=1}^{m}\sum_{j=1}^{n}(l_{ij}^* - h_{ij}Q_{ij}^*) \times (\tau_{ij} - \tau_{ij}^*) \geq 0, \quad \forall \tau \in R_+^{mn}, \tag{13.31}$$

which implies equilibrium conditions (13.14).

Now we recover the equilibrium conditions for the licenses. Let $s = s^*, Q = Q^*, d = d^*$, $t = t^*, \tau = \tau^*$, $\lambda = \lambda^*$, and $p_s = p_s^*, p_Q = p_Q^*$, and substitute these values into (13.19) provides:

$$\sum_{i=1}^{m}(p_s^* - \tau_i^*) \times (l_i - l_i^*) \geq 0, \quad \forall l \in R_+^m, \tag{13.32}$$

which implies equilibrium conditions (13.13).

Making the same substitutions into (13.19) as immediately preceding except letting now $l = l^*$ and keeping λ free, we obtain:

$$\sum_{i=1}^{m}\sum_{j=1}^{n}(p_Q^* - \tau_{ij}^*) \times (l_{ij} - l_{ij}^*) \geq 0, \quad \forall \lambda \in R_+^{mn}, \tag{13.33}$$

which implies equilibrium conditions (13.15).

As for the price equilibrium conditions, we set $s = s^*, Q = Q^*, d = d^*$, $t = t^*, \tau = \tau^*$, and $l = l^*, \lambda = \lambda^*$, and $p_Q = p_Q^*$, and substitute these values into (13.19), which yields

$$\sum_{i=1}^{m}(l_i^0 - l_i^*) \times (p_s - p_s^*) \geq 0, \quad \forall p_s \geq 0, \tag{13.34}$$

which implies equilibrium condition (13.16).

Making now the same substitutions as immediately preceding, except setting now $p_s = p_s^*$, and keeping p_Q free, we obtain

$$\sum_{i=1}^{m}\sum_{j=1}^{n}(l_{ij}^0 - l_{ij}^*) \times (p_Q - p_Q^*) \geq 0, \quad \forall p_Q \geq 0, \tag{13.35}$$

which implies equilibrium condition (13.17).

Now we turn to the commodity shipment equilibrium conditions. We set $t = t^*, \tau = t^*, l = l^*, \lambda = \lambda^*$, and $p_s = p_s^*, p_Q = p_Q^*$, and make these substitutions into variational inequality (13.19), which yields:

$$\sum_{i=1}^{m}(\pi_i(s^*) + h_i\tau_i^*) \times (s_i - s_i^*) + \sum_{i=1}^{m}\sum_{j=1}^{n}(c_{ij}(Q^*) + h_{ij}\tau_{ij}^*) \times (Q_{ij} - Q_{ij}^*)$$
$$- \sum_{j=1}^{n}\rho_j(d^*) \times (d_j - d_j^*) \geq 0,$$
$$\forall (s, Q, d) \in K, \tag{13.36}$$

which, of course, in view of feasibility constraints (13.1) and (13.2), is equivalent to:

$$\sum_{i=1}^{m}\sum_{j=1}^{n}(\pi_i(s^*) + h_i\tau_i^* + c_{ij}(Q^*) + h_{ij}\tau_{ij}^* - \rho_j(d^*)) \times (Q_{ij} - Q_{ij}^*) \geq 0, \quad \forall Q \in R_+^{mn}. \tag{13.37}$$

Letting now $Q_{kl} = Q_{kl}^*$ for all $kl \neq ij$, and making this substitution into (13.37), we obtain

$$(\pi_i(s^*) + h_i\tau_i^* + c_{ij}(Q^*) + h_{ij}\tau_{ij}^* - \rho_j(d^*)) \times (Q_{ij} - Q_{ij}^*) \geq 0, \quad \forall Q_{ij} \geq 0, \tag{13.38}$$

which implies equilibrium condition (13.18) since inequality (13.38) must hold for any (i,j). □

We now put variational inequality (13.19) into standard form. We define the column vector $X \equiv (s, Q, d, t, \tau, l, \lambda, p_s, p_Q) \in \mathcal{K}$ and the column vector

$$F(X) \equiv (\Pi(X), C(X), \mathcal{P}(X), T(X), \mathcal{T}(X), L(X), \Lambda(X), P_s(X), P_Q(X)),$$

where $\Pi(X)$, $T(X)$, and $L(X)$ are each m-dimensional column vectors with component i given, respectively, by

$$\Pi_i(X) : \pi_i(s) + h_i\tau_i$$
$$T_i(X) : l_i - h_i s_i$$
$$L_i(X) : p_s - \tau_i.$$

$C(X)$, $\mathcal{T}(X)$, and $\Lambda(X)$ are mn-dimensional column vectors with component i,j given, respectively, by:

$$C_{ij}(X) : c_{ij}(Q) + h_{ij}\tau_{ij}$$
$$\mathcal{T}_{ij}(X) : l_{ij} - h_{ij}Q_{ij}$$
$$\Lambda_{ij}(X) : p_Q - \tau_{ij}.$$

$\mathcal{P}$ is the n-dimensional vector with component j given by:

$$\mathcal{P}_j(X) = \rho_j(d),$$

$P_s(X)$ consists of the single element:

$$P_s(X) = \sum_{i=1}^{m}(l_i^0 - l_i)$$

and $P_Q(X)$ consists of the single element:

$$P_Q(X) = \sum_{i=1}^{m}\sum_{j=1}^{n}(l_{ij}^0 - l_{ij}).$$

Therefore, variational inequality (13.19) may now be expressed as:

$$\langle F(X^*), X - X^* \rangle \geq 0, \quad \forall X \in \mathcal{K}.$$

We now establish the relationships between the equilibrium marginal costs of emission abatement and the equilibrium license prices.

Proposition 1 *If $s_i^* > 0$, then*

$$\tau_i^* = p_s^*.$$

Proof: From equilibrium condition (13.12), we see that if $s_i^* > 0$, then $l_i^* > 0$, and, hence, from equilibrium condition (13.13) it follows that $\tau_i^* = p_s^*$. □

Proposition 2 *If $Q_{ij}^* > 0$, then*

$$\tau_{ij}^* = p_Q^*.$$

Proof: If $Q_{ij}^* > 0$, from equilibrium condition (13.14) it follows that $l_{ij}^* > 0$, and, thus, from equilibrium condition (13.15) that $\tau_{ij}^* = p_Q^*$. □

We now study whether the equilibrium pattern is independent of the initial license allocation and how to guarantee that the environmental emission standards imposed by the regulatory body are met in equilibrium. We first answer the question, in the subsequent proposition, whether the initial allocation of licenses affects the equilibrium pattern.

Proposition 3 (Equilibrium Pattern Independence from Initial License Allocation)
If $l_i^0 \geq 0$, for all supply markets i, and $\sum_{i=1}^m l_i^0 = \bar{E}_s$, with $\bar{E}_s$ fixed and positive and $l_{ij}^0 \geq 0$, for all transportation links (i,j), and $\sum_{i=1}^m \sum_{j=1}^n l_{ij}^0 = \bar{E}_Q$, with $\bar{E}_Q$ fixed and positive, then the equilibrium pattern is independent of the initial license allocation $\{l_i^0\}$.

Proof: The terms in variational inequality (13.19) are either independent of l_i^0 for all i, and l_{ij}^0, for all i and j, or depend only on the sums $\sum_{i=1}^m l_i^0$ and $\sum_{i=1}^m \sum_{j=1}^n l_{ij}^0$. □

Since we have assumed that the emissions generated by the supply markets are distinct from the emissions generated by the commodity shipment process, we now let $\bar{E}_s$ denote the environmental quality standard for the emissions generated by the supply markets, and we let $\bar{E}_Q$ denote the environmental quality standard for the emissions generated by the transportation links.

In the following proposition we show that the environmental standards are met by the equilibrium pattern, provided that the sum of the initial license allocation for the supply markets is equal to the environmental emission standard $\bar{E}_s$, and to the environmental emission standard $\bar{E}_Q$ for the transportation links.

Proposition 4 (Achievement of Environmental Standards)
If $\sum_{i=1}^m l_i^0 = \bar{E}_s$, and $\sum_{i=1}^m \sum_{j=1}^n l_{ij}^0 = \bar{E}_Q$, then the equilibrium vector achieves the environmental standards: $\bar{E}_s$ and $\bar{E}_Q$.

Proof: We have from equilibrium conditions (13.12) and (13.16) that

$$\sum_{i=1}^m h_i s_i^* \leq \sum_{i=1}^m l_i^* \leq \sum_{i=1}^m l_i^0 = \bar{E}_s,$$

and from equilibrium conditions (13.14) and (13.17) that

$$\sum_{i=1}^{m}\sum_{j=1}^{n} h_{ij}Q_{ij}^* \leq \sum_{i=1}^{m}\sum_{j=1}^{n} l_{ij}^* \leq \sum_{i=1}^{m}\sum_{j=1}^{n} l_{ij}^0 = \bar{E}_Q.$$

Hence, the environmental standards are met by the equilibrium pattern. □

Analogous results to Propositions 3 and 4 can be found in Nagurney and Dhanda (1996) for the case of firms and in Nagurney, Ramanujam, and Dhanda (1998) in the case of transportation networks.

13.3 QUALITATIVE PROPERTIES

In this section, we present qualitative properties of the equilibrium pattern, including an existence result. We also establish properties of the function $F(\cdot)$ which enters the variational inequality model for our problem that are needed to establish the convergence of the algorithmic scheme in the subsequent section.

Theorem 3 (Existence)
If $(s^*, Q^*, d^*, t^*, \tau^*, l^*, \lambda^*, p_s^*, p_Q^*) \in \mathcal{K}$ satisfies variational inequality (13.19) *then the equilibrium commodity supply, shipment, and demand pattern $(s^*, Q^*, d^*) \in K$ is a solution to the variational inequality problem:*

$$\sum_{i=1}^{m} \pi_i(s^*) \times (s_i - s_i^*) + \sum_{i=1}^{m}\sum_{j=1}^{n} c_{ij}(Q^*) \times (Q_{ij} - Q_{ij}^*) - \sum_{j=1}^{n} \rho_j(d^*) \times (d_j - d_j^*) \geq 0,$$
$$\forall (s, Q, d) \in K^1, \tag{13.39}$$

where

$$K^1 \equiv \{Q \geq 0 |\, (13.1), (13.2), (13.8) - -(13.11) \textit{ are satisfied}\}.$$

A solution to (13.39) *is guaranteed to exist since K^1 is compact and π, c, and ρ are assumed to be continuous. Moreover, there exist vectors: $t^* \in R_+^m$, $\tau^* \in R_+^{mn}$, $l^* \in R_+^m$, $\lambda^* \in R_+^{mn}$, p_s^*, and p_Q^* with*

$$(s^*, Q^*, d^*, t^*, \tau^*, l^*, \lambda^*, p_s^*, p_Q^*) \in \mathcal{K}$$

a solution to variational inequality (13.19).

Proof: Assume that (13.39) is not true, that is, there exists an $(s, Q, d) \in K^1$, such that

$$\sum_{i=1}^{m} \pi_i(s^*) \times (s_i - s_i^*) + \sum_{i=1}^{m}\sum_{j=1}^{n} c_{ij}(Q^*) \times (Q_{ij} - Q_{ij}^*) - \sum_{j=1}^{n} \rho_j(d^*) \times (d_j - d_j^*) < 0,$$
$$\forall (s, Q, d) \in K^1. \tag{13.40}$$

According to (13.40) and (13.19) we then have

$$\sum_{i=1}^{m} h_i t_i^* \times (s_i - s_i^*) + \sum_{i=1}^{m}\sum_{j=1}^{n} h_{ij}\tau_{ij}^* \times (Q_{ij} - Q_{ij}^*)$$
$$+ \sum_{i=1}^{m}(l_i^* - h_i s_i^*) \times (\tau_i - \tau_i^*) + \sum_{i=1}^{m}\sum_{j=1}^{n}(l_{ij}^* - h_{ij}Q_{ij}^*) \times (\tau_{ij} - \tau_{ij}^*)$$
$$+ \sum_{i=1}^{m}(p_s^* - \tau_i^*) \times (l_i - l_i^*) + \sum_{i=1}^{m}\sum_{j=1}^{n}(p_Q^* - \tau_{ij}^*) \times (l_{ij} - l_{ij}^*)$$
$$+ \sum_{i=1}^{m}(l_i^0 - l_i^*) \times (p_s - p_s^*) + \sum_{i=1}^{m}\sum_{j=1}^{n}(l_{ij}^0 - l_{ij}^*) \times (p_Q - p_Q^*)$$
$$\geq - \sum_{i=1}^{m}\pi_i(s^*) \times (s_i - s_i^*) - \sum_{i=1}^{m}\sum_{j=1}^{n} c_{ij}(Q^*) \times (Q_{ij} - Q_{ij}^*)$$
$$+ \sum_{j=1}^{n}\rho_j(d^*) \times (d_j - d_j^*) > 0. \tag{13.41}$$

Letting now $s_i, Q_{ij}, \tau_i, \tau_{ij}, l_i, \lambda_{ij}$, p_s, and p_Q equal to zero for all i and j, and substituting these values into the first inequality in (13.41), after some algebraic simplifications, yields:

$$\sum_{i=1}^{m} l_i^0(-p_s^*) + \sum_{i=1}^{m}\sum_{j=1}^{n} l_{ij}^0(-p_Q^*) \geq 0. \tag{13.42}$$

But the left-hand side of (13.42) is clearly nonpositive and, hence, a contradiction has been obtained and therefore variational inequality (13.39) must be satisfied.

Moreover, K^1 is compact, since

$$\sum_{i=1}^{m} l_i \leq \sum_{i=1}^{m} l_i^0 \quad \text{and} \quad \sum_{i=1}^{m}\sum_{j=1}^{n} l_{ij} \leq \sum_{i=1}^{m}\sum_{j=1}^{n} l_{ij}^0$$

and, consequently, from (13.8) and (13.9) it follows then that $\sum_{i=1}^{m} h_i s_i \leq \sum_{i=1}^{m} l_i^0$ and $\sum_{i=1}^{m}\sum_{j=1}^{n} h_{ij}Q_{ij} \leq \sum_{i=1}^{m}\sum_{j=1}^{n} l_{ij}^0$. Therefore, s, l, and Q, λ lie in a compact set and through (13.2) d also lies in a compact set. Hence, the existence of the vector $(s^*, Q^*, d^*, l^*, \lambda^*)$ satisfying (13.39) is guaranteed by the standard existence theory of variational inequalities (cf. Kinderlehrer and Stampacchia (1980)).

Also, according to the Lagrange Multiplier Theorem, there exist multipliers t^*, τ^*, p_s^*, p_Q^* associated with inequalities (13.12), (13.14), and (13.16), and (13.17), respectively, and these, together with the vector $(s^*, Q^*, d^*, l^*, \lambda^*)$ above must satisfy variational inequality (13.19). □

Theorem 4 (Uniqueness)
Assume that the supply price functions $\pi(s)$, the transportation cost functions $c(Q)$, and minus the demand price functions $-\rho(d)$ are strictly monotone in s, Q, and d,

respectively, that is,

$$\begin{aligned}&\langle \pi(s^1)-\pi(s^2), s^1-s^2\rangle + \langle c(Q^1)-c(Q^2), Q^1-Q^2\rangle\\&\quad-\langle \rho(d^1)-\rho(d^2), d^1-d^2\rangle > 0,\\&\qquad\forall (s^1,Q^1,d^1),(s^2,Q^2,d^2)\in\mathcal{K},\quad (s^1,Q^1,d^1)\neq(s^2,Q^2,d^2).\end{aligned}\tag{13.43}$$

Then the equilibrium supply, commodity shipment, and demand pattern is unique.

Proof: Assume, on the contrary that there are two distinct equilibrium patterns, denoted by X^1 and X^2. Then both must satisfy variational inequality (13.19), that is,

$$\langle F(X^1), X-X^1\rangle \geq 0, \quad \forall X\in\mathcal{K}\tag{13.44}$$

and

$$\langle F(X^2), X-X^2\rangle \geq 0, \quad \forall X\in\mathcal{K}.\tag{13.45}$$

Substituting X^2 for X into (13.44) and X^1 for X into (13.45), and adding the resultants, after algebraic simplifications, yields:

$$\begin{aligned}&\sum_{i=1}^{m}(\pi_i(s^1)-\pi_i(s^2))\times(s_i^1-s_i^2)+\sum_{i=1}^{m}\sum_{j=1}^{n}(c_{ij}(Q^1)-c_{ij}(Q^2))\times(Q_{ij}^1-Q_{ij}^2)\\&\quad-\sum_{j=1}^{n}(\rho_j(d^1)-\rho_j(d^2))\times(d_j^1-d_j^2)\leq 0,\end{aligned}\tag{13.46}$$

but due to the assumption of strict monotonicity this can only hold as an equality and, hence, $s^1=s^2$, $Q^1=Q^2$, and $d^1=d^2$. □

We now establish certain properties of the function $F(\cdot)$ which are used to prove convergence of the algorithmic scheme in Section 13.4.

Lemma 1 (Monotonicity)
Assume that the supply price functions are monotonically increasing in the supplies, the transportation cost functions are monotonically increasing in the commodity shipments, and the demand prices are monotonically decreasing in the demands, that is, for every $(s^1,Q^2,d^2),(s^2,Q^2,d^2)\in K$, we have that

$$\begin{aligned}&\langle \pi(s^1)-\pi(s^2), s^1-s^2\rangle + \langle c(Q^1)-c(Q^2), Q^1-Q^2\rangle\\&\quad-\langle \rho(d^1)-\rho(d^2), d^1-d^2\rangle \geq 0,\quad \forall (s^1,Q^1,d^1),(s^2,Q^2,d^2)\in K,\end{aligned}\tag{13.47}$$

Then $F(X)$ is monotone, that is,

$$\langle F(X^1)-F(X^2), X^1-X^2\rangle \geq 0, \quad \forall X^1,X^2\in\mathcal{K}.\tag{13.48}$$

Proof: From the definition of $F(X)$ for this model the left-hand side of inequality (13.48) is given by:

$$\sum_{i=1}^{m}(\pi_i(s^1)-\pi_i(s^2))\times(s_i^1-s_i^2)+\sum_{i=1}^{m}\sum_{j=1}^{n}(c_{ij}(Q^1)-c_{ij}(Q^2))\times(Q_{ij}^1-Q_{ij}^2)$$
$$-\sum_{j=1}^{n}(\rho_j(d^1)-\rho_j(d^2))\times(d_j^1-d_j^2)$$
$$+\sum_{i=1}^{m}(h_i\tau_i^1-h_i\tau_i^2)\times(s_i^1-s_i^2)+\sum_{i=1}^{m}\sum_{j=1}^{n}(h_{ij}\tau_{ij}^1-h_{ij}\tau_{ij}^2)\times(Q_{ij}^1-Q_{ij}^2)$$
$$+\sum_{i=1}^{m}((l_i^1-h_is_i^1)-(l_i^2-h_is_i^2))\times(\tau_i^1-\tau_i^2)$$
$$+\sum_{i=1}^{m}\sum_{j=1}^{n}((l_{ij}^1-h_{ij}Q_{ij}^1)-(l_{ij}^2-h_{ij}Q_{ij}^2))\times(\tau_{ij}^1-\tau_{ij}^2)$$
$$+\sum_{i=1}^{m}((p_s^1-\tau_i^1)-(p_s^2-\tau_i^2))\times(l_i^1-l_i^2)$$
$$+\sum_{i=1}^{m}\sum_{j=1}^{n}((p_Q^1-\tau_{ij}^1)-(p_Q^2-\tau_{ij}^2))\times(l_{ij}^1-l_{ij}^2)$$
$$+\sum_{i=1}^{m}(-l_i^1+l_i^2)\times(p_s^1-p_s^2)+\sum_{i=1}^{m}\sum_{j=1}^{n}(-l_{ij}^1+l_{ij}^2)\times(p_Q^1-p_Q^2). \tag{13.49}$$

After rearranging and simplifying, (13.49) reduces to:

$$\sum_{i=1}^{m}(\pi_i(s^1)-\pi_i(s^2))\times(s_i^1-s_i^2)+\sum_{i=1}^{m}\sum_{j=1}^{n}(c_{ij}(Q^1)-c_{ij}(Q^2))\times(Q_{ij}^1-Q_{ij}^2)$$
$$-\sum_{j=1}^{n}(\rho_j(d^1)-\rho_j(d^2))\times(d_j^1-d_j^2), \tag{13.50}$$

which, by the assumption of monotonicity, is greater than or equal to zero. □

Lemma 2 (Lipschitz Continuity)
If the supply price functions, the transportation cost functions, and the demand price functions have bounded first order derivatives, then the function $F(X)$ is Lipschitz continuous.

Proof: Follows along the lines of the arguments used in the proof of Lemma 3 in Nagurney (1994). □

13.4 THE ALGORITHM

In this section, the modified projection method of Korpelevich (1977) is proposed for the solution of variational inequality (13.19).

Modified Projection Method

Step 0: Initialization:
Set $X^0 \in \mathcal{K}$. Let $\mathcal{T} = 1$ and let α be a scalar such that $0 < \alpha < \frac{1}{L}$, where L is the Lipschitz continuity constant.

Step 1: Computation:
Compute $\bar{X}^{\mathcal{T}}$ by solving the variational inequality subproblem:

$$\langle [\bar{X}^{\mathcal{T}} + \alpha F(X^{\mathcal{T}-1}) - X^{\mathcal{T}-1}], X - \bar{X}^{\mathcal{T}} \rangle \geq 0, \quad \forall X \in \mathcal{K}. \tag{13.51}$$

Step 2: Adaptation:
Compute $X^{\mathcal{T}}$ by solving the variational inequality subproblem:

$$\langle [X^{\mathcal{T}} + \alpha F(\bar{X}^{\mathcal{T}}) - X^{\mathcal{T}-1}], X - X^{\mathcal{T}} \rangle \geq 0, \quad \forall X \in \mathcal{K}. \tag{13.52}$$

Step 3: Convergence Verification:
If $|X^{\mathcal{T}} - X^{\mathcal{T}-1}| \leq \epsilon$, for $\epsilon > 0$, a prespecified tolerance, then stop; otherwise, set $\mathcal{T} = \mathcal{T} + 1$ and go to Step 1.

We now discuss the solution of the variational inequality subproblems (13.51) and (13.52).

First, we note the simple form of the decomposed subproblems which can be computed very efficiently. In particular, recall that the feasible set $\mathcal{K}$ is a Cartesian product, where K has the network structure of the spatial price network equilibrium problem, whereas the other sets consist solely of the nonnegative orthant in different dimensions and, hence, yield subproblems of box-type constraints, each of which can be solved explicitly and in closed form using a simple formula. Moreover, even the resulting spatial price equilibrium problem over K is actually a quadratic programming problem akin to the classical spatial price equilibrium models of Samuelson (1952) and Takayama and Judge (1971) and, consequently, amenable to solution by many quadratic programming algorithms, including special-purpose ones (cf. Nagurney (1999) and the references therein).

Convergence for the algorithm is given in the following theorem.

Theorem 5 (Convergence)
If the supply price functions, transportation cost functions, and minus demand price functions are assumed to be monotone and to have bounded first order derivatives, then the modified projection method converges to the solution of variational inequality (13.19).

Proof: From Lemma 1, we conclude that $F(X)$ is monotone and from Lemma 2 that $F(X)$ is Lipschitz continuous. Hence, according to Korpelevich (1977), the modified projection method is guaranteed to converge to a solution of the variational inequality problem. □

13.5 NUMERICAL EXAMPLES

In this section, we present numerical results to illustrate the behavior of the model as well as the algorithm. We implemented the modified projection method in FORTRAN and utilized the IBM SP2. For the solution of the spatial price network equilibrium problem encountered in both the computation and the adaptation steps of the modified projection method, we utilized the Euler method (cf. Nagurney, Takayama, and Zhang (1995)) due to its simplicity.

The convergence tolerance ϵ was set to .0001 for all the examples. We initialized the modified projection method by setting all the variables equal to zero and we set $\alpha = .3$. We report the CPU time, exclusive of input/output times, as well as the number of iterations.

We solved three examples consisting of two supply markets and two demand markets. The data differed in the emissions factors. The common data for all three examples were as follows: The supply price functions were:

$$\pi_1(s) = 5s_1 + 1s_2 + 2, \quad \pi_2(s) = 2s_2 + 1.5s_1 + 1.5,$$

the transportation cost functions were:

$$c_{11}(Q) = Q_{11}, \quad c_{12}(Q) = 2Q_{12} + .035,$$

$$c_{21}(Q) = 3Q_{21} + .162, \quad c_{22}(Q) = 2Q_{22} + .115,$$

and the demand price functions were:

$$\rho_1(d) = -2d_1 - 1.5d_2 + 28.75, \quad \rho_2(d) = -4d_2 - d_1 + 41.$$

Example 1 In the first example, we let the emission parameters and the initial license allocations be as follows: $h_i{=}i$ and $l_i^0 = i$, for all supply markets i, $h_{ij} = j$ and $l_{ij}^0 = j$, for all transportation links (i, j).

The algorithm converged in .01 second of CPU time and required 180 iterations for convergence. It yielded the following equilibrium pattern:

$$\begin{aligned}
s_1^* &= 2.2931 \quad s_2^* = 0.3538,\\
Q_{11}^* &= 0.00 \quad Q_{12}^* = 2.2931 \quad Q_{21}^* = 0.00 \quad Q_{22}^* = 0.3538,\\
d_1^* &= 0.00 \quad d_2^* = 2.6469,\\
\tau_1^* &= 11.9718 \quad \tau_2^* = 11.9716,\\
\tau_{11}^* &= 0.0000 \quad \tau_{12}^* = 0.0000 \quad \tau_{21}^* = 0.0000 \quad \tau_{22}^* = 0.0000,\\
l_1^* &= 2.2929 \quad l_2^* = 0.7074,\\
l_{11}^* &= 0.0945 \quad l_{12}^* = 4.5976 \quad l_{21}^* = 0.0934 \quad l_{22}^* = 1.1208,\\
p_s^* &= 11.9717 \quad p_Q^* = 0.0000.
\end{aligned}$$

The market for supply market emission licenses cleared whereas there was an excess supply of transportation emission licenses equal to .0937.

Example 2 In the second example, the data were as in Example 1, except that we now modified the emission parameters as follows: $h_i = .5 \times i$, for all supply markets i, $h_{ij} = .5 \times j$ for all transportation links (i, j). Note that such a change could represent that there were improvements in technology both at the supply markets and on the transportation links in that the emission parameters were reduced.

The algorithm converged in .03 second of CPU time and required 393 iterations for convergence. It yielded the following equilibrium pattern:

$$
\begin{aligned}
s_1^* &= 2.2277 \quad s_2^* = 1.8622,\\
Q_{11}^* &= .3871 \quad Q_{12}^* = 1.8905 \quad Q_{21}^* = 0.0496 \quad Q_{22}^* = 1.8126,\\
d_1^* &= 0.4367 \quad d_2^* = 3.7031,\\
\tau_1^* &= 13.3703 \quad \tau_2^* = 13.3703,\\
\tau_{11}^* &= 0.0000 \quad \tau_{12}^* = 0.0000 \quad \tau_{21}^* = 0.0000 \quad \tau_{22}^* = 0.0000,\\
l_1^* &= 1.1384 \quad l_2^* = 1.8619,\\
l_{11}^* &= 0.5513 \quad l_{12}^* = 1.8905 \quad l_{21}^* = 0.8098 \quad l_{22}^* = 2.7485,\\
p_s^* &= 13.3703 \quad p_Q^* = 0.0000.
\end{aligned}
$$

The market for supply market emission licenses cleared as well as the market for the transportation emission licenses.

Example 3 In the third example, the data were as in Example 1, except that we now modified the emission parameters as follows: $h_i = .1 \times i$, for all supply markets i, $h_{ij} = .1 \times j$ for all transportation links (i, j). Note that such a change could represent that there were further improvements in technology both at the supply markets and on the transportation links in that the emission parameters were reduced.

The modified projection method converged in .01 second of CPU time and required 54 iterations for convergence. It yielded the following equilibrium pattern:

$$
\begin{aligned}
s_1^* &= 2.1353 \quad s_2^* = 4.5093,\\
Q_{11}^* &= 0.6273 \quad Q_{12}^* = 1.5080 \quad Q_{21}^* = 1.3105 \quad Q_{22}^* = 3.1988,\\
d_1^* &= 1.9377 \quad d_2^* = 4.7068,\\
\tau_1^* &= 0.0000 \quad \tau_2^* = 0.0000,\\
\tau_{11}^* &= 0.0000 \quad \tau_{12}^* = 0.0000 \quad \tau_{21}^* = 0.0000 \quad \tau_{22}^* = 0.0000,\\
l_1^* &= 0.4098 \quad l_2^* = 1.3894,\\
l_{11}^* &= 0.1260 \quad l_{12}^* = 0.5532 \quad l_{21}^* = 0.2215 \quad l_{22}^* = 0.9243,\\
p_s^* &= 0.0000 \quad p_Q^* = 0.0000.
\end{aligned}
$$

In this example, there was an excess supply of supply market emission permits at a level of 1.2007 and also an excess supply of transportation emission permits at a level of 4.1750.

Note also the above commodity production, shipment, and consumption pattern is also a spatial price equilibrium pattern and satisfies equilibrium conditions (13.6).

It is worth noting that as the emission parameters were decreased in the numerical examples, the demand for the commodity at the markets increased.

Acknowledgments This research was supported by the John F. Smith Memorial Fund at the Isenberg School of Management. This support is gratefully acknowledged.

The author is grateful to the anonymous referees for helpful comments and suggestions on the earlier version of the manuscript.

References

Allen, W. G. (1996), "Model Improvements for Evaluating Pricing Strategies," *Transportation Research Record* **1498**, 75-81.

Anderson, W. P., Kanaroglu, P. S., Miller, E. J., and Buliung, R. N. (1996), " Simulating Automobile Emissions in an Integrated Urban Model," *Transportation Research Record* **1520**, 71-80.

Crocker, T. D. (1966), "The Structuring of Atmospheric Pollution Control Systems," in *The Economics of Air Pollution*, pp. 61-86, H. Wolozin, editor, W. W. Norton, New York.

Dafermos, S., and Nagurney, A. (1987), "Oligopolistic and Competitive Behavior of Spatially Separated Markets," *Regional Science and Urban Economics* **17**, 245-254.

Dales, J. H. (1968), *Pollution, Property, and Prices*, University of Toronoto, Toronto, Ontario, Canada.

DeCorla-Souza, P., Everett, J., Cosby, J., and Lim, P. (1985), "Trip-Based Approach to Estimate Emissions with Environmental Protection Agency's MOBILE Model," *Transportation Research Record* **1444**, 118-125.

Florian, M., and Los, M. (1982), "A New Look at Static Spatial Price Equilibrium Models," *Regional Science and Urban Economics* **12**, 579-597.

Glover, E. L., and Brzezinski, D. J. (1989), "MOBILE4 Exhaust Emission Factors and Inspection/Maintenance Benefits for Passenger Cars," Technical Report, *EPA-AA-TSS-I/M-89-3.*

Kinderlehrer, D., and Stampacchia, G. (1980), *An Introduction to Variational Inequalities and Their Applications*, Academic Press, New York.

Korpelevich, G. M. (1977), "The Extragradient Method for Finding Saddle Points and Other Problems," *Matekon* **13**, 35-49.

Montgomery, W. D. (1972), "Markets in Licenses and Efficient Pollution Control Programs," *Journal of Economic Theory* **5**, 395-418.

Nagurney, A. (1987), "Computational Comparisons of Spatial Price Equilibrium Methods," *Journal of Regional Science* **27**, 55-76.

Nagurney, A. (1999), *Network Economics: A Variational Inequality Approach*, second and revised edition, Kluwer Academic Publishers, Boston, Massachusetts.

Nagurney, A. (1994), "Variational Inequalities in the Analysis and Computation of Multi-Sector, Multi-Instrument Financial Equilibria," *Journal of Economic Dynamics and Control* **18**, 161-184.

Nagurney, A., and Dhanda, K. (1996) "A Variational Inequality Approach for Marketable Pollution Permits," *Computational Economics* **9**, 363-384.

Nagurney, A., and Dhanda, K. K. (1997a), "Marketable Pollution Permits in Oligopolistic Markets with Transaction Costs," *Operations Research*, in press.

Nagurney, A., and Dhanda, K. K. (1997b), "Variational Inequalities for Marketable Pollution Permits with Technological Investment Opportunities: The Case of Oligopolistic Markets," *Mathematical and Computer Modelling* **26**, 1-25.

Nagurney, A., Ramanujam, P., and Dhanda, K. K. (1998), "A Multimodal Traffic Network Equilibrium Model with Emission Pollution Permits: Compliance Versus Noncompliance," *Transportation Research D* **3**, 349-374.

Nagurney, A., Takayama, T., and Zhang, D. (1995), "Massively Parallel Computation of Spatial Price Equilibrium Problems as Dynamical Systems," *Journal of Economic Dynamics and Control* **18**, 3-37.

Samuelson, P. A. (1952), "Spatial Price Equilibrium and Linear Programming," *American Economic Review* **42**, 283-303.

Takayama, T., and Judge, G. G. (1971), *Spatial and Temporal Price and Allocation Models*, North Holland, Amsterdam, The Netherlands.

Tietenberg, T. H. (1985), *Emissions Trading: An Exercise in Reforming Pollution Policy*, Resources for the Future, Inc., Washington, DC.

14 A NEW DUAL ALGORITHM FOR SHORTEST PATH REOPTIMIZATION

Sang Nguyen

Stefano Pallottino

Maria Grazia Scutellà

Abstract: Shortest path problems are among the most studied network flow problems, with interesting applications in various fields. In large scale transportation systems, a sequence of shortest path problems must often be solved, where the $(k+1)^{st}$ problem differs only slightly from the kth one. Significant reduction in computational time may be obtained from an efficient reoptimization procedure that exploits the useful information available after each shortest path computation in the sequence. Such reduction in computational time is essential in many on-line applications. This work is devoted to the development of such reoptimization algorithm. We shall focus on the sequence of shortest path problems to be solved for which problems differ by the origin node of the path set. After reviewing the classical algorithms described in the literature so far, which essentially show a Dijkstra-like behavior, a new dual approach will be proposed, which could be particularly promising in practice.

14.1 REOPTIMIZING SHORTEST PATHS COMPUTATIONS

In large scale transportation systems, shortest path problems must often be solved sequentially, where the $(k+1)^{st}$ problem differs only slightly from the kth one.

Quite often in practice, each problem in the sequence is solved from the beginning and independently of the previous one. The first attempts at exploiting reoptimization techniques were made by Murchland (1970) and Dionne (1978), who investigated the problem of finding the shortest cost matrix when the cost of one arc of the input graph changes. Although of some theoretical interest, the proposed approaches required an impractical large amount of memory for transportation applications. Another more

M. Gendreau and P. Marcotte (eds.), Transportation and Network Analysis: Current Trends, 221–235.

efficient approach based on the updating of the shortest cost matrix using Dijkstra's method (Dijkstra, 1959) is due to Fujishige (1981).

The first efficient reoptimization strategy can be found in Gallo (1980), where two general situations arising in transportation applications are considered and clever reoptimization procedures are proposed for their solution. The general problem considered is as follows: assume that a shortest path tree relative to a given origin r, say T_r^*, has been determined; the problem is that of finding a new shortest path tree when either the origin is changed, or exactly one arc is given a new cost. Note that most applications in transportation can be reduced either to one of these two cases, or to a combination of them. Based on Gallo's reoptimization strategy, an efficient algorithm is proposed in Gallo and Pallottino (1982) for the case of varying origin nodes that fully exploits the arc *reduced cost* property (see for instance Nemhauser (1972) and Lawler (1976)). Following a different direction, a dual simplex algorithm is suggested in Florian, Nguyen and Pallottino (1981) for the same problem. These two algorithms are considered to be the two most efficient ones in the literature.

In the following, we will first formally state the problem of finding a new shortest path tree when the origin node is changed, then review the reduced cost algorithm (Gallo and Pallottino, 1982) and the dual simplex algorithm (Florian, Nguyen and Pallottino, 1981) in terms of dual ascent methods[1], and finally propose a new dual approach. Unlike the other two algorithms found in the literature, which follow a Dijkstra-like node inserting order, our approach adopts a different (and, hopefully, more efficient) insertion policy.

14.2 THE SHORTEST PATH REOPTIMIZATION PROBLEM

14.2.1 The shortest path tree problem

Let $G = (N, A)$ be a directed graph, where N is the set of *nodes*, of cardinality n, and A is the set of *arcs*, of cardinality m. Let $c : A \to R$ be a function which assigns a cost c_{ij} to each arc $(i,j) \in A$. For simplicity, we shall assume that G is strongly connected and does not have parallel arcs.

Given an origin node or *root* $r \in N$, the *shortest path problem* (SP_r) consists in finding a directed tree T_r^* such that the path from r to i in T_r^* is a minimum cost path from r to i in G. Observe that, by our previous assumption, T_r^* is always a spanning tree. In order to guarantee a finite solution to the problem, we shall assume that no directed cycle with negative cost exists in G.

Using a linear programming notation, the problem and its dual can be formulated as follows:

$$
\begin{aligned}
(SP_r) \quad min \quad & \sum_{(i,j)\in A} c_{ij}x_{ij} & \\
& \sum_{(j,i)\in BS(i)} x_{ji} - \sum_{(i,j)\in FS(i)} x_{ij} = b_i & \forall i \in N, \\
& x_{ij} \geq 0 & \forall (i,j) \in A
\end{aligned}
$$

$$
\begin{aligned}
(DSP_r) \quad max \quad & (1-n)\pi_r + \sum_{j\neq r} \pi_j & \\
& \pi_j - \pi_i \leq c_{ij} & \forall (i,j) \in A
\end{aligned}
$$

where $FS(i)$ and $BS(i)$ denote the *forward star* and the *backward star* of i, respectively:

$$FS(i) = \{(u,v) \in A : u = i\}, \qquad BS(i) = \{(u,v) \in A : v = i\};$$

and $b_r = -n+1$, $b_i = 1$, $\forall i \neq r$, whereas π_i denotes the dual variable associated with i (usually referred to as the *price* or *potential* of i).

Given a potential vector π, using the *reduced cost* $\bar{c}_{ij} = c_{ij} + \pi_i - \pi_j$ of an arc $(i,j) \in A$, the dual feasibility can be equivalently stated as

$$\bar{c}_{ij} \geq 0, \qquad \forall (i,j) \in A.$$

14.2.2 The reoptimization problem

Assume that a shortest path tree T_r^* has been determined. The reoptimization problem under consideration consists of finding the shortest path tree T_s^* when the origin changes to a node $s \neq r$. Two main algorithmic approaches for this problem have been proposed in the literature in the early 80's, and we will review these in Sections 14.2.3 and 14.2.4. Both algorithms rely on the observation that the optimum potentials relative to T_r^* are feasible to the dual problem (DSP_s), and thus dual-like approaches seem to be promising avenues for computing the required shortest path tree rooted at s.

These algorithms (as well as the new one presented in this work) are instances of a general *dual ascent reoptimization paradigm*, which maintains during the algorithmic execution:

- a (partial) shortest path tree rooted at the new origin s, $T_s = (N_s, A_s)$,
- a partial tree $T_r = (N_r, A_r)$ of T_r^* that still has to be moved to T_s,
- a feasible dual vector $\pi^{(s)}$, which satisfies the complementarity slackness conditions associated with problem (SP_s) and its dual (DSP_s) (at the beginning, $\pi^{(s)}$ coincides with the optimum dual vector $\pi^{(r)}$ relative to the previous origin r):

  $$\bar{c}_{ij} \geq 0, \quad \forall (i,j) \in A, \quad \text{dual feasibility;}$$

  $$\bar{c}_{ij} = 0, \quad \forall (i,j) \in A_s, \quad \text{complementary slackness conditions (CSC),}$$

- the dual vector $\pi^{(s)}$ also satisfies the CSC for the arcs of A_r if the potentials of the nodes belonging to N_r either remained unchanged, or changed by the same value:

  $$\bar{c}_{ij} = 0, \qquad \forall (i,j) \in A_r. \tag{14.1}$$

In the following, the nodes of N_r will be referred to as *external nodes*.

Given a subset of external nodes $S \subseteq N_r$, we denote respectively by $BS_B(S)$ and $BS_E(S)$ the set of arcs incident into S originating from N_s, called *border arcs*, and those originating from the complement of S in N_r, called *external arcs*:

$$BS_B(S) = \{(i,j) \in A : i \in N_s, j \in S\};$$

$$BS_E(S) = \{(i,j) \in A : i \in N_r \backslash S, j \in S\}.$$

Based on this partitioning, the following critical values for each subset $S \subseteq N_r$ may be defined:

$$\delta_S = min\{\bar{c}_{ij} : (i,j) \in BS_B(S)\};$$

$$\eta_S = min\{\bar{c}_{ij} : (i,j) \in BS_E(S)\};$$

$$\gamma_S = min\{\delta_S, \eta_S\};$$

where $\delta_S = \infty$ if $BS_B(S) = \emptyset$ and similarly $\eta_S = \infty$ if $BS_E(S) = \emptyset$.

Consider the *cut* $(\bar{S}, S)$ separating S from its complement $\bar{S} = N \backslash S$, and the cutset $A(S) = \{(i,j) \in A : i \in \bar{S}, j \in S\}$, then γ_S, called the *dual gap* of S, is the minimum reduced cost of the cutset $A(S)$. Therefore, given a feasible dual solution, γ_S represents the maximum increment that can be added to all potentials in S while maintaining dual feasibility and guaranteeing that at least one new reduced cost decreases to 0 after the updating. In particular, when $S = \{j\}$ we have:

$$\delta_j = min\{\bar{c}_{ij} : (i,j) \in BS_B(j)\};$$

$$\eta_j = min\{\bar{c}_{ij} : (i,j) \in BS_E(j)\};$$

$$\gamma_j = min\{\delta_j, \eta_j\}.$$

When finite, γ_j and δ_j provide bounds for the optimal solution of (DSP_s), i.e., for the optimum potential associated with node j; in fact, it is possible to prove that $\pi_j^{(s)} - \pi_s^{(s)} + \gamma_j$ is a lower bound, while $\pi_j^{(s)} - \pi_s^{(s)} + \delta_j$ is an upper bound (Pallottino and Scutellà, 1997).

Observe that, by definition:

$$\delta_S = min\{\delta_j : j \in S\}, \qquad \forall S \subseteq N_r,$$

and, in the special case where $S = N_r$, we have:

$$\delta_{N_r} = \Delta = min\{\delta_j : j \in N_r\}. \tag{14.2}$$

According to the previous definition, Δ is the *dual gap* associated with the set of external nodes N_r. Therefore, it is the maximum increment that can be added to the potentials of all the external nodes while maintaining the dual feasibility and ensuring that the reduced cost of at least one border arc reaches 0 after the updating.

The typical dual ascent iteration. A dual ascent operation is performed at each iteration in an attempt to augment T_s. It works as follows: a subset $S \subseteq N_r$ is selected, and the potential of each node in S is increased as much as possible while maintaining dual feasibility. The objective is to enlarge the set of the arcs with zero reduced costs that can be added to T_s; these arcs satisfy the CSC.

With the previously introduced definitions, the dual ascent operation consists then of increasing the potential of each node in S by the amount γ_S:

$$\pi_i^{(s)} = \begin{cases} \pi_i^{(s)} + \gamma_S, & i \in S, \\ \pi_i^{(s)}, & \text{otherwise.} \end{cases}$$

Each arc $(i,j) \in BS_B(S)$ which has a zero reduced cost after the updating, if any, is then moved to T_s. [2]

At termination, when the optimal potentials $\pi^{(s)}$ for root s have been computed (i.e., when all the external nodes have been moved to T_s), the minimum path costs are given by:

$$C_i = \pi_i^{(s)} - \pi_s^{(s)}, \qquad \forall i \in N.$$

A particular case happens when $|S| = 1$, say $S = \{j\}$ for some external node j; in such a case, the dual ascent operation, referred to as a *local operation at node* j, corresponds to increasing the potential of j by the amount γ_j:

$$\pi_j^{(s)} = \pi_j^{(s)} + \gamma_j. \tag{14.3}$$

Another special case is when $S = N_r$. In this case, $\gamma_S = \Delta$, where Δ is the dual gap associated with N_r; the dual ascent operation, referred to as a *global operation*, becomes:

$$\pi_i^{(s)} = \pi_i^{(s)} + \Delta, \qquad \forall i \in N_r. \tag{14.4}$$

Observe that Δ is a threshold gap for the potentials of the external nodes, in the sense that it is the minimum increment that warrants that at least one external node, say $w = \text{argmin}\{\delta_j : j \in N_r\}$, called in the following a *closest node*, can be added to the partial tree T_s.

When $S = N_r$ at each iteration, we have a dual ascent interpretation of the classical Dijkstra algorithm in the reoptimization context. In fact, the resulting algorithm is equivalent to solving the shortest path tree problem with root s, by Dijkstra's approach, using the optimum reduced costs obtained from the previous shortest path tree computation with root r as the arc costs. Observe that in this case, the CSC hold for the arcs of A_r, as pointed out in (14.1).

An enhanced dual ascent approach. The above dual ascent operation, and thus the entire reoptimization paradigm, can be further specialized as follows. Let $w \in N_r$ be a closest node and $T(w) = (N(w), A(w))$ be the subtree of T_r rooted at w; moreover, let (v,w) denote a border arc such that $\bar{c}_{vw} = \Delta$. At the beginning, all the arcs in T_r have a zero reduced cost, and consequently all the arcs belonging to $A(w)$ have a zero reduced cost too. By this property, if one performs a global operation then both w and the entire tree $T(w)$ may be transferred to T_s. In other words, a path from s to any node $i \in N(w)$ in the updated tree $T_s \cup \{(v,w)\} \cup T(w)$ is a shortest path from s to i. However, the global operation (14.4) implies the updating of the potential of all nodes in N_r. One efficient way to avoid the repetitive updating of potentials is to consider the partial graph (hereafter called the *reduced graph*) obtained by removing all the external arcs (i,j) such that $i \in N_r \backslash N(w)$ and

$j \in N(w)$ (as implied earlier and shown in (Bertsekas, Pallottino and Scutellà, 1995), a shortest path in the reduced graph is also a shortest path in the original graph):

$$A := A \backslash BS_E(N(w)). \tag{14.5}$$

Now, if we set $S = N(w)$, and apply the dual ascent operation in the reduced graph, we obtain a potentially more efficient operation, called the *enhanced dual ascent operation.* The above arc reduction (14.5) leads to $BS_E(N(w)) = \emptyset$ and $\gamma_N(w) = \delta_N(w) = \Delta$. Thus, only the node potentials of $N(w)$ increase by the threshold gap Δ:

$$\pi_i^{(s)} = \begin{cases} \pi_i^{(s)} + \Delta, & i \in N(w), \\ \pi_i^{(s)}, & \text{otherwise.} \end{cases} \tag{14.6}$$

In addition, as the potentials of the nodes remaining in N_r do not change, each arc in the remaining T_r maintains a zero reduced cost as pointed out in (14.1), and allows to embed the enhanced dual ascent operation in an iterative process.

The following property can be easily established:

Property 1: The enhanced dual ascent paradigm moves nodes to T_s exactly in the same order as Dijkstra's algorithm (using the arc reduced costs).

From an algorithmic point of view, the enhanced paradigm is thus equivalent to Dijkstra's approach in the reoptimization context; however, it can be computationally more efficient since a smaller set of potentials must be updated at each iteration and each node potential is updated at most once.

The two main reoptimization algorithms from the literature (Gallo and Pallottino, 1982, and Florian, Nguyen and Pallottino, 1981), described in the following, are both derived from the enhanced dual ascent paradigm.

14.2.3 The Reduced cost algorithm

Although presented as a primal approach, the reduced cost algorithm (Gallo and Pallottino, 1982) is indeed a dual algorithm. In fact, it performs the same operations as the enhanced dual ascent paradigm, where the graph reductions are not performed explicitly (i.e., there is no physical removal of the external arcs).

For efficiency purposes, the key operation of finding the border arc (v, w) entering a closest node w is performed on the set Q of nodes $j \in N_r$ such that their border backward stars $BS_B(j)$ are not empty. So, the computation of the dual gap Δ becomes:

$$\Delta = min\{\delta_j : j \in Q\}.$$

For each $j \in Q$, only one border arc (i, j) is maintained, i.e., an arc (i, j) such that $\bar{c}_{ij} = \delta_j$; the other border arcs entering j are considered removed from the graph; therefore, the search of a closest node w in Q and the computation of Δ can be performed in $O(n)$ time.

Moreover, the computation of values $\delta_j, \forall j \in N_r$, can be performed globally in $O(m)$ time. Indeed, when a node i is moved to T_s, by examining its forward star it is possible to update the value δ_j of each node j in T_r such that an arc (i, j) exists in $FS(i)$. Finally, in order to move the subtree $T(w)$ to T_s, the set Q is implemented as two lists Q' and Q'', such that the former contains all the nodes with minimum δ-value and the latter contains the remaining nodes. Once a closest node w is selected from Q', the inspection of its forward star allows one to set the δ-value of all the sons of w in T_r to a value equal to $\delta_w = \Delta$; these sons are then inserted into Q' and thus all the nodes of $T(w)$ will be moved into Q' and subsequently into T_s. The search for a node with minimum δ-value among those in Q'' is only performed when $Q' = \emptyset$.

Since the reduced cost algorithm performs the same operations as the enhanced dual ascent approach, it can be considered as an efficient "ad hoc" implementation of Dijkstra's algorithm in the reoptimization context, according to Property 1.

14.2.4 The Dual simplex algorithm

Based on the observation that the optimum tree T_r is a primal unfeasible but dual feasible basis for the new problem when the root changes to s, this algorithm specializes the classical dual simplex algorithm to determining the new optimum tree T_s.

Given the current dual feasible basis, the most negative valued basic arc is determined without actually computing the primal basic solution: indeed, it is the first reverse arc along the basic tree path from s to r. The removal of this arc from the basis divides the tree into two subtrees $T_r = (N_r, A_r)$ and $T_s = (N_s, A_s)$. The new basic arc is a border arc $(v, w) \in (N_s, N_r)$ oriented from N_s to N_r that provides a minimum reduced cost, such that $\bar{c}_{vw} = \delta_w = \Delta$ (w is a closest node). From a practical point of view, as in the reduced cost approach, such a minimum reduced cost arc may be identified by maintaining only one border arc for each node j in Q. The updating of the δ-values is carried out by scanning the forward stars of the nodes entering T_s.

Similarly to the reduced cost approach, the arc reduction (14.5) is implicitly performed, and the entire subtree $T(w) = (N(w), A(w))$, rooted at w, is moved to T_s. However, unlike the former approach, a more sophisticated data structure describing T_r is needed to explicitly move the entire $T(w)$, and to traverse it in order to update the node potentials as in (14.6).

Observe that, as in the reduced cost case, the potential updating is exactly the enhanced dual ascent operation. Therefore, the dual simplex algorithm can also be viewed as an efficient implementation of Dijkstra's approach in the reoptimization context.

14.3 A NEW DUAL ALGORITHM FOR SHORTEST PATH REOPTIMIZATION

The new reoptimization algorithm is closely related to a recently introduced family of dual ascent shortest path algorithms, called the *hanging family* (Pallottino and Scutellà, 1997).[3]

Although algorithms of the hanging family seem to be suitable for the kind of reoptimization under consideration, they can be essentially reduced to the two previously reviewed reoptimization algorithms and, by Property 1, to a Dijkstra-like procedure, as shown in Pallottino and Scutellà, 1998. Ad hoc dual ascent operations are thus necessary to devise a novel reoptimization approach.

In the following, several basic properties that allow to efficiently pruning the graph and discovering subtrees which can be moved simultaneously to T_s will be presented. These properties extend the enhanced dual ascent paradigm previously described and embody the key ingredients of the new reoptimization approach.

14.3.1 Basic properties

Consider the typical dual ascent iteration in the reoptimization context, where both $T_s = (N_s, A_s)$ and $T_r = (N_r, A_r)$ are described by a unique predecessor function p, such that $i = p(j)$ indicates that arc (i,j) belongs either to A_s or to A_r. For each external node $j \in N_r$, $j \neq r$, let $T(j) = (N(j), A(j))$ denote the subtree of T_r rooted at j.

Given the current feasible dual vector $\pi^{(s)}$, for each external node $j \neq r$ we define:

$$\alpha_j = min\{\bar{c}_{ij} : (i,j) \in BS_E(j), i \neq p(j)\}, \tag{14.7}$$

where $\alpha_j = \infty$ if $BS_E(j)\backslash\{(p(j),j)\} = \emptyset$. When finite, α_j is the minimum increment to be added to the potential of j, due to the external arcs, if we ignore the critical arc $(p(j),j)$, which by complementarity slackness with respect to tree T_r, has a zero reduced cost. In the case of the root r, let $\alpha_r = \eta_r$.

Furthermore, for each external node $j \neq r$, let

$$\delta_j^* = min\{\delta_j, \delta_{p(j)}^*\}; \tag{14.8}$$

$$\alpha_j^* = min\{\alpha_j, \alpha_{p(j)}^*\}, \tag{14.9}$$

where $\delta_r^* = \delta_r$ and $\alpha_r^* = \alpha_r$.

A finite δ_j^* is the minimum increment to be added to the potential of j due to the border arcs entering j (δ_j) and to the border arcs entering the ancestors of j in T_r (recall that the paths from such ancestors to j have a zero reduced cost). On the other hand, a finite α_j^* is the minimum increment to be added to the potential of j due to the external arcs entering j and the ancestors of j in T_r.

To give added insights to the values introduced, consider the graph and the shortest path tree rooted at $r = 1$ in Figure 14.1, where the optimal potentials are associated with the nodes.

Let $s = 4$ be the new root; the trees T_r and T_s are depicted in Figure 14.2, together with the border arcs, their reduced costs and the values $\delta_j, \forall j \in N_r$. It is $\Delta = min\{\delta_j : j \in N_r\} = 1$.

The values $\alpha_j, j \in N_r$, computed according to (14.7), are shown in Figure 14.3. Furthermore, the values δ_j^* and $\alpha_j^*, j \in N_r$, defined according to (14.8) and (14.9), are respectively: $\delta_1^* = 5$, $\delta_2^* = 4$, $\delta_3^* = 3$ and $\delta_6^* = 1$; $\alpha_1^* = 4$, $\alpha_2^* = 4$, $\alpha_3^* = 2$ and $\alpha_6^* = 2$.

Based on the above defined values, the following properties can be readily proved.

Property 2: $\delta_j^* + \pi_j^{(s)} - \pi_s^{(s)}$ is an upper bound to the shortest path length from s to j; when α_j is finite, $\alpha_j + \Delta + \pi_j^{(s)} - \pi_s^{(s)}$ is a lower bound to the length of the paths from s to j which enter node j through an external arc other than $(p(j),j)$.

Such a property follows easily from the definitions of δ_j^*, α_j and Δ.

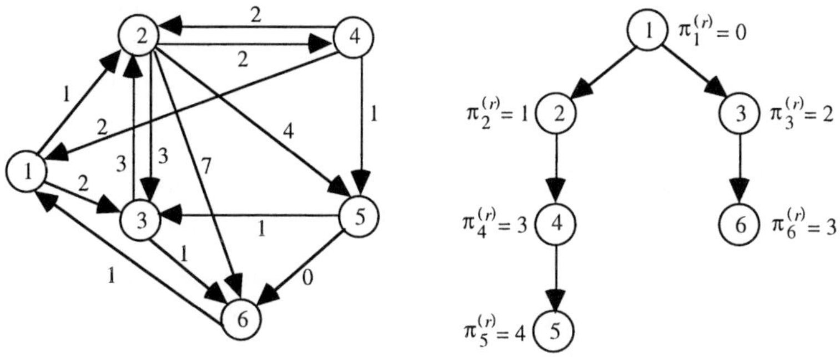

Figure 14.1 The graph G and the shortest path tree T_r

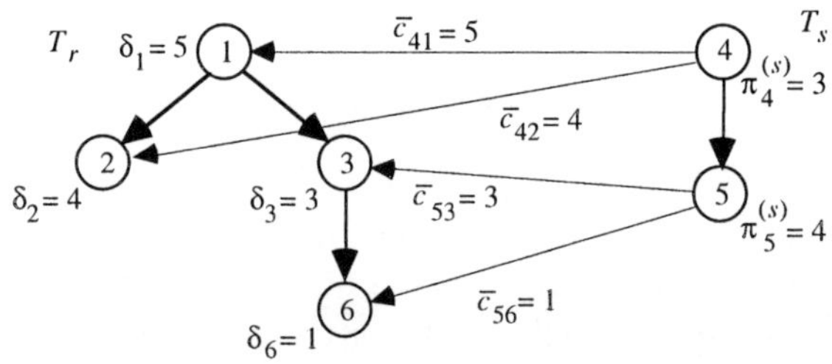

Figure 14.2 The trees T_r and T_s and the border arcs

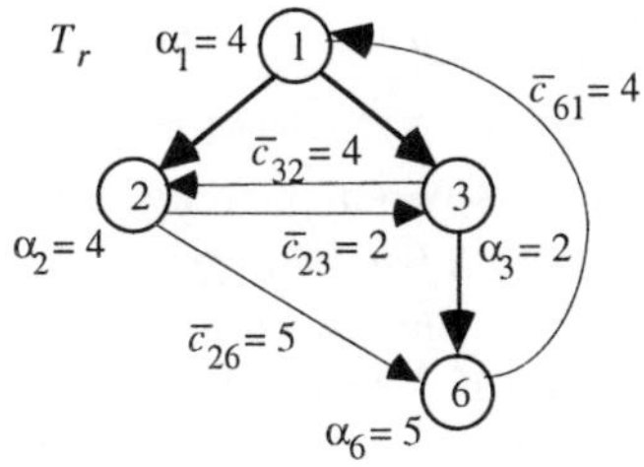

Figure 14.3 The values $\alpha_j, j \in N_r$

Property 3: If $\delta_j^* < \delta_j$ for some external node j, then all the border arcs entering j can be (implicitly) removed from G.

Proof: By Property 2, $\delta_j^* + \pi_j^{(s)} - \pi_s^{(s)}$ is an upper bound to the shortest path length from s to j; indeed, it is the length of a path of G from s to j going through an ancestor of j in T_r. Note that when $\delta_j^* < \delta_j$, the path in G, from s to j, terminating with the border arc that provides the value δ_j, has a length greater than the one having length $\delta_j^* + \pi_j^{(s)} - \pi_s^{(s)}$. All the border arcs entering j can thus be removed since there is at least a shortest path to j not terminating with any of these arcs.

Due to Property 3, one needs to maintain a border arc only for the external nodes j such that $\delta_j = \delta_j^*$, hereafter denoted as *effective candidate nodes.* One practical consequence is that, if one needs to compute the value Δ, it is sufficient to consider the set $Q^* = \{j \in N_r : \delta_j^* = \delta_j\} \subseteq Q$ instead of Q.

Property 4: If $\alpha_j + \Delta \geq \delta_j^*$ for some node $j \in N_r$, then all the external arcs (i, j), $i \neq p(j)$ (if $j \neq r$), can be (implicitly) removed from G.

Proof: By Property 1, when α_j is finite, $\alpha_j + \Delta + \pi_j^{(s)} - \pi_s^{(s)}$ is a lower bound to the length of the paths from s to j which enter node j through an external arc other than $(p(j), j)$; on the other hand, $\delta_j^* + \pi_j^{(s)} - \pi_s^{(s)}$ is an upper bound to the shortest path length from s to j. Condition $\alpha_j + \Delta \geq \delta_j^*$ thus implies that all the external arcs other than $(p(j), j)$ can be removed from G.

We now have all the ingredients to introduce an additional graph reduction, involving the arcs of T_r.

Property 5: If $\alpha_j^* + \Delta \geq \delta_j$ for an effective candidate node $j \in Q^* \backslash \{r\}$, then $(p(j), j)$ can be removed from G.

Proof: By definition, α_j^* is the minimum increment to be added to the potential of j due to the external arcs entering j and the other ancestors of j in T_r. Therefore, when α_j^* is finite, $\alpha_j^* + \Delta + \pi_j^{(s)} - \pi_s^{(s)}$ is a lower bound to the length of all paths from s to j which enter node j through an external arc, including the tree arc $(p(j), j)$. By Property 2, $\delta_j^* + \pi_j^{(s)} - \pi_s^{(s)}$ is an upper bound to the shortest path length from s to j; moreover, since j is an effective node, $\delta_j + \pi_j^{(s)} - \pi_s^{(s)}$ is the length of a path of G having the border arc entering j as its last arc. By combining all these properties, $\alpha_j^* + \Delta \geq \delta_j$ implies that all the external arcs entering j can be removed from G, including $(p(j), j)$.

Observe that removing tree arcs from T_r may transform T_r into a forest. An interesting consequence of Property 5 is that any effective candidate node j satisfying Property 5 can be hanged on T_s via the border arc (i, j), which is indeed the only arc entering j. This may happen even if $\delta_j > \Delta$. Unfortunately, at least in general, the descendants of j in T_r, i.e., the entire subtree $T(j)$, cannot be moved to T_s, since it is possible that the minimum cost path from s to some descendants of j does not pass through node j. For example, consider the previous graph where all the external nodes are effective candidate nodes, i.e., $Q^* = \{1, 2, 3, 6\}$. Since $\alpha_3^* + \Delta = 2 + 1 = 3 = \delta_3$, then all the external arcs entering node 3 can be removed. It follows that the optimal potential of node 3 is $\pi_3^{(s)} = \pi_3^{(s)} + \delta_3 = 2+3 = 5$. However, $\pi_6^{(s)} + \delta_6 = 3+1 < \pi_3^{(s)} = 5$, and consequently $T(3)$ cannot be moved to T_s.

Stronger conditions have thus to be discovered to allow the transfer of entire subtrees.

Definition 1: An effective candidate node j is said to be *independent* if $\delta_i \geq \delta_j$, for each descendant $i \in N(j)$; equivalently, if we define

$$\bar{\delta}_j = min\{\delta_i : i \in N(j)\},$$

we must have that $\bar{\delta}_j = \delta_j$.

Definition 2: A subtree $T(j)$ rooted at an independent node $j \in N_r$ is said to be *hangable* if j satisfies Property 5 and if $\alpha_i + \Delta \geq \delta_j$ for each proper descendant $i \in N(j)$; equivalently, if we define

$$\bar{\alpha}_j = min\{\alpha_i : i \in N(j)\},$$

we must have that $\bar{\alpha}_j + \Delta \geq \delta_j$.

Theorem 1: Let $T(j)$ be a hangable subtree; then the optimum potential of each node $i \in N(j)$ is $\pi_i^{(s)} = \pi_i^{(s)} + \delta_j$.

Proof: Since Property 5 is satisfied, all the external arcs entering j can be removed: therefore $\pi_j^{(s)} + \delta_j$ is the optimum potential of node j.

Since j is independent, i.e., $\bar{\delta}_j = \delta_j$, the path going through the (unique) border arc entering j is the best one among the paths of G, ending with a border arc, which go to a proper descendant of j (recall that all the arcs of T_r have a zero reduced cost). All the border arcs entering proper descendants of j can thus be removed from G.

Moreover, since $\alpha_i + \Delta \geq \delta_j$ for each proper descendant $i \in N(j)$, also all the external arcs entering the proper descendants of j, with the exception of the arcs of $T(j)$, can be removed from G.

It follows that the border arc entering j is the only arc linking the entire subtree $T(j)$ to T_s. The result thus follows.

Corollary 1: All the hangable subtrees $T(j)$ can be simultaneously moved to T_s; the optimum potential of each node $i \in N(j)$ is obtained by increasing the current potential value $\pi_i^{(s)}$ by the amount δ_j:

$$\pi_i^{(s)} = \pi_i^{(s)} + \delta_j, \qquad \forall i \in N(j). \tag{14.10}$$

14.3.2 The algorithm

The properties stated in Section 14.3.1 suggest considering a subset $\bar{Q}$ of nodes $j \in Q^*$ such that:

$$\delta_j \leq min\{\bar{\delta}_j, \alpha_j^* + \Delta, \bar{\alpha}_j + \Delta\}. \tag{14.11}$$

$\bar{Q}$ defines in fact the *roots of the hangable subtrees*, i.e., it specifies subtrees which can be simultaneously moved to T_s.

The typical iteration of the new reoptimization algorithm is based on the discovering of such a set $\bar{Q}$.

Typical iteration

- compute all the values δ_j, δ_j^*, $\bar{\delta}_j$, α_j, α_j^*, $\bar{\alpha}_j$, for each node $j \in N_r$;
- compute the set $Q^* = \{j \in N_r : \delta_j = \delta_j^*\}$ and the value $\Delta = min\{\delta_j : j \in Q^*\}$;
- for each node j satisfying conditions (14.11), i.e., for each root of a hangable subtree, move the subtree $T(j)$ to T_s and update the potentials as in (14.10).

Observe that at least a closest node w satisfies conditions (14.11); so, $T(w)$ is hangable by construction and can be moved to T_s. In the worst case, exactly one subtree may be moved to T_s at each iteration, i.e., $T(w)$. In such a case, the proposed algorithm behaves like the two previously described reoptimization algorithms and shows a Dijkstra-like behavior.

However, it may happen that, during some algorithmic iterations, more subtrees can be moved in parallel to T_s. In such a case, the new reoptimization algorithm is different from the classical reoptimization algorithms from the literature. In particular, subtrees $T(j)$ that require an increase of their potential by an amount $\delta_j > \Delta$ can all be moved to T_s. When this happens, the new reoptimization algorithm extends the enhanced dual ascent operation on a collection of disjoint node sets.

To illustrate the behavior of the proposed approach, consider the example discussed previously, where the cost of the arc $(5,6)$ changes from 0 to 3. The values associated with the external nodes are reported in the following table. We have $Q^* = \{1,2,3\}$ and $\Delta = 3$. Since $\delta_1^* = 5 > 3 = \bar{\delta}_1$, node 1 is not the root of a hangable subtree.

Table 14.1

external nodes	1	2	3	6
δ_j	5	4	3	4
α_j	4	4	2	5
δ_j^*	5	4	3	3
α_j^*	4	4	2	2
$\bar{\delta}_j$	3	4	3	4
$\bar{\alpha}_j$	2	4	2	5

On the other hand $T(3)$ is hangable, since $\delta_3 = \Delta$. In this example $T(2)$ is hangable too, since $\delta_2 \leq min\{\bar{\delta}_2, \alpha_2^* + \Delta\}$, although it is $\delta_2 = 4 > 3 = \Delta$. $T(2)$ and $T(3)$ can thus be moved simultaneously to T_s.

The trees T_r and T_s, after moving $T(2)$ and $T(3)$ to T_s, are illustrated in Figure 14.4.

At the next iteration, $(4,1)$ and 1 enter T_s, and $\pi_1^{(s)}$ is set to value 5. The shortest path tree rooted at s is shown in Figure 14.5 together with the minimum costs $C_i = \pi_i^{(s)} - \pi_s^{(s)}$ associated with each node i in T_s.

In order to complete the description of the reoptimization algorithm, we will now specify how to efficiently discover hangable subtrees at each iteration.

Consider a typical iteration. At the beginning, compute the values δ_j and α_j, $\forall j \in N_r$. Moreover, set $\Delta = \delta_r$, $Q^* = \emptyset$, and $\bar{\alpha}_j = \alpha_j$ and $\bar{\delta}_j = \delta_j$, $\forall j \in N_r$; finally, set $\alpha_r^* = \alpha_r$ and $\delta_r^* = \delta_r$ (*initialization step*).

After the initialization step, visit T_r in a depth-first way, starting from r, and perform the following operations for each visited node $j \neq r$:

- if j is visited in the descending phase, then set $\alpha_j^* = min\{\alpha_j, \alpha_{p(j)}^*\}$ and $\delta_j^* = min\{\delta_j, \delta_{p(j)}^*\}$;

  if $\delta_j = \delta_j^*$, then $Q^* = Q^* \cup \{j\}$ and $\Delta = min\{\Delta, \delta_j\}$;

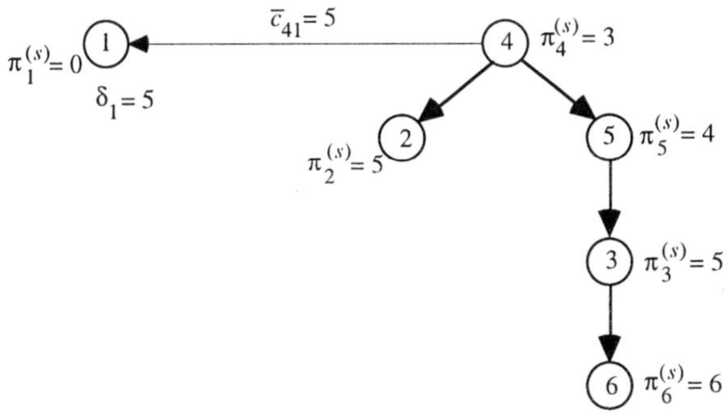

Figure 14.4 The new trees T_r and T_s

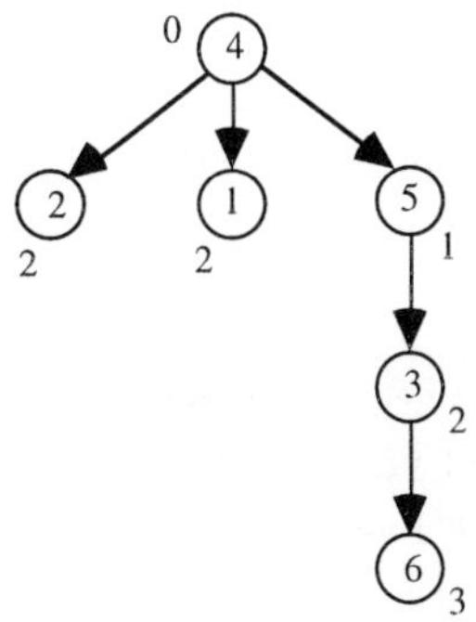

Figure 14.5 The shortest path tree rooted at s

- if j is visited in the ascending phase, then set $\bar{\alpha}_{p(j)} = min\{\bar{\alpha}_{p(j)}, \bar{\alpha}_j\}$ and $\bar{\delta}_{p(j)} = min\{\bar{\delta}_{p(j)}, \bar{\delta}_j\}$.

At the end of the visit of T_r, if $\delta_r = \Delta$ then the entire forest T_r can be moved to T_s, and the computation ends; otherwise, repeat the following operations until $Q^* = \emptyset$:

- select and remove a node $j \in Q^*$;
- if j still belongs to T_r and if j satisfies conditions (14.11), then move $T(j)$ to T_s; update the potentials as in (14.10).

Theorem 2: (Algorithm correctness and time complexity) The reoptimization algorithm correctly finds the shortest path tree T_s in $O(mn)$ time.

Proof: The algorithm correctness follows immediately from Theorem 1, since only hangable subtrees are moved during each algorithm iteration.

As far as the time complexity is concerned, observe that the algorithm performs at most n iterations, since at each iteration at least a closest node w is moved to T_s together with the corresponding subtree $T(w)$.

Consider now a typical iteration. During all the initialization steps, the computation of values $\delta_j, \forall j \in N_r$, can be performed globally in $O(m)$ time, as shown for the reduced cost algorithm. On the other hand, the computation of values $\alpha_j, \forall j \in N_r$, requires $O(m)$ time at each algorithm iteration, since in the worst case the backward star of each node in G has to be analyzed.

The other initialization operations can be performed in $O(n)$ time. The depth-first visit of T_r and the construction of Q^* require $O(n)$ time, too.

Finally, the movement of the hangable subtrees requires globally $O(n)$ time. The stated complexity follows.

Due to the computational cost required by each algorithm iteration, and in particular by the computation of values $\alpha_j, \forall j \in N_r$, it is possible to devise a relaxed version of the reoptimization approach, using lower bounds of $\alpha_j, \forall j \in N_r$, instead of the correct values (observe that, if lower bounds are used, and if a node $j \in Q^*$ is discovered during an iteration which satisfies conditions (14.11), then j is certainly the root of a hangable subtree; in other words, the algorithm correctness is preserved). In particular, if the values $\alpha_j, \forall j \in N_r$, are updated only in k pre-specified iterations, and the old values are maintained in the other iterations, then the algorithm time complexity becomes $O(n^2 + km)$. By suitably choosing k, better time complexity values can thus be derived.

We suggest to consider $k = 1$, i.e., to compute the α's only in the first initialization phase, and set $k = \lceil n^2/m \rceil$: in both cases the algorithm time complexity reduces to $O(n^2)$. From a practical point of view the reduced effort in computing the α's could lead to very efficient implementations.

Finally, to obtain a non-Dijkstra-like behavior, a necessary condition is that $\alpha_j^* > 0$ for at least one external node j. Such a condition requires, in particular, that value $\alpha_r > 0$.

Observe that $\alpha_r = 0$ implies that there exists an external arc entering r, say (i, r), which has a zero reduced cost. In other words, there exists a zero cost cycle in G, say C, which goes through the old root node r. When this happens, we suggest to shrink C into a new node, say R, which takes the place of the old root r. If the α-value associated with the new root R is again 0, i.e. $\alpha_R = 0$, then the shrinking operation can be iterated, until the tree to be hanged has a root with a nonzero α-value.

It is easy to show that the proposed reoptimization algorithm is still correct when the shrink reduction operations are performed. Furthermore, the time complexity values stated in Theorem 2 and the ensuing observations remain valid: indeed, all the shrink reduction operations can be performed in $O(n)$ time, and the reverse operations, i.e., the cycle unfolding after the algorithm termination, can be performed in $O(n)$ time too (we refer the reader to Ahuja, Magnanti and Orlin, 1993, and Tarjan, 1983 for details about cycle shrinking and unfolding operations).

We strongly believe that the proposed new dual approach is a promising avenue for the reoptimization problems considered. Numerical experimentations are presently carried out and the results will be reported in a subsequent paper.

Acknowledgments Research grants by M.U.R.S.T. progetto cofinanziato 1997, INDAM-GNIM, and NSERC Canada.

Notes

1. For the sake of completeness, the more recent auction algorithms (Bertsekas, Pallottino and Scutellà, 1995) have to be mentioned as well. These algorithms, which belong to the family of the dual methods for standard shortest path computations, might in fact be suitable for the reoptimization problem under consideration. However, auction methods will not be described in this work, since no specialized reoptimization algorithm has been proposed so far.

2. The dual ascent operation can be also performed by selecting disjoint sets S_1, S_2, ..., S_h of external nodes, and by increasing the node potentials in each subset S_i by the corresponding dual gap γ_{S_i}, $i = 1, \ldots, h$.

3. In the hanging family, two basic dual ascent operations are performed. These operations can be viewed as particular cases of the dual ascent operation introduced for the shortest path reoptimization problem (more sophisticated operations, that mix the two basic ones in various ways, have been proposed): the *local operation* (14.3), performed at a single external node, and the *classic operation*, that corresponds to (14.4).

References

Ahuja, R. K., Magnanti, T. L. and Orlin, J. B. (1993). *Network flows: Theory, algorithms, and applications.* Prentice Hall, Englewood Cliffs, NJ.

Bertsekas, D. P., Pallottino, S. and Scutellà, M. G. (1995). "Polynomial Auction algorithms for shortest paths". *Computational Optimization and Applications* 4, 99–125.

Dijkstra, E. W. (1959). "A note on two problems in connexion with graphs". *Numerische Mathematik* 1, 269–271.

Dionne, R. (1978). "Étude et extension d'un algorithme de Murchland". *INFOR* 16, 132–146.

Florian, M., Nguyen, S. and Pallottino, S. (1981). "A dual simplex algorithm for finding all shortest paths" *Networks* 11, 367–378.

Fujishige, S. (1981). "A note on the problem of updating shortest paths". *Networks* 11, 317–319.

Gallo, G. (1980). "Reoptimization procedures in shortest path problems". *Rivista di Matematica per le Scienze Economiche e Sociali* 3, 3–13.

Gallo, G. and Pallottino, S. (1982). "A new algorithm to find the shortest paths between all pairs of nodes". *Discrete Applied Mathematics* 4, 23–35.

Lawler, E. L. (1976). *Combinatorial optimization: Networks and matroids.* Holt, Rinehart and Winston, New York.

Murchland, J. D. (1970). *A fixed matrix method for all shortest distances in a directed graph and for the inverse problem.* Ph.D. Thesis, Univ. of Karlsruhe.

Nemhauser, G. (1972). "A generalized permanent label setting algorithm for the shortest path between specified nodes". *Journal of Mathematical Analysis and Applications* 38, 328–334.

Pallottino, S. and Scutellà, M. G. (1997). "Dual algorithms for the shortest path tree problem". *Networks* 29, 125–133.

Pallottino, S. and Scutellà, M. G. (1998). "Shortest path algorithms in transportation models: classical and innovative aspects". in (Marcotte, P. and Nguyen, S., Eds.) *Equilibrium and advanced transportation modelling*, Kluwer, Boston, 245–281.

Tarjan, R. E. (1983). *Data structures and network algorithms.* SIAM, Philadelphia, PA.

15 BIPROPORTIONAL MATRIX BALANCING WITH UPPER BOUNDS

Heinz Spiess

Abstract: The purpose of the note is to look at the problem of biproportional matrix balancing when upper bounds are imposed on the matrix elements. This problem can be formulated as a convex minimization problem. Using the Kuhn-Tucker optimality conditions the functional form of the resulting model is derived. The dual formulation of the problem is derived and it is shown how it can be solved by a cyclic coordinate descent method. This leads to the proposal of an efficient solution algorithm.

15.1 INTRODUCTION

The purpose of this note is to analyze an extension of the standard biproportional matrix balancing (Murchland, 1977), also known as "Fratar" (Lamond and Stewart, 1981), "Furness method" (Furness 1970), "entropy transportation problem" or "entropy spatial interaction model" (Chabini and Florian, 1995), and referred to as "two-dimensional matrix balancing" in the EMME/2 transportation planning package (Spiess, 1984; INRO, 1996).

As in the standard two-dimensional balancing, a given matrix (containing synthetic friction factors or "prior" matrix data) is "factored up" to fit a given set of productions and attractions. But in addition to satisfying the given productions and attractions, upper bounds are now imposed on the elements of the resulting matrix. Such bounds can be thought to represent transport capacities, but they can also be simply used to limit the growth of individual matrix elements with respect to the prior matrix.

In the next section we briefly introduce the standard biproportional matrix balancing model and its optimization formulation as a maximum entropy model. In Section 15.3 the extended model is formulated and analyzed. In Section 15.4, finally, an efficient solution algorithm is formulated, which is based on the coordinate descent method applied to the dual problem.

M. Gendreau and P. Marcotte (eds.), Transportation and Network Analysis: Current Trends, 237–243.

15.2 STANDARD BIPROPORTIONAL MATRIX BALANCING

The problem of two-dimensional matrix balancing consists in finding factors a_p and b_q for each origin $p \in P$ and destination $q \in Q$ so that, when multiplied to the corresponding rows and columns of the prior matrix G_{pq}, the marginal totals of resulting matrix g_{pq} correspond to the given productions O_p and attractions D_q. This can be expressed with the following set of equations:

$$g_{pq} = a_p b_q G_{pq}, \qquad pq \in PQ, \tag{15.1}$$

$$\sum_{q \in Q} g_{pq} = O_p, \qquad p \in P, \tag{15.2}$$

$$\sum_{p \in P} g_{pq} = D_q, \qquad q \in Q. \tag{15.3}$$

It is easy to show that the above problem can also be expressed as the following convex optimization problem (also referred to as maximum entropy transportation problem):

$$\min \sum_{pq \in PQ} g_{pq}(\log g_{pq} - \log G_{pq} - 1) \tag{15.4}$$

subject to (15.2) and (15.3).

Defining α_p and β_q as the dual variables associated with constraints (15.2) and (15.3), we obtain from the Kuhn-Tucker optimality conditions the functional form

$$g_{pq} = e^{-\alpha_p - \beta_q} G_{pq}, \qquad pq \in PQ, \tag{15.5}$$

which corresponds exactly to (15.1) when using the substitutions

$$a_p = e^{-\alpha_p}, \qquad p \in P, \tag{15.6}$$

$$b_q = e^{-\beta_q}, \qquad q \in Q. \tag{15.7}$$

The dual formulation of problem (15.4) can be written as the following unconstrained optimization problem:

$$\min_{\alpha\beta} \sum_{pq \in PQ} G_{pq} e^{-\alpha_p - \beta_q} + \sum_{p \in P} \alpha_p O_p + \sum_{q \in Q} \beta_q D_q. \tag{15.8}$$

The standard biproportional matrix balancing model is known to be always feasible when $\sum_{p \in P} O_p = \sum_{q \in Q} D_q$ and $G_{pq} > 0$ for all $pq \in PQ$.

Problems which have $G_{pq} = 0$ for some O-D pairs can also be handled in this framework. But in this case, the O-D pairs with $G_{pq} = 0$ are not to be considered part of the set PQ and $g_{pq} = 0$ is used for these O-D pairs where necessary, e.g. where it appears in a sum over origins or destinations.

Note, however, that problems with $G_{pq} = 0$ for some O-D pairs are not always feasible. In this case, there is no easy method to determine the feasibility of a problem a priori, so that the primal feasibility of a solution found by a dual algorithm always has to be checked explicitly.

15.3 BIPROPORTIONAL MATRIX BALANCING WITH UPPER BOUNDS

We now turn our attention to an extension of the model defined in the previous section in which explicit upper bounds U_{pq} are imposed on the resulting matrix elements g_{pq}. Since imposing upper bounds will have a direct impact on the functional form of the resulting model, it is not possible to formulate the extended model by simply adding a further constraint to problem (15.1)-(15.3).

However, the minimization problem (15.4) lends itself very well to an extension by adding upper bound constraints. This leads to the following extended problem formulation:

$$\min \sum_{pq \in PQ} g_{pq}(\log g_{pq} - \log G_{pq} - 1) \tag{15.9}$$

subject to

$$\sum_{q \in Q} g_{pq} = O_p, \qquad p \in P, \tag{15.10}$$

$$\sum_{p \in P} g_{pq} = D_q, \qquad q \in Q, \tag{15.11}$$

$$g_{pq} \le U_{pq}, \qquad pq \in PQ. \tag{15.12}$$

Note that this model is a "single" entropy model which associates an entropy term term only with actual trips, as is the case for the uncapacitated balancing model. This is different from the model proposed by Chabini and Florian (1995), who formulate the same type of problem using an objective function based on "double entropy", i.e. they associate an entropy term not only with the actual trips g_{pq}, but also with the resulting slacks $U_{pq} - g_{pq}$.

Using again α_p and β_q as the dual variables associated with the production (15.10) and attraction constraints (15.11) and the new dual variables $\mu_{pq} \ge 0$ for the upper bound constraints (15.12), we can state the Kuhn-Tucker optimality conditions for the above problem as:

$$\log g_{pq} - \log G_{pq} + \alpha_p + \beta_q + \mu_{pq} = 0, \qquad pq \in PQ, \tag{15.13}$$

$$\mu_{pq}(U_{pq} - g_{pq}) = 0, \qquad pq \in PQ. \tag{15.14}$$

The optimality condition (15.13) can be rewritten to yield the following functional form for the resulting model:

$$g_{pq} = G_{pq} e^{-\alpha_p - \beta_q - \mu_{pq}}, \qquad pq \in PQ. \tag{15.15}$$

By applying the complementary slackness conditions (15.14) to the dual variable μ_{pq} and distinguishing between the cases $\mu_{pq} = 0$ and $\mu_{pq} > 0$, (15.15) becomes

$$g_{pq} = \begin{cases} G_{pq} e^{-\alpha_p - \beta_q} = a_p b_q G_{pq} & \text{for } \mu_{pq} = 0 \\ U_{pq} & \text{for } \mu_{pq} > 0 \end{cases} \tag{15.16}$$

which can be written more concisely as

$$g_{pq} = \min\{G_{pq} e^{-\alpha_p - \beta_q}, U_{pq}\} = \min\{a_p b_q G_{pq}, U_{pq}\}, \qquad pq \in PQ. \tag{15.17}$$

Note that the above model formulation does no longer explicitly contain the dual variables μ_{pq}.

Without loss of generality we can assume that $U_{pq} > 0$ for all $pq \in PQ$, as matrix elements g_{pq} can always be forced to zero by setting $G_{pq} = 0$ and removing the corresponding O-D pairs from the set PQ, as explained in Section 15.2.

The next step is to formulate the Lagrangian dual for problem (15.9):

$$\min_{\alpha,\beta\ \mu\geq 0} -L(\alpha,\beta,\mu) = \sum_{pq\in PQ} G_{pq}e^{-\alpha_p-\beta_q-\mu_{pq}} + \sum_{p\in P}\alpha_p O_p + \sum_{q\in Q}\beta_q D_q + \sum_{pq\in PQ}\mu_{pq}U_{pq} \tag{15.18}$$

Note that, except for the non-negativity of μ_{pq}, the dual problem is unconstrained. For this type of problem a simple method of successive coordinate descent is known to converge to the optimal solution (see e.g. Luenberger, 1984). Thus, it can be used to solve the dual problem (15.18) and find the optimal values of the dual variables. If the primal problem (15.9) is feasible, the optimal values of the dual variables α_p and β_q can be inserted into (15.17) to find the optimal solution of the primal problem.

Applying the coordinate descent method to the dual problem (15.18) amounts to solving cyclically the first order optimality conditions for the corresponding dual variables. This implies finding the zeroes of the partial derivatives of the dual objective function with respect to all α_p and β_q, and all non-zero μ_{pq}:

$$\begin{aligned} \partial L(\alpha,\beta,\mu)/\partial\alpha_p &= -\textstyle\sum_{q\in Q} G_{pq}e^{-\alpha_p-\beta_q-\mu_{pq}} + O_p &&= 0, \quad p\in P, && (15.19)\\ \partial L(\alpha,\beta,\mu)/\partial\beta_q &= -\textstyle\sum_{p\in P} G_{pq}e^{-\alpha_p-\beta_q-\mu_{pq}} + D_q &&= 0, \quad q\in Q, && (15.20)\\ \partial L(\alpha,\beta,\mu)/\partial\mu_{pq} &= -G_{pq}e^{-\alpha_p-\beta_q-\mu_{pq}} + U_{pq} &&= 0, \quad \forall\mu_{pq}>0. && (15.21) \end{aligned}$$

By observing that (15.21) is equivalent to (15.17), it is easy to see that conditions (15.19) and (15.21) can be combined into the simpler production equations

$$\sum_{q\in Q}\min\{a_p b_q G_{pq}, U_{pq}\} = O_p, \qquad p\in P, \tag{15.22}$$

and the conditions (15.20) and (15.21) combine into the simpler attraction equations

$$\sum_{p\in P}\min\{a_p b_q G_{pq}, U_{pq}\} = D_q, \qquad q\in Q. \tag{15.23}$$

Solving (15.22) for α_p will simultaneously satisfy the first order conditions (15.19) for origin p and (15.21) for all $\mu_{pq} > 0$ related to origin p. In the same manner solving (15.23) for β_q will simultaneously satisfy the first order conditions (15.20) for destination q and (15.21) for all $\mu_{pq} > 0$ related to destination q.

Before turning our attention to the solution algorithm, let us briefly look at the question of feasibility. The introduction of upper bounds influences the feasibility in a similarly complex manner as do the zeroes in the matrix G_{pq}. While, of course, the conditions

$$\sum_{q\in Q} U_{pq} \geq O_p \qquad p\in P, \tag{15.24}$$

$$\sum_{p\in P} U_{pq} \geq D_q \qquad q\in Q \tag{15.25}$$

are necessary conditions for feasibility, they are by no means sufficient.

15.4 SOLUTION ALGORITHM

The proposed solution algorithm can be classified as a coordinate descent method applied to the dual problem formulation. As it is based on iteratively solving equations (15.22) and (15.23), it does not optimize along dual coordinates one by one, but always satisfies the first order conditions simultaneously for all dual variables associated with either one origin or one destination — which should result in an improved convergence compared to a one-by-one coordinate descent method.

Algorithm 1 describes the basic algorithm proposed to solve the biproportional matrix balancing problem with upper bounds (15.9). This algorithm assumes that equations of the type $\sum_i \min\{xf_i, u_i\} = T$ can be solved for x. How this can be done efficiently is shown in Algorithm 2.

Algorithm 1:

0. Initialization
Set $b_q^0 = 1$ for $q \in Q$ and set $k := 1$.

1. Balance origins
For each origin $p \in P$ solve the equation

$$\sum_{q \in Q} \min\{a_p^k b_q^{k-1} G_{pq}, U_{pq}\} = O_p$$

for variable a_p^k by applying Algorithm 2.

2. Balance destinations
For each destination $q \in Q$ solve the equation

$$\sum_{p \in P} \min\{a_p^k b_q^k G_{pq}, U_{pq}\} = D_q$$

for variable b_q^k by applying Algorithm 2.

3. Test stopping criteria
If convergence is reached for the the multipliers a_p and b_q, e.g. $||a^k - a^{k-1}|| + ||b^k - b^{k-1}|| < \epsilon_1$, then go to Step 4, otherwise set $k := k + 1$ and return to Step 1.

4. Compute primal solution
If the maximum production constraint violation for all $p \in P$ is smaller than some predefined tolerance value, i.e.

$$\left| \sum_{q \in Q} \min\{a_p^k b_q^k G_{pq}, U_{pq}\} - O_p \right| < \epsilon_2,$$

then compute the optimal primal solution

$$g_{pq} = \min\{a_p^k b_q^k G_{pq}, U_{pq}\},$$

otherwise the primal problem is infeasible. STOP.

Now let us consider the equation

$$\sum_i \min\{xf_i, u_i\} = T. \tag{15.26}$$

which occurs in Step 1 of Algorithm 1 with $i = q$, $f_i = b_q^{k-1} G_{pq}$, $u_i = U_{pq}$, and in Step 3 with $i = q$, $f_i = a_p^k G_{pq}$, $u_i = U_{pq}$. Without loss of generality we can assume that $f_i > 0$ and $u_i > 0$ for all $i \in I$. If there were any elements i with $f_i = 0$ and/or $u_i = 0$, these could simply be dropped, since they do not influence the solution of (15.26).

The problem can be solved by first sorting the elements i according to their ratios u_i/f_i and then scanning the elements in this order and checking if the optimal value x lies in the range between two consecutive u_i/f_i ratios. This leads to the following algorithm:

Algorithm 2: (Solve $\sum_i \min\{xf_i, u_i\} = T$)

0. Initialization
Set $F = \sum_i f_i$ and $U = T$.

1. Sort elements
Sort the elements i in increasing order of u_i/f_i ratios.

2. Scan elements
Using the order established in Step 1, do for each i:
If $u_i F \leq f_i U$ then
set $F := F - f_i$ and $U := U - u_i$,
else
go to Step 3.
If this point is reached after having scanned all elements, then the problem is infeasible. STOP.

3. Compute optimal solution
Set the solution to $x = U/F$ and STOP.

So far, the above algorithms have only been implemented in a "quick and dirty" manner purely for testing purposes. In the few examples tested so far (all based on data taken from the standard EMME/2 Winnipeg data bank using 154 traffic zones), the algorithm has been found to converge quite rapidly to the optimal solution (4-7 iterations).

15.5 CONCLUSIONS

The methodology of introducing upper bounds is presented in this note only for the biproportional matrix balancing. However, it can most likely be applied directly to other related problems, such as the multi-proportional matrix balancing or Evans and Kirby's three-dimensional matrix balancing.

References

Bregman, L. (1967). "The Relaxation Method of Finding the Common Point of Convex Sets and its Application to the Solution of Problems in Convex Programming," *U.S.S.R. Computational Math. Mathematical Phys.* 7, 200-217.

Chabini, I. and Florian, M. (1995). "An Entropy Based Primal-Dual Algorithm for Convex and Linear Cost Transportation Problems," Centre de recherche sur les transports, Université de Montréal, Publication 963 (CRT #95-17).

Evans, S.P. and Kirby, H.R. (1974). "A Three-Dimensional Furness Procedure for Calibrating Gravity Models," *Transportation Research*, 8, 105-122.

Furness, K.P. (1970). "Time Function Interaction," *Traffic Engineering and Control* Vol 7, No 7, 19-36.

INRO Consultants Inc. (1997). *EMME/2 User's Manual.*

Lamond, B. and Stewart, N.F. (1981). "Bregman's Balancing Method," *Transportation Research*, 15B, 239-248.

Luenberger, D.G. (1984). *Linear and Nonlinear Programming.* Second Edition, Addison-Wesley.

Murchland, J. (1977). *The Multi-proportional Problem.* Univerity College London, research note JDM 263.

Spiess, H. (1984). *Contributions à la théorie et aux outils de planification de réseaux de transport urbain.* Ph.D. thesis, Département d'informatique et de recherche opérationnelle, Publication 382, Centre de recherche sur les transports, Université de Montréal.